Hybrid Polymer Composite Materials

Related titles

Advances in Polymer Nanocomposites (ISBN: 978-1-84569-940-6)

Nanofibers and Nanotechnology in Textiles (ISBN: 978-1-84569-105-9)

Polymer Nanocomposites (ISBN: 978-1-85573-969-7)

Woodhead Publishing Series in Composites Science and Engineering

Hybrid Polymer Composite Materials

Processing

Edited by

Vijay Kumar Thakur

Manju Kumari Thakur

Raju Kumar Gupta

ELSEVIER

WP
WOODHEAD PUBLISHING
An imprint of Elsevier

Woodhead Publishing is an imprint of Elsevier
The Officers' Mess Business Centre, Royston Road, Duxford, CB22 4QH, United Kingdom
50 Hampshire Street, 5th Floor, Cambridge, MA 02139, United States
The Boulevard, Langford Lane, Kidlington, OX5 1GB, United Kingdom

British Library Cataloguing-in-Publication Data
A catalogue record for this book is available from the British Library

Library of Congress Cataloging-in-Publication Data
A catalog record for this book is available from the Library of Congress

ISBN: 978-0-08-100789-1 (print)
ISBN: 978-0-08-100790-7 (online)

For information on all Woodhead Publishing publications
visit our website at https://www.elsevier.com/books-and-journals

Publisher: Matthew Deans
Acquisition Editor: Gwen Jones
Editorial Project Manager: Charlotte Rowley
Production Project Manager: Debasish Ghosh
Cover Designer: Greg Harris

Typeset by MPS Limited, Chennai, India

Contents

List of Contributors

Mohammad Asim Laboratory of Biocomposite Technology, Institute of Tropical Forestry and Forest Products (INTROP), Universiti Putra Malaysia, 43400 Serdang, Selangor, Malaysia

Mohamed Bassyouni King Abdulaziz University, Rabigh, Saudi Arabia; Higher Technological Institute, Tenth of Ramdan City, Egypt

Juan-Ramón Campos-Cruz Instituto Tecnológico de Aguascalientes, Aguascalientes, Mexico

Diego Cazorla-Amorós University of Alicante, Alicante, Spain

Sibele P. Cestari Universidade Federal do Rio de Janeiro (UFRJ), Rio de Janeiro, RJ, Brazil

Mihai C. Corobea National Institute for Research & Development in Chemistry and Petrochemistry – ICECHIM Bucharest, Bucharest, Romania

Daniela de França da Silva Freitas Universidade Federal do Rio de Janeiro (UFRJ), Rio de Janeiro, RJ, Brazil

Saptarshi Dhibar Materials Science Centre, Indian Institute of Technology Kharagpur, Kharagpur, India

Dan Donescu National Institute for Research & Development in Chemistry and Petrochemistry – ICECHIM Bucharest, Bucharest, Romania

Atsushi Gabe University of Alicante, Alicante, Spain

Signorino Galvagno University of Messina, Messina, Italy

Johnsy George Defence Food Research Laboratory, Mysore, India

Marius Ghiurea National Institute for Research & Development in Chemistry and Petrochemistry – ICECHIM Bucharest, Bucharest, Romania

Pei Sean Goh Universiti Teknologi Malaysia (UTM), Skudai, Johor, Malaysia

Daniela Iannazzo University of Messina, Messina, Italy

Ahamad Fauzi Ismail Universiti Teknologi Malaysia (UTM), Skudai, Johor, Malaysia

José-Enrique Jaime-Leal Instituto Tecnológico de Aguascalientes, Aguascalientes, Mexico

Umair Javaid King Abdulaziz University, Rabigh, Saudi Arabia; National University of Science and Technology, NUST, Islamabad, Pakistan

Mohammad Jawaid Laboratory of Biocomposite Technology, Institute of Tropical Forestry and Forest Products (INTROP), Universiti Putra Malaysia, 43400 Serdang, Selangor, Malaysia; Chemical Engineering Department, College of Engineering, King Saud University, Riyadh, Saudi Arabia

Jun-ichi Kadokawa Kagoshima University, Kagoshima, Japan

Luis C. Mendes Universidade Federal do Rio de Janeiro (UFRJ), Rio de Janeiro, RJ, Brazil

Emilia Morallón University of Alicante, Alicante, Spain

María José Mostazo-López University of Alicante, Alicante, Spain

Prabir K. Mukherjee Government College of Engineering and Textile Technology, Serampore, India

Mohammad Nasir School of Industrial Technology, Universiti Sains Malaysia, 11800 Penang, Malaysia

Alessandro Pistone University of Messina, Messina, Italy

Muklesur Rahman Aliah University, Kolkata, India

Ramengmawii Laboratory of Biocomposite Technology, Institute of Tropical Forestry and Forest Products (INTROP), Universiti Putra Malaysia, 43400 Serdang, Selangor, Malaysia

Norma-Aurea Rangel-Vázquez Instituto Tecnológico de Aguascalientes, Aguascalientes, Mexico

Ricardo Rangel-Vázquez PCC. Real de Haciendas S/N, Aguascalientes, Mexico

Naheed Saba Laboratory of Biocomposite Technology, Institute of Tropical Forestry and Forest Products (INTROP), Universiti Putra Malaysia, 43400 Serdang, Selangor, Malaysia

S.N. Sabapathi Defence Food Research Laboratory, Mysore, India

Marina Salamò University of Messina, Messina, Italy

David Salinas-Torres University of Liège, Liège, Belgium

Santiranjan Shannigrahi Institute of Materials Research and Engineering, Singapore, Singapore

Himani Sharma Doon University Dehradun, Dehradun, Uttarakhand, India

Mohit Sharma Institute of Materials Research and Engineering, Singapore, Singapore

Siddaramaiah Sri Jayachamarajendra College of Engineering, Mysore, India

Catalin I. Spataru National Institute for Research & Development in Chemistry and Petrochemistry – ICECHIM Bucharest, Bucharest, Romania

Mohamed Thariq Hameed Sultan Aerospace Manufacturing Research Centre, Faculty of Engineering, Universiti Putra Malaysia, 43400 Serdang, Selangor, Malaysia

Syed W. ul Hasan King Abdulaziz University, Rabigh, Saudi Arabia; University of Engineering and Technology, Lahore, Pakistan

Kar Chun Wong Universiti Teknologi Malaysia (UTM), Skudai, Johor, Malaysia

Processing of hybrid polymer composites—a review

1

Mohammad Asim[1], *Mohammad Jawaid*[1,2], *Naheed Saba*[1], *Ramengmawii*[1], *Mohammad Nasir*[3] *and*
Mohamed Thariq Hameed Sultan[4]
[1]Laboratory of Biocomposite Technology, Institute of Tropical Forestry and Forest Products (INTROP), Universiti Putra Malaysia, 43400 Serdang, Selangor, Malaysia, [2]Chemical Engineering Department, College of Engineering, King Saud University, Riyadh, Saudi Arabia, [3]School of Industrial Technology, Universiti Sains Malaysia, 11800 Penang, Malaysia, [4]Aerospace Manufacturing Research Centre, Faculty of Engineering, Universiti Putra Malaysia, 43400 Serdang, Selangor, Malaysia

Chapter Outline

Hybrid Polymer Composite Materials: Processing. DOI: http://dx.doi.org/10.1016/B978-0-08-100789-1.00001-0

1.1 Introduction

The final cost and process of construction materials are rising every year due to a scarcity of unprocessed materials and the high cost of energy. Alternative constituents in wood materials for construction and nondegradable materials are now a global concern for energy saving, conservation of natural resources, and eco-systems. Worldwide efforts are being made to achieve sustainable economic growth and development of high-performance products with good properties [1]. A global shortage of solid wood materials due to environmental concerns are promoting consumption of fibrous materials; an improvement of composites and other various materials are also being considered [2,3]. Environmental pollution and shortage of energy resources are global challenges for finding substitute sources of renewable and sustainable biomass energy resources. Lignocellulosic materials are the most suitable and abundant bioresource in the world with annual production reaching up to 170–200 billion tons [4]. Due to its light weight and eco-friendly nature, natural fiber composites are the center of attraction for industries over traditional composites [5]. There are many sources of natural fibers like plant fibers, animal fibers, fruit fibers, and stalk fibers. Natural fibers, as a cemented material, have been used for several decades in many countries [6]. These natural fibers are now a major component in making of bio-composite materials like boards, paper, and many structures [7]. Nowadays, polymer technology is considered to be an advanced technology that is used because of environmental concerns; it is easy to process and is ideal for recycling products [8]. Polymers have unique characteristics (thermoset and thermoplastic) that make them compatibile with materials in all conditions such as various temperatures, varying densities, mechanical, and physical and thermal properties [9]. These polymers are very reliable and suitable for reinforcement with natural and synthetic (glass fiber, carbon, and aramid) fibers. These polymer composites are not used as component of automobiles, ships, and structural applications, but they do provide desirable properties such as high-strength to weight ratio, ease of fabrication, complex shapes, low cost, and good resistance to corrosion and marine fouling [10,11]. Synthetic fiber-reinforced composites are not degradable due to its high molecular mass and hydrophobic character, while natural fiber-reinforced polymer composites are hydrophilic low molecular mass products. Natural fiber-reinforced polymer composites have some disadvantages such as low strength, poor interfacial bonding between fiber and matrix, and moisture uptake compared to artificial fiber-reinforced composites. For long-term performance, natural fiber-reinforced polymer composites are very responsive to moisture absorption, and it affects fiber/matrix bonding [12]. Surface treatment of natural fibers is a method to reduce its hydrophilic character and enhance adhesive property of fiber/matrix.

Chemical treatments change natural fiber's surface morphology and eliminate some unwanted chemicals that attract to hydroxyl groups, a hurdle in fiber/matrix bonding [13,14]. Mechanical properties of treated fiber polymer composites and hybrid composites improved strength [15] over untreated fiber composites [16,17]. Various types of manufacturing processes for natural fiber-reinforced composites have been invented. Specifications for polymers and fibers are a deciding factor for the manufacturing process. Advance technologies, manufacturing processes, and innovative researches are used to get the high-strength engineering composites related to new applications area [18]. There is an increasing demand from many industries, such as automotive, building and construction, electrical and electronic industries, to promote natural fiber polymer composites thereby leading to a very competitive market [19,20].

1.2 Fibers

Natural fiber-reinforced polymer composites are attractive for advanced technologies because of their various advantages over conventional fibers. These advantages include low density, acceptable specific strength, less wear during processing, low cost, low energy, renewability, and biodegradability.

Synthetic composite materials such as aramid, carbon, and glass fiber-reinforced plastics, dominate the aerospace, leisure, automobile, construction, and sporting industries. Synthetic fibers are the most widely accepted reinforcement material due to their low cost and fairly good mechanical properties.

1.2.1 Natural fibers

The mainstream of the world's energy is provided by natural resources like petrochemical sources, coal, and natural gases. All of these energy sources and its products are finite [4] and uncertain, which necessitate replacing them with easily available raw materials and cheap sustainability. Currently, several types of natural fibers reinforced in thermoset and thermoplastic polymers have revealed several advantages: renewable resources which are light in weight and available at low cost- which is widly accepted in aerospace all of which are appealing from a marketing perspective. Many natural materials and wastes such as shell flour, wood flour, and pulp have been used as fillers in polymers [21]. Natural fibers can be extracted from many parts of plants such as bast stem, leaf, and seeds from the plants in a bundle form. Fiber classification is shown in Fig. 1.1 [22].

Many countries produce agricultural plants and fruits not only for food but also for generating raw materials for bio-composite industries. Table 1.1 shows the annual different types of natural fibers and its annual production from various sources [23]. Natural fibers are produced at a rate of about 30 million tons every year. Natural fibers revealed moderately high mechanical strength, and stiffness and can replaced with metals and used as reinforcing materials in polymeric matrices to make useful structural composites material.

1.2.1.1 Fiber treatment

The major hurdle in fiber composites is the compatibility of a fiber-matrix. Natural fibers show poor compatibility and relatively high moisture absorption during reinforcement

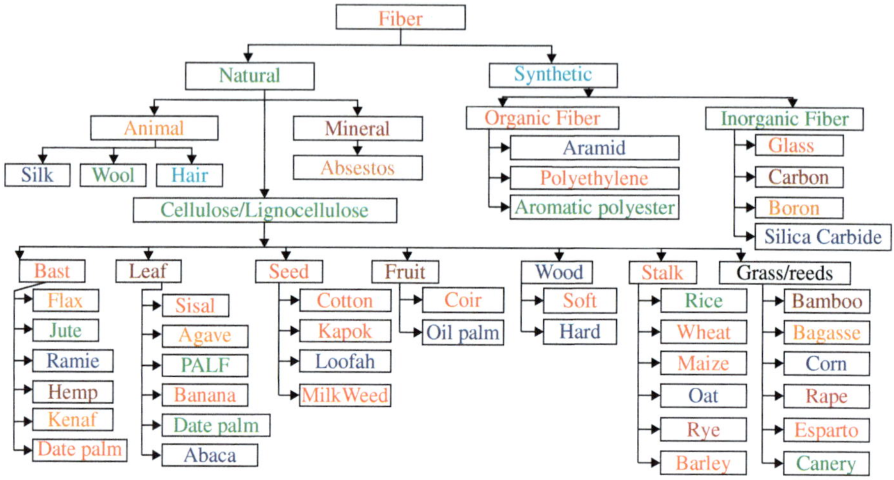

Figure 1.1 Classification and subclassification of fiber [22].

Table 1.1 Annual production of natural fibers and sources

Fiber source	World production (10^3 tons)	Origin	Fiber source	World production (10^3 tons)	Origin
Abaca	70	Stem	Nettles	Abundant	Stem
Bamboo	10,000	Stem	Oil palm fruit	Abundant	Fruit
Banana	200	Fruit	Palm rah	Abundant	Stem
Broom	Abundant	Stem	Ramie	100	Stem
Coir	100	Stem	Roselle	250	Stem
Cotton lint	18,500	Stem	Rice husk	Abundant	Fruit/ grain
Elephant grass	Abundant	Stem	Rice straw	Abundant	Stem
Flax	810	Stem	Sisal	380	Stem
Hemp	215	Stem	Sun hemp	70	Stem
Jute	2500	Stem	Wheat straw	Abundant	Stem
Kenaf	770	Stem	Wood	1,750,000	Stem
Linseed	Abundant	Fruit	Sugarcane bagasse	75,000	Stem
Pineapple	Abundant	Leaf	Cantala	–	Leaf
Caroa	–	Leaf	China jute	–	Stem

into polymer composites. Treatments of natural fibers modify the properties of a fiber surface to improve its adhesion with a different matrix. The mechanical strength of composites could be achieved with a strong interface, and the efficiency of stress transfer from the matrix to the fiber could be reduced with a weaker interface [24,18].

The hydrophilic nature of natural fiber is a main cause of weak bonding between fibres and matrix directly affect mechanical strength of polymer composites. Many other physical impurities and hydroxyl groups on the fiber surface are obstacles to making good enforcement materials [25]. Much research has been carried out on surface modification of natural fibers and its composites [26–29]. The mechanical properties of composites depend on the orientation and characteristics of fibers and matrix, and adhesion between fiber and matrix [30]. Weak bonding between fibers and matrix directly affect to mechanical strength [25].

Many types of chemical treatments effect the fiber surface to make better interfacial adhesion and better mechanical properties of composites. Some chemical surface treatments include sodium chlorite [31,32], metha acrylate [33], isocyanate, silane treatment [28], acetylation, mercerization, etherification, enzymatic treatment, peroxide treatments, benzoylation, dicumyl peroxide treatment, and plasma treatment [34].

Pineapple leaf fibers (PALF) and kenaf fibers have been treated with alkali, silane, and combined alkali and silane to study the effect of mechanical, morphological, and fiber/matrix adhesion. PALF and kenaf fiber treated with silane revealed excellent mechanical and fiber/matrix bonding with compare of untreated and alkali-treated fibers [28]. Surface treatment of hemp fibers were investigated as a means of improving interfacial shear strength (IFSS) of hemp fiber-reinforced polylactide (PLA) and unsaturated polyester (UPE) composites. Fibers were treated with sodium hydroxide, acetic anhydride, maleic anhydride, and silane. A combined treatment using sodium hydroxide and silane was also carried out. IFSS of PLA/hemp fiber samples increased after treatment, except in the case of maleic anhydride treatment. Increased IFSS could be explained by better bonding of PLA with treated fibers and increased PLA trans-crystallinity [35]. Alkali-treated and untreated natural rubber tested mechanical and morphological properties of composites. Treated composites showed higher mechanical, morphological properties and better interfacial interaction [36,37]. Hemp fibers treated with sodium hydroxide, acetic anhydride, maleic anhydride, and silane to improve interfacial bonding and to compare the effect of chemical treatments on hemp fiber. All chemical treatments effected fiber surface and improved interfacial strength of hemp fiber and matrix [35]. Banana fibers are treated with alkali, benzoyl chloride, $KMnO_4$, and triethoxy octyl silane (TEOS) to investigate thermal conductivity, thermal diffusivity, and specific heat. Treated banana fiber/polypropylene (PP) composites enhance in the thermo-physical properties while the benzoylated banana fiber composites showed the highest values of thermal conductivity and thermal diffusivity, and NaOH treatment has an influence on the thermo-physical properties [38].

1.2.2 Synthetic fiber

Synthetic fibers such as glass fibers, carbon fibers, and boron fiber are gaining more attention over last two decades. Natural fiber composites are having healthy competition with synthetic composites such as glass–polypropylene and glass–epoxies. Synthetic fiber-reinforced polymer composites provide high-strength and stiffness materials that are widely accepted as aerospace components and automotive industries [39]. Nowadays, natural fibers are used to replace synthetic fibers because of environmental concern as a reinforced material in polymer composites for engineering materials [40]. Hybrid composites of natural and synthetic fibers are often used to enhance the mechanical strength of polymer composites [21].

1.3 Polymer

Polymers are a combination of many smaller molecules. Units of these smaller molecules is a monomer, and a combination of monomers converts into polymer, a word taken from Greek, which means "many members." Cellulose, lignin, starch, and natural rubber are best examples of natural polymers. Natural polymers began to chemically modify into many other products such as gun cotton, vulcanized rubber, and celluloid. The chemical reaction by which polymers are synthesized from monomers is termed polymerization; however, this is a generic term because there are a number of chemical mechanisms involved in different polymerization reactions. Synthetic polymers, as modern polymers, helped improve technology and applied sciences during the early 20th century. Synthetic polymers are developed by chemical reactions and applied in building constructions in addition to other uses. The usefulness of synthetic polymers, which have different behaviors, characteristics, and applications, are endless and are only restricted by some chemicals and thermodynamic laws—and by the creativity of chemists.

1.3.1 Thermoset

Thermoset reins are very flexible for desired properties industries such as their high-modulus, strength, durability, and thermal and chemical resistances because of high cross-linking density [41,42]. As a result, thermosetting resins have a very low impact resistance and cannot be reshaped after curing/polymerization [43]. There are several thermosets available such as polyesters, phenol-formaldehyde resins, epoxy resins, and polyurethanes [44]. In addition, thermosets are used to enhance high performance and impact resistance for industrial applications, such as building materials and transportation [45]. Various types of fillers such as natural fiber and synthetic fibers are often added to the resin to form composite materials. Composite preparation of thermoset polymers are usually based on manual lay-up and spray-up as techniques [9].

1.3.2 Thermoplastic

Thermoplastics are synthesized from plants in large amounts and transformed through chemical processing. Some of the most important thermoplastics are polyethylene [low density polyethylene (LDPE) and high density polyethylene

(HDPE)], polypropylene (PP), poly(vinyl chloride) (PVC) and polystyrene [44]. These polymers can be used in many possible applications structural purposes such as wire and light duty utilities. Thermoplastic polymers are also being used as a matrix for natural and synthetic fibers [46]. Thermoplastic polymers can melt at specific temperatures and can be shaped and reshaped (through reheating) according to the mold. Reprocessing thermoplastic polymers can lose its physical properties due to a breakage of polymeric chains; it is best to not recycle thermoplastics.

1.4 Polymer composites

In recent decades, composite materials have increased their popularity in many sectors because of their versatile uses, ease of preparation, availability of raw materials, eco-friendliness, improved mechanical for desired products, low density, and low cost. Most composite materials are composed of a polymeric matrix. Varieties of research based on reinforcement of natural fibers such as date palm, rubber wood [47], coir, oil palm empty fruit bunch (OPEFB), ramie, sugar palm, bamboo, hemp, flax, sisal, jute, pineapple leaves fibers, kenaf, etc., have been investigated [23,48]. Sugar palm fibers/phenolic composites have been studied to investigate mechanical properties and compatibility of fiber-matrix bonding. The composites exhibit very good mechanical properties at a 30 percent volume and fiber/matrix adhesion [49]. Carbon fiber/phenolic resin composites were investigated and an improvement in thermal conductivity, after adding highly crystalline multi-walled carbon nanotubes, was found. Phenolic resin worked as a thermal connection between adjacent carbon fibers and resulted in an improvement of the thermal conductivity [50]. Silane-treated, cellulose fiber-reinforced phenolic composites improved its fiber-matrix adhesion and mechanical properties such as tensile and flexural property [51]. Jute fiber/epoxy composite treated with alkali and silane showed a very good fiber/matrix interfacial strength. Single fiber pull-out tests were examined with scanning electron microscopy (SEM) and atomic force microscopes (AFM) characterization of the fracture samples [52]. An alkali-treated woven flax fiber/epoxy composite revealed improved fibers/matrix adhesion [48]. Hemp fiber-reinforced UPE composites were studied to evaluate physical and mechanical properties. The percentage of moisture content was increased as the fiber volume fraction enhanced and mechanical strength was decreased due to moisture content [53]. A treated bagasse fiber-reinforced UPE composite revealed good physical, mechanical properties, storage modulus, and fiber-matrix interaction [54]. Chen et al. [55]. studied the interface of the bamboo/vinyl ester composite and effect of water absorption on IFSS of composite. The carbon/vinyl ester composites have superior properties, higher specific strength, and specific modulus than the marine steel whereas glass/vinyl ester composites have a higher specific strength but lower specific modulus than marine steel [56]. Bamboo fiber-reinforced polyester composites developed a composite material of high-strength and light weight applications [57]. Different properties such as water absorption and mechanical properties of bamboo fiber-reinforced polyester composites have been reported [58]. Alkali-treated ramie/poly (lactic acid) laminated composites enhance mechanical properties [16]. Modified TiO_2 grafted flax fiber-reinforced PLA composites were investigated, and a significant improvement in mechanical, physical, and dynamic mechanical analysis of treated composites [59] was found. A composite of kenaf fibers/PLA showed very good mechanical strength and also investigated biodegradability in moist condition [60].

1.5 Hybrid composites

The reinforcement of two or more natural fibers into a single matrix develops a hybrid composite. Many researchers have studied to select the best combination of natural fibers to achieve the best outcome for utilization and to minimize the negative aspects [21,61]. A combination of different types of fibers in a single matrix can generate high-valued hybrid biocomposites. Basically, three types of reinforcement methods have been incorporated: (1) a proper mixture of two types of short fibers before adding a matrix or adding fibers in to polymer alternatively [62]; (2) layering of fiber mat or fabric and matrix [63], and (3) in the case of glass fiber-LC fiber composite systems, the addition of nonwoven and woven fabric as reinforcements [64,22].

Carbon nanotube and carbon fiber/phenolic hybrid composites were investigated for mechanical, ablation properties, and thermal conductivity. Carbon materials as reinforcements improved properties significantly [65]. Research [66] shows the merits of combining high-modulus glass fibers with banana fibers in phenolic resoles to develop high-performance, cost-effective, light weight hybrid composites; the hybrid composites showed high mechanical and physical performance. Jute and oil palm fibers reinforced epoxy hybrid composite prepared by hand lay-up technique. A hybrid composite was examined for its fiber/matrix interface, mechanical strength, and dynamic mechanical analysis [61]. Kenaf/aramid/epoxy hybrid composites were examined for its mechanical properties for automobile application. It showed very good tensile strength and impact resistance properties [67]. Glass–carbon/epoxy composites were investigated for the range of strain during impact loadings and mechanical strength for aircraft [68]. Nano filler/kenaf/epoxy based hybrid nanocomposites was prepared by using a hand lay-up technique. Three percent nano fibers improved a hybrid composite of mechanical and morphological properties [69]. Jute/glass fiber-reinforced UPE hybrid composites were studied for its mechanical and physical properties [70]. Flax, hemp, and glass fiber-reinforced polyurethane (PU) hybrid composites were investigated to achieve a better mechanical composite. Physical and mechanical properties of hybrid composites were very high compared to its composite alone [71]. Montmorillonite/rice husk hybrid filler–filled low-density polyethylene nanocomposite films were prepared by extrusion blown film. Morphological, mechanical, oxygen (O_2) barrier, and thermal properties [72] were investigated. A kenaf fiber/coir fiber/polypropylene/

montmorillonite nanoclay hybrid composite was prepared through a hot compression method, and physical, mechanical and biodegradable properties were investigated [73].

1.6 Parameters of processing methods

Manufacturing processes of composites are reported to be used in the development of biodegradable products from natural fiber composites. Ho et al. [74] discussed manufacturing process and methods of hybrid biocomposites related to natural fiber composites. Most of the processing methods of natural fiber composites are related to conventional processing methods such as pultrusion, injection molding, compression molding, vacuum infusion molding, hot press processes, and resin transfer molding (RTM).

1.6.1 Pultrusion

A combination of pull and extrusion process is known as a pultrusion process. The mechanism of a pultrusion process is very much similar to an extrusion process, but in pultrusion, materials are poured into molds for developing products. Thermosetting polymers can also be used in this process. Impregnated fiber composites pass through a polymer bath chamber and cross many types of mold. The shape of the final product depends on the final mold; it can be either shape circular, rectangular, square, and I-shaped or H-shaped. Final bio-composite products are used to cure inside the mold.

1.6.2 Filament winding

Filament winding is an open mold process, and the mold is a rotating mandrel. It is composite manufacturing process that produces its products circular in shape. In this procedure, natural fibers first pass through the resin bath chamber then the impregnated composite move towards the rotating mandrel. Equal distribution in a mandrel can be achieved. In this process, winding patterns consist of three types: hoop, helical, and polar.

1.6.3 Hand lay-up

Hand lay-up method is basically known for open molding processing to develop polymer composite products. It requires a high-skilled operator to operate the fabrication task skillfully. Before fabrication, molds are required to spray some mold-release agent for ease of handling composites and to produce a smooth surface. The density of the acquired composite should be calculated before mixing the fiber and matrix; it helps to use an optimum amount of raw material. Small-size fibers, fiber mats, and fabric can be use in the technique. Thermosets and thermoplastic are both suitable for hand lay-up techniques such as UPE, polyurethane, phenolic, vinyl ester, PLA, and epoxy. A hand lay-up is also known as a wet lay-up. Automobile companies resort to a hand lay-up technique when using natural fiber composites for components.

1.6.4 Resin transfer molding

RTM is a well-established composite manufacturing technique that is very much suitable for fabricating composites for automotive and aircraft components. In this technique, fiber should be cut into precise sizes by a knife or scissors. In this process the resin is injected separately onto a bed of stationary platform. This process is very popular due to the capacity of high production and it is cheap to produce. Many studies have been done using RTM.

In the RTM process, oven-dry fiber is placed in a mold, and two mold plates are clamped tightly to avoid leakage of resin during injection process. By using dispensing-equipment resin or a polymer injected into the mold using single or multiple inlet ports, the mold depends on the complexity of the shape of a final product until the mold is filled with resin. Once the resin is cooled, composites are used to remove it from the mold. Post-curing is also required to ensure the resin is fully cured.

1.6.5 Vacuum bagging

Prepreg molding is classified in to two process: vacuum bagging and autoclave molding. The difference between these two processes concerns the method of matrix curing: first in vacuumed bag and second in an autoclave. As a composite manufacturing process, vacuum bagging provides compaction pressure and consolidation of plies within the laminate. It is also known as an extended version of hand lay-up process and vacuum bag molding. In this process, the mold base is kept horizontally, and the prepreg composites are placed in equal distribution into the mold base. It is then horizontally layered and placed over the composites. Finally all the composite laminates are covered by a vacuum bag and sealed properly using sealers. Many types of polymer materials can be used including epoxy, phenolic, and polyimide [75].

1.6.6 Compression molding

Compression molding is the most common thermoset and thermoplastic polymer composite manufacturing process. It is normally used to produce composite components in high production volume such as automotive components. Basically, there are two types of compression molding processes: cold and hot.

For thermoset polymers, intermediate semi-cured composite material called a molding compound is used. Bulk molding compound (BMC) and sheet molding compound (SMC) are two widely used thermosetting-based molding compounds.

For thermoplastics polymers, a glass mat thermoplastic (GMT) is normally used as molding compound. In the laboratory scale compression molding, composite materials are mixed in an internal mixer and twin screw extruder for homogeneous distribution.

For the two types of the compression molding processes, only pressure is applied for the cold press technique, but in the hot press technique pressure and temperature both are required. Cold pressure technique requires room temperature for curing process, while a hot press technique applies temperature on the mold; it then transfers the heat to the composite and initiates the curing process [76].

1.6.7 Injection molding

Injection molding is one of the most famous and widely used manufacturing technique to produce plastic components. Injection molding can also produce natural fiber composites but the size of the fiber should be very small: in particle or in powder form. An injection molding compound is required to prepare in granules form by using a twin screw extruder. The granules then go to mold through a hopper and a heated barrel. Granules are heated in a barrel. All the melted materials are then injected into the mold to make the composite.

1.7 Advantage and disadvantage of processing methods

There are several important advantages and disadvantages in processing natural fiber-reinforced composites during composite manufacturing process. This mainly has to do with the physical, mechanical, and thermal properties. These properties also depend on fiber selection, surface modification of fiber, polymer selection, and the processing method [77].

1.7.1 Resin transfer molding (RTM)

The RTM manufacturing process and fabrication plays a very important role in the quality and performance of products. A composite having a good surface provides better mechanical properties. Some composites that have foreign materials such as dust, dirt, or particles intensify the micro-cracks, and poor fiber-matrix bonding causes lower mechanical properties. The process and response of materials decide the connection of process and implementation such as visco-elastic behavior, permeability, and impregnation [78].

This process is efficient for providing very large and complex shapes and design flexibility. It saves on labor costs, and fast production and products are smooth and dimension tolerant. A disadvantage of this technique is that it requires a professional for tooling design and construction. Reinforcement of materials loading can be difficult with complex parts. The tools are very complex and costly, so it can be difficult to understand properly.

1.7.2 Compression molding

Compression molding is very high-pressure, high-volume, plastic molding technique designed for high-strength products. Most industries have chosen this process because of the many advantages during the process such as low cost, a short cycle time, high-volume production, dimensional accuracy, improved impact strength, and uniform shrinkage. It provides uniform flow and uniform density. Natural fiber-reinforced polymer composites with compression molding application have been studied [79,80].

1.7.3 Injection molding

Machining activities such as screw speed and the volumetric flow rate can be calculated using an injection molding technique. Wastage of materials is very low: a negligible leakage flow from the valve as well as screw speed are constant throughout the process. The screw, installed inside the electric injection molding machine, is empowered, and the servomotor controls it precisely. The servomotor also provides accuracy in rheological measurements [74].

Some problems related to dimension accuracy and its mechanical properties and optical properties can be connected with the residual stress in a part [81,82]. These problems come due to some residual stress that can be flow induced or thermally induced. Warpage can take place due to thermally induced residual stresses. Environmentally, stress crack and shrinkage take place due to inhomogeneous cooling of a part, which may lead to stress unevenly distributed in the part and in turn may lead to part warpage [83]. These problems generally occur due to a thin wall, so this problem can be minimized through the proper design of parts and innovation in cooling and molding technologies [84,85].

1.7.4 Hand lay-up

Hand lay-up technique is a very simple and open mold low volume production method that used is by many automobile and aircraft industries. The hand-layout technique required is a one-sided, cost-effective infiltration of resin through the reinforcement fabric over the surface of the mold by a hand lay-up technique that is quite similar to other manufacturing techniques of composites. The drawback of the hand lay-up process is interfacial bonding of fiber and matrix that can affect the mechanical properties of composites.

1.7.5 Common disadvantage of natural fiber composites

Natural fibers are hydrophilic in nature due to having hydroxyl groups that also have a microbial degradation character. These fibers are sometimes not compatible with a hydrophobic matrix and can cause a bad interfacial bonding. There are some suggestions to reduce the hydrophilic character of fiber through surface modification [86–88]. Interfacial bonding in a composite plays a very important role to transferring stress through the fibers to matrix. Treated surface fiber with many different functionalized monomers can enhance the interfacial bonding.

Homogeneous distribution of natural fiber provides mechanical strength to the composites; bad dispersion of fiber can destroy the structure and its strength. The size of natural fiber and surface treatment also plays a very important role in dispersion. Thermal degradation is a very important concern for the strength of composites. Lignocellulosic materials change its chemical and mechanical behavior when it comes in contact to temperature range of 100°C–250°C [77].

1.8 Applications

Hybrid polymer composites have been commercialized by using a wide range of applications and developments of new products. There are many applications of polymer composite based on products and components. Even today, after polymer composites have nearly completed its century of development, the technical potential of thermosets and thermoplastic development is far from exhausted.

1.8.1 Application of hybrid polymer composites

There's a developing trend to use natural fiber-reinforced composites in automobile components for both interior and exterior purposes [89]. These composites are very light in weight ratio and having very good mechanical strength. Lightweight vehicles enhance fuel efficiency and reduce greenhouse gas emission, reduction in vehicle weight up to 25% save 250 million barrels of crude oil and in CO_2 emissions of 220 billion pounds per year [90,91]. Among the polymers, poly propylene used for automotive industries components due to its lesser density, better mechanical properties, and easier in manufacturing process. Other polymers such as polyethylene, polystyrene, and polyamides were also used in automotive industries. Hybrid polymer composites are very useful in aircraft, furniture, and construction purposes, as shown in Fig. 1.2. Glass and kenaf fiber–reinforced, polypropylene-hybrid composites are manufactured for automotive brake levers. The design of automotive brake levers was made by the cooperation of automotive component vendors, automotive component manufacturers, and composite material experts. Hybrid glass-kenaf fiber–reinforced polymer composite are also designed for composite automotive bumper beams [92].

Glass–sugar palm fiber-reinforced UPE composites are used for small boats. Boat designers work in a team comprising a designer, material experts, and boat manufacturers. Kenaf and glass-reinforced UPE resin-based hybrid composites were developed and characterized successfully for structural applications [93]. Natural fiber hybrid composites have also been used as fire-resistant materials. Porous microstructure of natural fibers provides fire-resistant properties [94,95].

1.8.2 Application of each processing method

1.8.2.1 Hand lay-up

Hand lay-up is one of the presently proven and established low cost and effective processes for widely constructing turbine blades composites and further extensive

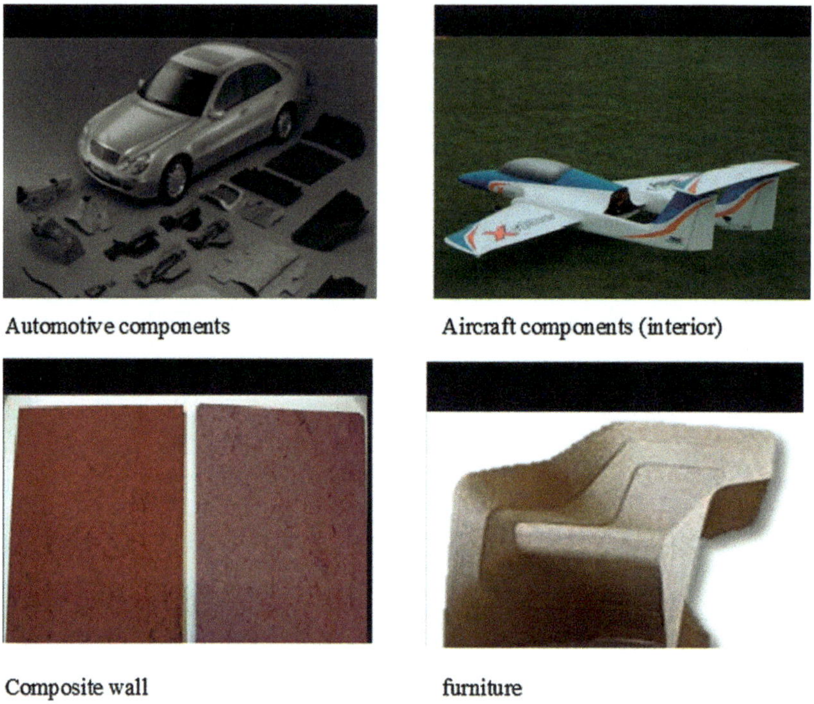

Figure 1.2 Various applications of hybrid polymer composites [23].

structural products such as airframe components, boats, truck and car body components, tanks, bath wares, housing, swimming pools, ducts decks and hulls for leisure boats, wind turbine blades, architectural moldings, etc. [96].

1.8.2.2 Compression molding

The compression molding process is one of the supreme techniques used for molding the thermoset and thermoplastic polymers. It helps to develop variety of useful plastic products such as table top, door stopper, tray, plastic rubber bands, construction materials, popular cap, and jar cover designs of different sizes. Compression moldings are widely used in the development of automotive applications through high-strength sheet molding [97].

1.8.2.3 Injection molding

Injection molding has been undertaken by worldwide researchers to expand industrial applications and to continuously improve the production of high-quality natural fiber composites [92]. A mobile case made by using inject molding by NEC Corporation [98]. Inject molding is very suitable to make electronic gadgets such as thin-walled technology, which is expected to expand for electronic applications. One smart phone company called Xiaomi explored electronic packing applications with thin-walled casings produced for manufacturing its ultra-slim Mi2A device [85].

1.8.2.4 Solvent casting

The applications of natural fiber-reinforced composites are far beyond the automotive industries. There are many eco-friendly products trending in the marketplace. There are many applications especially in the housing sector such as roofing panels and doors and windows that have thermal and acoustic insulation properties [99].

1.9 Conclusion

From this study, it can be concluded that it is possible to carry out a manufacturing processing study for hybrid polymer composites. Natural fiber and synthetic fiber-reinforced polymer composites used to make hybrid composites with different manufacturing processes leads to the success of the manufacturing of products from natural fiber composites. Studies of thermoset and thermoplastic polymers and related fibers with different manufacturing processes found advantages and disadvantages the during production of hybrid composites. Manufacturing processes are supported to select for automotive components, aircraft applications, house appliances, and electronic gadgets. Hybrid polymer composites have resulted in very important contributions in modern technology and related developments.

References

[1] H.P.S. Abdul Khalid, H.M. Fizree, M. Jawaid, O.S. Alattas, Preparation and characterization of nano- structured materials from oil palm ash: a bio-agricultural waste from oil palm mill, BioResources 6 (2011) 4537–4546.

[2] A. Ashori, A. Nourbakhsh, Bio-based composites from waste agricultural residues, Waste Manag. 30 (2010) 680–684.

[3] C.K. Abdullah, M. Jawaid, H.P.S. Abdul Khalil, A. Zaidon, A. Hadiyane, Oil palm trunk polymer composite: morphology, water absorption, and thickness swelling behaviours, BioResources 7 (2012) 2948–2959.

[4] N. Saba, M. Jawaid, K.R. Hakeem, M.T. Paridah, A. Khalina, O.Y. Alothman, Potential of bioenergy production from industrial kenaf (Hibiscus cannabinus L.) based on Malaysian perspective, Renew. Sustain. Energy Rev. 42 (2015) 446–459. Available from: http://dx.doi.org/10.1016/j.rser.2014.10.029.

[5] L.T. Drzal, A.K. Mohanty, M. Misra, Bio-composite materials as alternatives to petroleum-based composites for automotive applications, Magnesium 40 (2001) 1–3.

[6] H. Savastano, P.G. Warden, R.S.P. Coutts, Brazilian waste fibre as reinforcement for cement-based composites, Cem. Concr. Compos. 22 (2000) 379–384, <http://dx.doi.org/10.1016/S0958-9465(00)00034-2>.

[7] V.D. Shinde, U.S. Bongarde, Review on natural fiber reinforcement polymer composites, Int. J. Eng. Sci. Innov. Technol. 3 (2014) 431–436.

[8] A.K. Mohanty, M. Misra, G. Hinrichsen, Biofibre, biodegradable polymers and biocomposites: an overview, Macromol. Mater. Eng. 276–277 (2000) 1–24, <http://dx.doi.org/10.1002/(SICI)1439-2054(20000301)276:1 < 1::AID-MAME1 > 3.0.CO;2-W>.
[9] J.M. Raquez, M. Deléglise, M.F. Lacrampe, P. Krawczak, Thermosetting (bio)materials derived from renewable resources: a critical review, Prog. Polym. Sci. 35 (2010) 487–509. Available from: http://dx.doi.org/10.1016/j.progpolymsci.2010.01.001.
[10] L. Yan, N. Chouw, K. Jayaraman, Flax fibre and its composites – a review, Compos, Part B Eng 56 (2014) 296–317. Available from: http://dx.doi.org/10.1016/j.compositesb.2013.08.014.
[11] I. Pillin, A. Kervoelen, A. Bourmaud, J. Goimard, N. Montrelay, C. Baley, Could oleaginous flax fibers be used as reinforcement for polymers? Ind. Crops Prod. 34 (2011) 1556–1563. Available from: http://dx.doi.org/10.1016/j.indcrop.2011.05.016.
[12] A. Le Duigou, P. Davies, C. Baley, Exploring durability of interfaces in flax fibrefiber/epoxy micro-composites, Compos. Part A Appl. Sci. Manuf 48 (2013) 121–128. Available from: http://dx.doi.org/10.1016/j.compositesa.2013.01.010.
[13] M. Rashid, S.A. Samad, M.A. Gafur, R. Qadir, A.M.S. Chowdhury, Effect of Reinforcement of Hydrophobic Grade Banana (Musa ornata) Bark Fiber on the Physicomechanical Properties of Isotactic Polypropylene, 2016 (2016).
[14] M.M. Kabir, H. Wang, K.T. Lau, F. Cardona, Chemical treatments on plant-based natural fibrefiber reinforced polymer composites: an overview, Compos. Part B Eng. 43 (2012) 2883–2892. Available from: http://dx.doi.org/10.1016/j.compositesb.2012.04.053.
[15] Y. Li, K.L. Pickering, R.L. Farrell, Analysis of green hemp fibrefiber reinforced composites using bag retting and white rot fungal treatments, Ind. Crops Prod. 29 (2009) 420–426. Available from: http://dx.doi.org/10.1016/j.indcrop.2008.08.005.
[16] J. Yang, L. Zhu, Z. Yang, L. Yao, Y. Qiu, Improving mechanical properties of ramie/poly (lactic acid) composites by synergistic effect of fabric cyclic loading and alkali treatment, J. Ind. Text. (2016). Available from: http://dx.doi.org/10.1177/1528083716648763.
[17] N. Saba, M. Jawaid, Mechanical properties of kenaf fiber reinforced polymer composite: a review, Constr. Build. Mater. 76 (2015) 87–96.
[18] O. Faruk, A.K. Bledzki, H.-P.P. Fink, M. Sain, Biocomposites reinforced with natural fibers: 2000–2010, Prog. Polym. Sci. 37 (2012) 1552–1596. Available from: http://dx.doi.org/10.1016/j.progpolymsci.2012.04.003.
[19] L. Mohammed, M.N.M. Ansari, G. Pua, M. Jawaid, M.S. Islam, A review on natural fiber reinforced polymer composite and its applications, Int. J. Polym. Sci. 2015 (2015). Available from: http://dx.doi.org/10.1155/2015/243947.
[20] T. Väisänen, A. Haapala, R. Lappalainen, L. Tomppo, Utilization of agricultural and forest industry waste and residues in natural fiber-polymer composites: a review, Waste Manag. (2016). Available from: http://dx.doi.org/10.1016/j.wasman.2016.04.037.
[21] M. Jawaid, H.P.S. Abdul Khalil, Cellulosic/synthetic fibre reinforced polymer hybrid composites: a review, Carbohydr. Polym. 86 (2011) 1–18. Available from: http://dx.doi.org/10.1016/j.carbpol.2011.04.043.
[22] N. Saba, P. Tahir, M. Jawaid, A review on potentiality of nano filler/natural fiber filled polymer hybrid composites, Polymers (Basel), 2014, pp. 2247–2273. Available from: http://dx.doi.org/10.3390/polym6082247.
[23] M. Asim, K. Abdan, M. Jawaid, M. Nasir, Z. Dashtizadeh, M.R. Ishak, et al., A review on pineapple leaves fibre and its composites, Int. J. Polym. Sci. (2015) 1–17.

[24] M. Nasir, A. Gupta, M.D.H. Beg, G.K. Chua, M. Jawaid, A. Kumar, et al., Fabricating eco-friendly binderless fiberboard from laccase-treated rubber wood fiber, Bioresources. 8 (2013) 3599–3608.

[25] E. Terpáková, L. Kidalová, A. Eštoková, J. Čigášová, N. Števulová, Chemical modification of hemp shives and their characterization, Proc. Eng 42 (2012) 931–941. Available from: http://dx.doi.org/10.1016/j.proeng.2012.07.486.

[26] M.S. Meon, M.F. Othman, H. Husain, M.F. Remeli, M.S.M. Syawal, Improving tensile properties of kenaf fibers treated with sodium hydroxide, Proc. Eng 41 (2012) 1587–1592. Available from: http://dx.doi.org/10.1016/j.proeng.2012.07.354.

[27] D. Puglia, M. Monti, C. Santulli, F. Sarasini, I.M. De Rosa, J.M. Kenny, Effect of alkali and silane treatments on mechanical and thermal behavior of Phormium tenax fibers, Fibers Polym. 14 (2013) 423–427. Available from: http://dx.doi.org/10.1007/s12221-013-0423-x.

[28] Mohammad Asim, Mohammad Jawaid, Khalina Abdan, M.R. Ishak, Effect of alkali and silane treatments on mechanical and fibre-matrix bond strength of kenaf and pineapple leaf fibre, J. Bionic Eng. 13 (2016) 426–435.

[29] R. Yahaya, S.M. Sapuan, M. Jawaid, Z. Leman, E.S. Zainudin, Effect of layering sequence and chemical treatment on the mechanical properties of woven kenaf–aramid hybrid laminated composites, Mater. Des. 67 (2015) 173–179. Available from: http://dx.doi.org/10.1016/j.matdes.2014.11.024.

[30] S. Huo, a Thapa, C. a Ulven, Effect of surface treatments on interfacial properties of flax fiber-reinforced composites, Adv. Compos. Mater. 22 (2013) 109–121. Available from: http://dx.doi.org/10.1080/09243046.2013.777996.

[31] G.M. Arifuzzaman Khan, M. Shaheruzzaman, M.H. Rahman, S.M. Abdur Razzaque, M.S. Islam, M.S. Alam, Surface modification of okra bast fiber and its physico-chemical characteristics, Fibers Polym. 10 (2009) 65–70. Available from: http://dx.doi.org/10.1007/s12221-009-0065-1.

[32] A.M.M. Edeerozey, H.M. Akil, A.B. Azhar, M.I.Z. Ariffin, Chemical modification of kenaf fibers, Mater. Lett. 61 (2007) 2023–2025. Available from: http://dx.doi.org/10.1016/j.matlet.2006.08.006.

[33] G.S. Solnit, The effect of methyl methacrylate reinforcement with silane-treated and untreated glass fibers, J. Prosthet. Dent. 66 (1991) 310–314, <http://dx.doi.org/10.1016/0022-3913(91)90255-U>.

[34] A. Azam, S. Khubab, Y. Nawab, J. Madiha, T. Hussain, Hydrophobic treatment of natural fibers and their composites – a review, J. Ind. Text. Accepted (2016) 1–31. Available from: http://dx.doi.org/10.1177/1528083716654468.

[35] M.A. Sawpan, K.L. Pickering, A. Fernyhough, Effect of fibre treatments on interfacial shear strength of hemp fibre reinforced polylactide and unsaturated polyester composites, Compos. Part A Appl. Sci. Manuf. 42 (2011) 1189–1196. Available from: http://dx.doi.org/10.1016/j.compositesa.2011.05.003.

[36] M. Nasir, R. Hashim, O. Sulaiman, A. Nordin, J. Lamaming, M. Asim, Laccase, an emerging tool to fabricate green composites: a review, BioResources 10 (2015) 1–23.

[37] H. Ismail, A.M. Norjulia, Z. Ahmad, The effects of untreated and treated kenaf loading on the properties of kenaf fibre-filled natural rubber compounds, Polym. Plast. Technol. Eng. 49 (2010) 519–524. Available from: http://dx.doi.org/10.1080/03602550903283117.

[38] S. Annie Paul, A. Boudenne, L. Ibos, Y. Candau, K. Joseph, S. Thomas, Effect of fiber loading and chemical treatments on thermophysical properties of banana fiber/polypropylene commingled composite materials, Compos. Part A Appl. Sci. Manuf. 39 (2008) 1582–1588. Available from: http://dx.doi.org/10.1016/j.compositesa.2008.06.004.

[39] L. Yan, N. Chouw, X. Yuan, Improving the mechanical properties of natural fibrefiber fabric reinforced epoxy composites by alkali treatment, J. Reinf. Plast. Compos. 31 (2012) 425–437. Available from: http://dx.doi.org/10.1177/0731684412439494.

[40] M. Assarar, D. Scida, A. El Mahi, C. Poilâne, R. Ayad, Influence of water ageing on mechanical properties and damage events of two reinforced composite materials: flax-fibrefibers and glass-fibre, Mater. Des. 32 (2011) 788–795. Available from: http://dx.doi.org/10.1016/j.matdes.2010.07.024.

[41] J. Guilleminot, S. Comas-Cardona, D. Kondo, C. Binetruy, P. Krawczak, Multiscale modelling of the composite reinforced foam core of a 3D sandwich structure, Compos. Sci. Technol. 68 (2008) 1777–1786. Available from: http://dx.doi.org/10.1016/j.compscitech.2008.02.005.

[42] B. Gourichon, M. Deléglise, C. Binetruy, P. Krawczak, Dynamic void content prediction during radial injection in liquid composite molding, Compos. Part A Appl. Sci. Manuf 39 (2008) 46–55. Available from: http://dx.doi.org/10.1016/j.compositesa.2007.09.008.

[43] S. Comas-Cardona, P. Groenenboom, C. Binetruy, P. Krawczak, A generic mixed FE-SPH method to address hydro-mechanical coupling in liquid composite moulding processes, Compos. Part A Appl. Sci. Manuf 36 (2005) 1004–1010. Available from: http://dx.doi.org/10.1016/j.compositesa.2004.11.009.

[44] E. Saldívar-Guerra, E. Vivaldo-Lima, Handbook of polymer synthesis, characterization, and processing, 2013. <http://dx.doi.org.10.1002/9781118480793>.

[45] M. Deléglise, C. Binétruy, P. Krawczak, Solution to filling time prediction issues for constant pressure driven injection in RTM, Compos. Part A Appl. Sci. Manuf. 36 (2005) 339–344. Available from: http://dx.doi.org/10.1016/j.compositesa.2004.07.001.

[46] K. Van De Velde, P. Kiekens, Thermoplastic polymers: overview of several properties and their consequences in flax fibre reinforced composites, Polym. Test. 20 (2001) 885–893, <http://dx.doi.org/10.1016/S0142-9418(01)00017-4>

[47] M. Nasir, A. Gupta, M.D. Hossen Beg, G.K. Chua, M. Asim, Laccase application in medium density fibreboard to prepare a bio-composite, RSC Adv. 4 (2014) 11520. Available from: http://dx.doi.org/10.1039/c3ra40593a.

[48] N. Saba, M. Jawaid, O.Y. Alothman, M. Paridah, A. Hassan, Recent advances in epoxy resin, natural fiber-reinforced epoxy composites and their applications, J. Reinf. Plast. Compos. (2015), 0731684415618459. <http://dx.doi.org/10.1177/0731684415618459>

[49] B. Rashid, Z. Leman, M. Jawaid, M.J. Ghazali, M.R. Ishak, The mechanical performance of sugar palm fibres (ijuk) reinforced phenolic composites, Int. J. Precis. Eng. Manuf. 17 (2016) 1001–1008. Available from: http://dx.doi.org/10.1007/s12541-016-0122-9.

[50] Y.A. Kim, S. Kamio, T. Tajiri, T. Hayashi, S.M. Song, M. Endo, et al., Enhanced thermal conductivity of carbon fiber/phenolic resin composites by the introduction of carbon nanotubes, Appl. Phys. Lett. 90 (2007). Available from: http://dx.doi.org/10.1063/1.2710778.

[51] E. Rojo, M.V. Alonso, M. Oliet, B. Del Saz-Orozco, F. Rodriguez, Effect of fiber loading on the properties of treated cellulose fiber-reinforced phenolic composites, Compos. Part B Eng. 68 (2015) 185–192. Available from: http://dx.doi.org/10.1016/j.compositesb.2014.08.047.

[52] T.T.L. Doan, H. Brodowsky, E. Mäder, Jute fibre/epoxy composites: surface properties and interfacial adhesion, Compos. Sci. Technol. 72 (2012) 1160–1166. Available from: http://dx.doi.org/10.1016/j.compscitech.2012.03.025.

[53] H.N. Dhakal, Z.Y. Zhang, M.O.W. Richardson, Effect of water absorption on the mechanical properties of hemp fibre reinforced unsaturated polyester composites, Compos. Sci. Technol. 67 (2007) 1674–1683. Available from: http://dx.doi.org/10.1016/j.compscitech.2006.06.019.
[54] V. Vilay, M. Mariatti, R. Mat Taib, M. Todo, Effect of fiber surface treatment and fiber loading on the properties of bagasse fiber-reinforced unsaturated polyester composites, Compos. Sci. Technol. 68 (2008) 631–638. Available from: http://dx.doi.org/10.1016/j.compscitech.2007.10.005.
[55] H. Chen, M. Miao, X. Ding, Influence of moisture absorption on the interfacial strength of bamboo/vinyl ester composites, Compos. Part A Appl. Sci. Manuf 40 (2009) 2013–2019. Available from: http://dx.doi.org/10.1016/j.compositesa.2009.09.003.
[56] K.N. Shivakumar, Carbon/vinyl ester composites for enhanced performance in marine applications, J. Reinf. Plast. Compos. 25 (2006) 1101–1116. Available from: http://dx.doi.org/10.1177/0731684406065194.
[57] A.V. Ratna Prasad, K. Mohana Rao, Mechanical properties of natural fibrefiber reinforced polyester composites: jowar, sisal and bamboo, Mater. Des. 32 (2011) 4658–4663. Available from: http://dx.doi.org/10.1016/j.matdes.2011.03.015.
[58] H.P.S. Abdul Khalil, I.U.H. Bhat, M. Jawaid, A. Zaidon, D. Hermawan, Y.S. Hadi, Bamboo fibre reinforced biocomposites: a review, Mater. Des. 42 (2012) 353–368. Available from: http://dx.doi.org/10.1016/j.matdes.2012.06.015.
[59] Mr Foruzanmehr, P.Y. Vuillaume, S. Elkoun, M. Robert, Physical and mechanical properties of PLA composites reinforced by TiO_2 grafted flax fibers, Mater. Des. 106 (2016) 295–304. Available from: http://dx.doi.org/10.1016/j.matdes.2016.05.103.
[60] J. Sahari, S.M. Sapuan, Natural fibre reinforced biodegradable polymer composites, Rev. Adv. Mater. Sci. 30 (2011) 166–174.
[61] M. Jawaid, H.P.S. Abdul Khalil, A. Hassan, R. Dungani, A. Hadiyane, Effect of jute fibre loading on tensile and dynamic mechanical properties of oil palm epoxy composites, Compos. Part B Eng. 45 (2013) 619–624. Available from: http://dx.doi.org/10.1016/j.compositesb.2012.04.068.
[62] M.S. Sreekala, M.G. Kumaran, M.L. Geethakumariamma, S. Thomas, Environmental effects in oil palm fiber reinforced phenol formaldehyde composites: studies on thermal, biological, moisture and high energy radiation effects, Adv. Compos. Mater. 13 (2004) 171–197. Available from: http://dx.doi.org/10.1163/1568551042580154.
[63] A.K. Mohanty, M. Misra, L.T. Drzal, Sustainable bio-composites from renewable resources: opportunities and challenges in the green materials world, J. Polym. Environ. 10 (2002) 19–26.
[64] M. Jacob, S. Thomas, K.T. Varughese, Mechanical properties of sisal/oil palm hybrid fiber reinforced natural rubber composites, Compos. Sci. Technol. 64 (2004) 955–965, <http://dx.doi.org/10.1016/S0266-3538(03)00261-6>
[65] J.M. Park, D.J. Kwon, Z.J. Wang, J.U. Roh, W. Il Lee, J.K. Park, et al., Effects of carbon nanotubes and carbon fiber reinforcements on thermal conductivity and ablation properties of carbon/phenolic composites, Compos. Part B Eng. 67 (2014) 22–29. Available from: http://dx.doi.org/10.1016/j.compositesb.2014.06.022.
[66] S.T. Seena Joseph, M.S. Sreekala Peter Koshy, Mechanical properties and water sorption behavior of phenol–formaldehyde hybrid composites reinforced with banana fiber and glass fiber, J. Appl. Polym. Sci. 109 (2008) 1439–1446. Available from: http://dx.doi.org/10.1002/app.
[67] R. Yahaya, S.M. Sapuan, M. Jawaid, Z. Leman, E.S. Zainudin, Effect of fibrefiber orientations on the mechanical properties of kenaf–aramid hybrid composites for

spall-liner application, Def. Technol. 12 (2015) 52–58. Available from: http://dx.doi.org/10.1016/j.dt.2015.08.005.
[68] K. Naresh, K. Shankar, B.S. Rao, R. Velmurugan, Effect of high strain rate on glass/carbon/hybrid fiber reinforced epoxy laminated composites, Compos. Part B Eng. 100 (2016) 125–135. Available from: http://dx.doi.org/10.1016/j.compositesb.2016.06.007.
[69] N. Saba, M.T. Paridah, K. Abdan, N.A. Ibrahim, Effect of oil palm nano filler on mechanical and morphological properties of kenaf reinforced epoxy composites, Constr. Build. Mater. 123 (2016) 15–26. Available from: http://dx.doi.org/10.1016/j.conbuildmat.2016.06.131.
[70] M.H. Zamri, H.M. Akil, A.A. Bakar, Z.A.M. Ishak, L.W. Cheng, Effect of water absorption on pultruded jute/glass fiber-reinforced unsaturated polyester hybrid composites, J. Compos. Mater. 46 (2012) 51–61. Available from: http://dx.doi.org/10.1177/0021998311410488.
[71] P. Pandey, D. Bajwa, C. Ulven, S. Bajwa, Influence of hybridizing flax and hemp-agave fibers with glass fiber as reinforcement in a polyurethane composite, Materials (Basel) 9 (2016) 390. Available from: http://dx.doi.org/10.3390/ma9050390.
[72] K. Majeed, A. Hassan, A.A. Bakar, M. Jawaid, Effect of montmorillonite (MMT) content on the mechanical, oxygen barrier, and thermal properties of rice husk/MMT hybrid filler-filled low-density polyethylene nanocomposite blown films, J. Thermoplast. Compos. Mater. (2014) 1–17. Available from: http://dx.doi.org/10.1177/0892705714554492.
[73] M.S. Islam, N.A.B. Hasbullah, M. Hasan, Z.A. Talib, M. Jawaid, M.K.M. Haafiz, Physical, mechanical and biodegradable properties of kenaf/coir hybrid fiber reinforced polymer nanocomposites, Mater. Today Commun. 4 (2015) 69–76, <http://dx.doi.org/10.1016/j.mtcomm.2015.05.001>.
[74] M. Ho, H. Wang, J. Lee, C. Ho, K. Lau, J. Leng, et al., Critical factors on manufacturing processes of natural fibre composites, Compos. Part B 8 (2012) 3549–3562. Available from: http://dx.doi.org/10.1016/j.compositesb.2011.10.001.
[75] S.M. Sapuan, N.B. Yusoff, The relationship between manufacturing and design for manufacturing in product development of natural fibre composites, in: M.S. Salit, M. Jawaid, N.B. Yusoff, M.E. Hoque (Eds.), Manufacturing of Natural Fibre Reinforced Polymer Composites, Springer, Berlin Heidelberg, 2015.
[76] R. Wirawan, S.M. Sapuan, R. Yunus, K. Abdan, Properties of sugarcane bagasse/ poly (vinyl chloride) composites after various treatments, J. Compos. Mater. 45 (2010) 1667–1674. Available from: http://dx.doi.org/10.1177/0021998310385030.
[77] M. Arifur Rahman, Fahmida Parvin, Mahbub Hasan, M.E. Hoque, Introduction to manufacturing of natural fiber-reinforced polymer composites, in: M.S. Salit, M. Jawaid, N.B. Yusoff, M.E. Hoque (Eds.), Manufacturing of Natural Fibre Reinforced Polymer Composites, Springer, Berlin Heidelberg, 2015.
[78] D. Verma, The manufacturing of natural fibre-reinforced composites by resin-transfer molding process, in: M.S. Salit, M. Jawaid, N.B. Yusoff, M.E. Hoque (Eds.), Manufacturing of Natural Fibre Reinforced Polymer Composites, Springer, Berlin Heidelberg, 2015, pp. 287–290.
[79] A. Memon, A. Nakai, Fabrication and mechanical properties of jute spun yarn/PLA unidirection composite by compression molding, Energy Proc. 34 (2013) 830–838. Available from: http://dx.doi.org/10.1016/j.egypro.2013.06.819.
[80] A. Memon, A. Nakai, Mechanical properties of jute spun Yarn/PLA tubular braided composite by pultrusion molding, Energy Proc. 34 (2013) 818–829. Available from: http://dx.doi.org/10.1016/j.egypro.2013.06.818.

[81] C. Weng, Modelling and simulation of residual stresses and birefringence in the precision injection moulding of microlens arrays. UMI Diss. Publ. (2010).

[82] X. Zhang, X. Cheng, K.A. Stelson, M. Bhattacharya, A. Sen, V.R. Voller, Approximate model of thermal residual stress in an injection molded part, J. Therm. Stress 25 (2002) 523–538. Available from: http://dx.doi.org/10.1080/01495730290074270.

[83] A.S. Maxwell, A. Turnbull, Measurement of residual stress in engineering plastics using the hole-drilling technique, Polym. Test. 22 (2003) 231–233. Available from: http://dx.doi.org/10.1016/S0142-9418(02)00087-9.

[84] X.M. Cheng, L. Zhou, N.Y. Sheng, Y.D. Wu, Injection molding and warpage of thin-walled parts based on simulated deformation, Proc. - 2009 Int. Conf. Comput. Intell. Softw. Eng., CiSE, 2009. Available from: http://dx.doi.org/10.1109/CISE.2009.5362816.

[85] M.D. Azaman, S.M. Sapuan, S. Sulaiman, A. Khalina, E.S. Zianudin, Processability of wood fibrefiber-filled thermoplastic composite thin-walled parts using injection moulding, in: M.S. Salit, M. Jawaid, N.B. Yusoff, M.E. Hoque (Eds.), Manufacturing of Natural Fibre Reinforced Polymer Composites, Springer, Berlin Heidelberg, 2015, pp. 351–368.

[86] M.J. John, R.D. Anandjiwala, Recent developments in chemical modification and characterization of natural fiber-reinforced composites, Polymer (Guildf) 29 (2008) 187–207. Available from: http://dx.doi.org/10.1002/pc.

[87] M.A. Khan, R.A. Khan, H.U. Zaman, M. Noor-A Alam, M.A. Hoque, Effect of surface modification of jute with acrylic monomers on the performance of polypropylene composites, J. Reinf. Plast. Compos. 29 (2009) 1195–1205. Available from: http://dx.doi.org/10.1177/0731684409103147.

[88] R.A. Khan, M.A. Khan, H.U. Zaman, F. Parvin, T. Islam, F. Nigar, et al., Fabrication and characterization of jute fabric-reinforced PVC-based composite, J. Thermoplast. Compos. Mater. 25 (2012) 45–58. Available from: http://dx.doi.org/10.1177/0892705711404726.

[89] O. Faruk, A.K. Bledzki, H.-P. Fink, M. Sain, Progress report on natural fiber reinforced composites, Macromol. Mater. Eng. 299 (2014) 9–26. Available from: http://dx.doi.org/10.1002/mame.201300008.

[90] G. Gejo, J. Kuruvilla, A. Boudenne, T. Sabu, Recent advances in green composites, Key Eng. Mater. 425 (2010) 107–166, <http://dx.doi.org/10.4028/www.scientific.net/KEM.425.107>.

[91] J.K. Pandey, S.H. Ahn, C.S. Lee, A.K. Mohanty, M. Misra, Recent advances in the application of natural fiber based composites, Macromol. Mater. Eng. 295 (2010) 975–989. Available from: http://dx.doi.org/10.1002/mame.201000095.

[92] M.M. Davoodi, S.M. Sapuan, D. Ahmad, A. Aidy, A. Khalina, M. Jonoobi, Concept selection of car bumper beam with developed hybrid bio-composite material, Mater. Des. 32 (2011) 4857–4865. Available from: http://dx.doi.org/10.1016/j.matdes.2011.06.011.

[93] A. Atiqah, M.A. Maleque, M. Jawaid, M. Iqbal, Development of kenaf-glass reinforced unsaturated polyester hybrid composite for structural applications, Compos. Part B Eng. 56 (2014) 68–73. Available from: http://dx.doi.org/10.1016/j.compositesb.2013.08.019.

[94] N. Saba, M. Jawaid, O.Y. Alothman, M.T. Paridah, A review on dynamic mechanical properties of natural fibrefiber reinforced polymer composites, Constr. Build. Mater. 106 (2016) 149–159. Available from: http://dx.doi.org/10.1016/j.conbuildmat.2015.12.075.

[95] S. Chapple, R. Anandjiwala, S.C. Happle, R.A. Nandjiwala, Flammability of natural fiber-reinforced composites and strategies for fire retardancy: a review, J. Thermoplast. Compos. Mater. 23 (2010) 871–893. Available from: http://dx.doi.org/10.1177/0892705709356338.

[96] R.D.S.T.C. Fong, N. Saba, C.K. Liew, K.L.G. M. Enamul Hoque, Yarn flax fibers for polymer-coated sutures and hand layup polymer composite laminates, in: M.S. Salit, M. Jawaid, N.B. Yusoff, M.E. Hoque (Eds.), Manufacturing of Natural Fibre Reinforced Polymer Composites, Springer, Berlin Heidelberg, 2015, pp. 155–176.

[97] S.S.M. Mahbub Hasan, M. Enamul Hoque, S.M. Sapuan, N. Saba, Manufacturing of Coir FibreFiber-Reinforced Polymer Composites by Hot Compression Technique, in: M.S. Salit, M. Jawaid, N.B. Yusoff, M.E. Hoque (Eds.), Manufacturing of Natural Fibre Reinforced Polymer Composites, Springer, Berlin Heidelberg, 2015, pp. 309–330.

[98] E. Zini, M. Scandola, Green composites: an overview, Polym. Compos. 32 (2011) 1905–1915. Available from: http://dx.doi.org/10.1002/pc.21224.

[99] I. Kong, K.Y. Tshai, M.E. Hoque, Manufacturing of natural fibrefiber-reinforced polymer composites by solvent casting method, in: M.S. Salit, M. Jawaid, N.B. Yusoff, M. E. Hoque (Eds.), Manufacturing of Natural Fibre Reinforced Polymer Composites, Springer, Berlin Heidelberg, 2015, pp. 331–350.

Bio-based hybrid polymer composites: a sustainable high performance material

2

Mohamed Bassyouni[1,2], *Umair Javaid*[1,3] *and Syed W. ul Hasan*[1,4]
[1]King Abdulaziz University, Rabigh, Saudi Arabia, [2]Higher Technological Institute, Tenth of Ramdan City, Egypt, [3]National University of Science and Technology, NUST, Islamabad, Pakistan, [4]University of Engineering and Technology, Lahore, Pakistan

Chapter Outline

Climate change has induced a sustainable and renewable approaches in the development of biodegradable materials, resulting in fewer carbon footprints. Natural fibers (NFs) are making their place as a worthy alternative to the synthetic fibers in reinforced polymer composites. Natural fiber reinforced plastics (NFRPs) are abundantly used in modern composite industry due to their high abundance, low cost, low density, and environment-friendly nature; NFs have several inexpensive, technical, and ecological advantages over synthetic fibers in the field of polymer composites. They are considered as prospective alternative to tradition reinforcements in polymer composites. In commercial applications, NFRPs are used in nonstructural automotive components i.e., beams, roof panels, boat hulls, tennis rackets, furniture, pipes, and tanks. There are more than 1000 species of bio-fibers which can be used as reinforcements in polymer composites, some

Hybrid Polymer Composite Materials: Processing. DOI: http://dx.doi.org/10.1016/B978-0-08-100789-1.00002-2

examples are: flax, kenaf, jute, coconut tree leaf, sisal, ramie, palm, sea husk, vakka, elephant grass, abaca leaf fiber, and sunflowers stalk flour. In the past, there has been intensive research in NFRPs where cellulosic fibers were found to have excellent stiffness and strength. However, there are some drawbacks like variability in fiber quality due to variation in harvesting and extraction of plants and interfacial adhesion. Moisture sensitivity, limited thermal resistance, long term durability in the presence of ultraviolet light and temperature, are also important factors. In this chapter, an overview of research in the area of NFs for fiber reinforced polymer composite applications are discussed in details. Fundamental properties of NFs/polymer matrix interfacial properties, surface modification, and manufacturing processes are described. Physical and mechanical properties of different bio-composites have been reviewed based on published data and predictive models using Cambridge Engineering Selector (CES) program.

2.1 Introduction

The scientific community is looking into alternative materials due to severe environmental problems in order to replace synthetic fibers such as Kevlar, glass, and carbon fiber in polymer composites [1–7]. In bio-composites polymer based natural fibers (NFs) such as jute, hemp, flax, coir, kenaf, and sisal are reinforced with synthetic or natural matrices [8]. NFs have some important properties such as abundance, renewable, processing flexibility, low fossil-fuel energy requirements, reliable mechanical properties, minimal health hazards, fracture resistance, greater energy recovery, low density and cost effectiveness as compared to synthetic fibers [9]. Fiber glass is more dense, tough to machine, and cannot be salvaged [10]. The energy consumption to produce flax fiber from cultivation to producing mat is 9.6 MJ/kg which is lesser as compared to 54.7 MJ/kg to produce glass fibers (GF) mat [11]. The inherent flexibility and easy delamination of NFs composites with respect to synthetic fiber composites enable them to struggle impact loads without smashing [12]. Some disadvantages are also presented by natural fiber polymer composites. All lingo-cellulosic fibers present heterogeneous characteristics as compared to synthetic fibers fabricated within particular calculated values. However, the interaction of these NFs with most of the synthetic and natural matrices is of considerable importance as poor fiber/matrix interface can lead to lesser load distribution from the matrix to the fiber [13]. The hydrophobicity of the NFs causes poor compatibility towards polymer matrix causing weak mechanical properties of the composites [14]. NFs contain packs of elementary fibers along with various irregularities and voids. Major constituents of NFs are cellulose and lignin in variable amounts. Another issue is hydrophilic (polar) nature of NFs, especially when used in hydrophobic (polar) polymer matrices. Cellulose which is mainly responsible for mechanical properties of NFs is ordered in microfibrils enclosed by the other two main components: hemicellulose, and lignin [15]. Lignin is an aromatic biopolymer, a vital cell wall component of all vascular plants. Hemicellulose is a large group of polysaccharides found in the primary and secondary cell walls of the plants [16,17]. The interfacial strength in NFs and polymer matrix can be increased by large number of physical or chemical surface treatments [18]. The physical treatments like corona, plasma are mostly used [19]. NFs are cellulose-rich and abundantly available materials. Their applications structural composites have been limited due to low price and improved performance of standard plastics [20]. Biomass wastes can be combusted or gasified into synthesis gas, carbon monoxide and hydrogen [21]. Among NFs, wood fibers are the most abundantly used cellulose fibers have extensive use in fiberboard, pulp, paper, and many other industrial applications. The use of wood composites date from centuries ago, currently 80% of natural fiber reinforced composites are made from wood fibers. Wood fibers consist of both live and dead cells in the wood, depending upon age and location of tree. They are obtained from timber by complex combination of chemical, biological, and mechanical processes. Wood fibers mainly exists within the layer of xylem tissue in the wood [22]. At the microscopic level (0.01–6 mm) there are two kinds of wood cells with different hierarchical structures, namely Tracheids (in softwoods and hardwoods) and Vessels (only in hardwoods). Dimensions of both are given in Table 2.1.

2.2 Nature and behavior of natural fibers

2.2.1 Properties of NFs

The surface property of NFs is one of the most important parameter influencing interfacial adhesion between matrix and surface of fibers. This property depends upon fiber morphology, chemical composition, extractive chemicals, and processing conditions [23]. Due to high polar character of NFs, they are less compatible with nonpolar resin, this leads to insufficient stress transfer in NFRPs. The use of various kinds of physical and chemical surface treatment methods can make surface of NFs compatible with organic resins. Chemical composition of common NFs are listed in Table 2.2. The mechanical properties of NFs can be analyzed using two methods namely macroscopic tests and indentation tests. The macroscopic tests focus on the performance of whole sample, the parameters include tensile, compression, impact, and flexural tests. Whereas, indentation test focus on measuring local properties of sample [24]. The mechanical performance depends upon growing parameters like growth, climate, and age of the plant [25]. Some mechanical and physical properties of common NFs in comparison with synthetic fibers are given in Table 2.3. Fig. 2.1 shows that ramie fiber has the highest specific strength (Tensile strength/Density) versus specific Young's modulus (Modulus of elasticity/Density) among the wide variety of NFs while coir and wool have the lowest values.

The values of physical and chemical properties of NFs are given in Table 2.4 derived by Cambridge Engineering Selector (CES). The high tensile strength of flax

Table 2.1 Types of wood fibers and dimensions

Type of wood fibers	Length (mm)	Width (μm)	Aspect ratio
Tracheids	2–6	20–40	50–200
Vessels	1–2	10–50	28–86

Table 2.2 **Chemical composition and microfibril angle of natural fibers**

Properties/fibers	Coir	Flax	Hemp	Jute	Ramie	Sisal
Diameter (μm)	100–450	100	25	60	40–50	100–300
Cellulose content (%)	36–43	62–72	67–75	59–71	68–76	74–75.2[a] 60–67[b]
Hemicellulose content (%)	0.2	16–18	16–18	12–13	13–14	10–13.9
Lignin content (%)	41–45	2–2.5	2.8–3.3	11.8–12.9	0.6–0.7	8–12[b] 7.6–7.98[a]
Microfibrillar angle (°)	30–45	10	6.2	7–9	7.5–12	10–20

[a]Brazilian.
[b]Indian.

Table 2.3 Mechanical properties of natural fibers and common synthetic fibers

Fibers	Density (g/cm^3)	Elongation (%)	Tensile strength (MPa)	Young's modulus (GPa)
Flax	1.5	1.2–3.2	345–2000	15–80
Hemp	1.48	1.6	550–900	26–80
Sisal	1.5	3.0–7.0	468–700	9.4–22
Coir	1.2	17–47	175	4.0–6.0
Softwood	1.5	–	600–1020	18–40
Hardwood	1.2	–	–	37.9
E-glass	2.5	2.5	2000–3500	70
S-glass	2.5	2.8	4570	86
Aramid	1.4	3.3–3.7	3000–3150	63.0–67.0

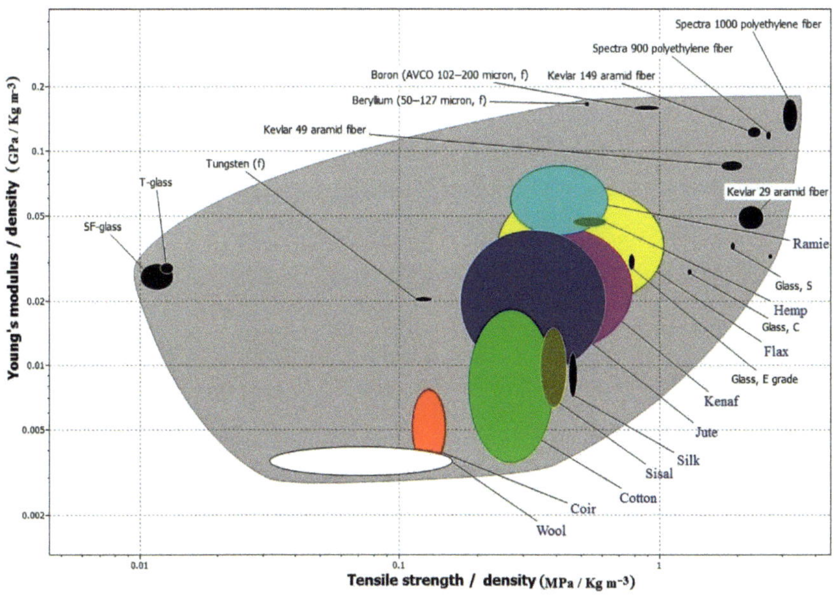

Figure 2.1 Specific tensile strength versus specific Young's modulus of NFs, metal, glass, and polymer fibers.

Table 2.4 **Physical and mechanical properties of NFs. (Using Cambridge Engineering Selector program)**

	Ramie	Flax	Kenaf	Hemp	Cotton	Sisal	Silk	Coir
Physical properties								
Density (kg/m^3)	1450–1550	1400–1500	1190–1200	1480–1500	1500–1600	1450–1500	1260–1350	1140–1200
Mechanical properties								
Young's modulus (GPa)	61.4–128	27.6–100	14–53	66.5–73.5	5.5–28	9.4–22	9.3–15	04-Sep
Yield strength (elastic limit) (MPa)	360–612	150–338	195–666	200–400	100–350	460–576	600–604	100–150
Tensile strength (MPa)	400–938	345–1500	240–930	690–921	287–597	511–64	600–604	131–175
Elongation (% strain)	1.2–3.8	2–3.2	1.6–2.9	1.52–1.68	07-8	02-7	4–6.5	15–40
Flexural modulus (GPa)	61.4–128	27.6–100	14–53	66.5–73.5	5.5–28	9.4–22	9.3–15	04-9
Flexural strength (modulus of rupture) (MPa)								135–240
Shear modulus (GPa)	14.1–16.3	1–2.1	22.3–24.6	1.2–2.4	1–2.1	3.67–9.17	2–3.6	1.45–2.17

Bulk modulus (GPa)		02-6		2.5–7	02-6		03-10	
Poisson's ratio	0.345–0.359	0.25–0.3	0.344–0.368	0.25–0.3	0.25–0.3	0.359–0.374	0.25–0.3	0.383–0.393
Fatigue strength at 10^7 cycles (MPa)	200–272		86.8–296			220–316		54–96
Mechanical loss coefficient (tan delta)	0.00279–0.00307	0.01–0.05	0.00212–0.00226	0.01–0.05	0.01–0.05	0.00407–0.00753	0.01–0.05	0.0106–0.0139
Impact and fracture properties								
Fracture toughness ($MPa \cdot m^{0.5}$)	9.2–19.7	01-2	1.16–14.8	01-2	01-2	19.6–101	01-2	3.33–11.3

fiber comparing with the other listed NFs can be attributed to the unidirectional cellulose microfibrils and tubular structure of flax while coir fiber (CF) has the lowest tensile strength as the fiber orientation, composition (low cellulose contents), and aspect ratio play a crucial role in the mechanical and physical properties of NFs. Ramie, kenaf and sisal fibers exhibit high service and glass temperature as tabulated in Table 2.5. NFs have superior electrical resistivity in the range 10^{14} to 10^{16} μohm.cm. The average water absorption (wt%) of NFs is around 2% as listed in Table 2.6. In addition, most of the NFs have a good UV radiation and organic solvent resistances. Primary production energy, CO_2 footprint and heat of combustion of NFs are listed in Table 2.7.

2.2.2 Processing of NFs

The main interests in NFs are due to their environmental friendly nature, low density, and reliable specific strength/stiffness [26]. Some of the drawbacks in using NFs in industrial application composites are their sensitive nature to moisture absorption, polar nature, and less thermal stability [27]. Insufficient fiber wetting can result in poor interfacial strength [28]. Interfacial strength can be increased by mechanical interlocking, chemical bonding or inter-diffusion bonding [29]. Mechanical properties are enhanced by making rough surface of fiber. Chemical bonding is done by introducing chemical groups in matrices and on the surface of fibers. This leads to formation of bond and resulting in strong interfacial strength. Chemical bonding can be achieved by using coupling agents. They act as a bridge between fibers and reinforcements. Some physical methods like plasma, heat, corona, ultraviolet treatments, electron radiation, and fiber heating are also available to increase interfacial strength between NFs and polymer matrices. Corona treatment uses high voltage generated plasma by quartz at low temperature and pressure [30].

In the grown form NFs have cellulose-rich core with external wall covered with cementing which comprises of waxes, fats, lignin, pectin, and hemicellulose. This cementing prevents formation of strong bond between hydrophilic fiber and hydrophobic matrix. A number of approaches have been used to render this problem, most common among them is the surface treatment of NFs with alkali to reduce cementing and increase cellulosic fractions. Surface treatment may improve the fiber reinforcement in polymer matrix; however, this will result in added cost and decreased economical compatibility of NF composites [31]. Alkali treatment (mercerization) is a method to enhance interfacial strength (IS) between NFs and polymer matrix [32]. It eliminates oils and waxes from the surface of NFs [33]. It reduces water absorption from 4.2 to 3.8%, surges the amorphous region via the dissolution, and leaches out fatty acids and some other lignin component from NFs [34,35]. In a study mercerized jute fibers were coated with oligomeric siloxane to improve IS, leading to improved mechanical properties [36]. Several researchers have studied effect of different wt% alkali solution along with different immersion times to treat NFs [37]. It swells the amorphous region of NFs and causes removal of approximately 40% of the hemicellulose in the fiber structure [38]. Mwaikambo and Ansell reported that the chemical composition of jute fibers (JFs): cellulose (51–84%), hemicellulose (12–20%), lignin (5–13%), and pectin (0.2%) [35]. The purpose of alkali treatment is to (1) remove impurity and hemicellulose on the fiber

Table 2.5 Thermal and electrical properties of NFs. (Cambridge Engineering Selector program)

	Ramie	Flax	Kenaf	Hemp	Cotton	Sisal	Silk	Coir
Glass temperature (°C)	390–400	110–130	380–390	100	110–130	380–390	77	
Maximum service temperature (°C)	400–420	110–130	400–420	110–130	110–130	400–420	77–87	
Minimum service temperature (°C)		−273		−273	−273		−273	
Thermal conductivity (W/m · °C)	0.25–0.35	0.2–0.3	0.25–0.35	0.2–0.3	0.2–0.3	0.25–0.35	0.2–0.3	
Specific heat capacity (J/kg · °C)	1200–1220	1200–1220	1200–1220	1200–1220	1200–1220	1200–1220	1360–1390	
Thermal expansion coefficient (μstrain/°C)	15–30	15–30	15–30	15–30	15–30	15–30	15–30	37.4–49.3
Electrical resistivity (μohm.cm)		10^{14}–10^{16}		10^{14}–10^{16}	10^{14}–10^{16}		10^{16}–10^{19}	

(Continued)

Table 2.5 (Continued)

	Ramie	Flax	Kenaf	Hemp	Cotton	Sisal	Silk	Coir
Dielectric constant (relative permittivity)		3–6		3–6	3–6		2–4	
Dissipation factor (dielectric loss tangent)		0.05–0.057		0.003–0.02	0.003–0.02		0.001–0.005	
Dielectric strength (dielectric breakdown) (MV/m)		6–8		6–8	6–8		6–8	

Table 2.6 Absorption and permeability properties of NFs. (Cambridge Engineering Selector program)

	Ramie	Flax	Kenaf	Hemp	Cotton	Sisal	Silk	Coir
Water absorption @ 24 h (%)	2.4–3.4					2–2.4		1.8–2.2
Water absorption @ sat (%)	12–17					10–12		9–11
Humidity absorption @ sat (%)	4–5.67					3.33–4		3–3.67
Water (fresh)	Excellent	Acceptable	Acceptable	Acceptable	Acceptable	Excellent	Acceptable	Excellent
Water (salt)	Excellent	Acceptable	Acceptable	Acceptable	Acceptable	Excellent	Acceptable	Excellent
Weak acids	Acceptable	Limited use	Limited use	Limited use	Limited use	Acceptable	Acceptable	Acceptable
Strong acids	Unacceptable	Unacceptable	Unacceptable	Unacceptable	Unacceptable	Unacceptable	Limited use	Unacceptable
Weak alkalis	Acceptable	Acceptable	Limited use	Limited use	Acceptable	Acceptable	Acceptable	Acceptable
Strong alkalis	Unacceptable	Limited use	Unacceptable	Unacceptable	Limited use	Unacceptable	Unacceptable	Unacceptable
Organic solvents	Acceptable	Acceptable	Acceptable	Acceptable	Acceptable	Acceptable	Acceptable	Acceptable
Oxidation at 500C	Unacceptable	Unacceptable	Unacceptable	Unacceptable	Unacceptable	Unacceptable	Unacceptable	Unacceptable
UV radiation (sunlight)	Good	Good	Good	Good	Fair	Good	Poor	Good
Wear resistance		Unacceptable		Unacceptable	Unacceptable		Unacceptable	
Flammability	Highly flammable	Highly flammable	Highly flammable	Highly flammable	Highly flammable	Highly flammable	Highly flammable	Highly flammable

Table 2.7 **Primary production energy, CO_2 footprint and heat of combustion of natural fibers (CES program)**

	Ramie	Flax	Kenaf	Hemp	Cotton	Sisal	Silk	Coir
Embodied energy, primary production (MJ/kg)	10.5–11.6	10.5–11.6	64.1–70.7	9.52–10.5	43.3–47.8	9.52–10.5	1750–1930	9.52–10.5
CO_2 footprint, primary production (kg/kg)	0.416–0.459	0.416–0.459	2.67–2.94	1.52–1.68	0.851–0.938	1.52–1.68	163–179	1.52–1.68
NOx creation (g/kg)	2.57–2.84	25.8–28.5	2.57–2.84	14.7–16.2	25.8–28.5	2.57–2.84	196–217	2.57–2.84
SOx creation (g/kg)	6.56–7.25	44.1–48.7	6.56–7.25	25.1–27.8	44.1–48.7	6.56–7.25	335–370	6.56–7.25
Water usage (l/kg)	3820–4220	2980–3290	500–1500	2380–2630	7380–8160	7880–8710	500–1500	2200–2440

Fabric production energy (MJ/kg)	2.48–2.73	2.48–2.73	2.48–2.73	2.48–2.73	2.48–2.73	2.48–2.73	2.48–2.73	2.48–2.73
Fabric production CO_2 (kg/kg)	0.198–0.218	0.198–0.218	0.198–0.218	0.198–0.218	0.198–0.218	0.198–0.218	0.198–0.218	0.198–0.218
Heat of combustion (net) (MJ/kg)	16.8–17.7	17–17.9	17–17.9	17.8–18.7	17–17.9	19.3–20.2	13.4–14.1	14.2–14.9
Combustion CO_2 (kg/kg)	1.39–1.46	1.39–1.46	1.39–1.46	1.54–1.62	1.39–1.46	1.5–1.58	1.18–1.24	1.39–1.46
Annual world production (tonne/year)	240,000–265,000	435,000–481,000	485,000–536,000	66,200–73,200	2.54e7–2.81e7	362,000–400,000	125,000–139,000	1.05e6–1.16e6

surface, and (2) imparting chemical accessibility to hydroxyl group of cellulose to promote silane coupling agent reactivity. In a study by Mahjoub et al. [37], kenaf fibers were mercerized using 5, 7, 10, and 15 wt% solution of NaOH between 3 and 24 h at room temperature. Results showed that using 5% alkali solution caused no tension on fiber structure. Fiore et al. reported increase in storage modulus and significant shift of tan(d) peaks to higher temperatures in alkali treated kenaf enriched polymer matrix [38].

2.2.3 Types and applications of NFs

The increasing awareness toward the global environment is demanding a shift of material selection in automotive, construction, and packing industry from organic polymeric material to green materials in order to avail biodegradability of products at the end of life. Due to high specific mechanical properties and CO_2 neutrality, low weight and reduced carbon footprints, NFs are replacing synthetic fiber reinforced petroleum based polymer composites in building products. NFs are used in door panels, package trays, hat racks, instrumental panels, internal engine covers, sun visors, boot liners, oil/air filters, seat backs, and underfloor paneling. Currently, most of the international automotive manufacturers use them and their use is expected to be increased in upcoming years [39]. A study by S. V. Joshi et al. [10] reveals that by replacing traditional acrylonitrile butadiene styrene (ABS) panels in Audi car door panels using hemp reinforced epoxy composites can decrease net energy consumption by 45% and considerably reduce emission of toxic gases. They have been extensively used in toys, packing materials, and electronic devices cases. The recycling and reuse of products are major issues of government policy around the globe. NFs made of renewable agricultural and forestry feedstock ensure greener and sustainable materials as compared to synthetic polymer composites. Moreover, biodegradability of NFs support their ultimate disposal by composting or incineration. The release of CO_2 due to combustion or atmospheric degradation of NFs is balanced by the biomaterial assimilation during the NFs biological growth. Thus there is a huge possibility to replace structural components with NFs. Moreover, NFs are a nonabrasive and cause low damage to tools and molding equipment in comparison to glass fiber. In sports, NFs are used in surfboards, one of the recent product in "Ecoboard" by Limitations Ltd, using bio-based resin and hemp fiber. Fishing rods are also being produced using NFs, a recent product is by CellulComp Ltd who extract nenocellulose from root vegetables. Bamdad Barari et al. found that the friction and wear resistance of bio-based epoxy can be improved in the presence of cellulose nano-fibers which form uniform tribo-layer composites as shown in Fig. 2.2 [40].

NFs have been found to be better replacement of wooden products in insulating structural panels due to their good mechanical properties. JFs are used in producing door panels, sanitary products, foot wear industry, home/garden furniture, and toys. Some applications of kenaf fibers are ropes, twines, engineering wood, insulated panels, oil and liquid absorbent materials, automotive interior lining, furniture [41], orthopedics and seed oil which is rich in omega polyunsaturated fatty acids, a well-known compound for reducing cholesterol.

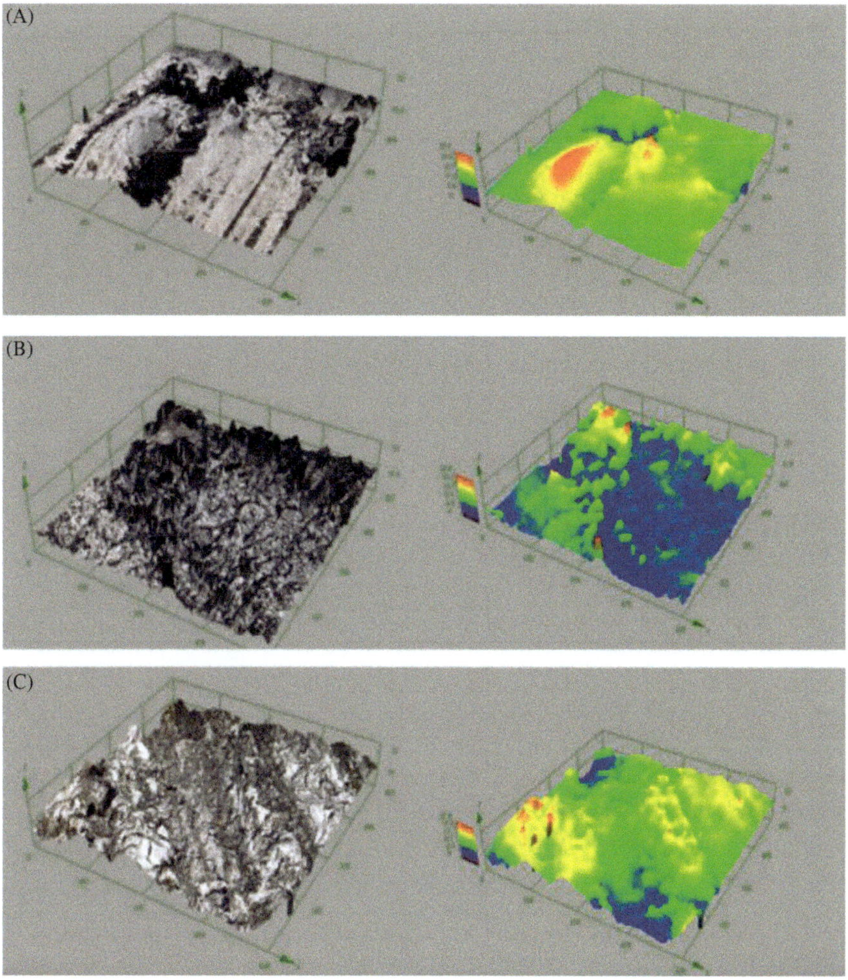

Figure 2.2 3D surface topography of (A) neat epoxy, (B) bio-epoxy reinforced by 0.9 vol.% nano-cellulose, and (C) bio-epoxy reinforced by 1.4 vol.% nano-cellulose at 7 N normal load and 0.15 m/s.
Reprinted from Bamdad Barari et al. 2016. Mechanical, physical and tribological characterization of nano-cellulose fibers reinforced bio-epoxy composites: an attempt to fabricate and scale the 'Green' composite, J. Carbohydr. Polym. 147 (2016) 282–293. With permission from Elsevier.

2.2.3.1 Flax fibers (FFs)

They have a complex structure with crystalline cellulose macro-molecules in microfibrils. These microfibrils are oriented along the fiber axis at an angle between 5 and 10 degree for flax fibers and surrounded by a matrix of noncellulosic components i.e., hemicellulose, lignin, pectins, and proteins [36]. The cell wall is comprised of inner concentric layers and outer primary walls. The fibrillar cells are elementary fibers which gather in the form of fiber bundles within the stem of higher plants. Fernández et al. have reported that flax stem surface in the epidermal region is comprises of pectins, adlipophilic component such as waxes [42]. Beneath the epidermis, there is internal zone which is rich in cellulose with even more amount of pectins and polysaccharides. The inner tissue of the stem contains lignin and noncellulosic polysaccharides. The cohesion of flax fibers bundles is due to pectins which are the main component in the primary cell walls and cell junctions. Marques et al. have analyzed the lipopholic extractives of flax fibers using gas chromatography. They found fatty alcohols and acids, aldehydes, and esters in flax fibers [43]. The production of flax fibers from bast fibers requires many steps like scotching and hackling as well as chemical treatments like alkali extraction, desizing, kiering or bleaching, and scouring. The main objective of all these steps is to remove impurities, separate and refine fiber bundles to arrange them in woven fabrics [43].

2.2.3.2 Kenaf fibers (KFs)

Kenaf (*Hibiscus cannabinus* L.) has been cultivated since 4000 BC for food and fiber extracting purposes with growth rate of 10 cm/day, rising to a height of 3 m in 3 months under optimum ambient conditions. It is acquired from processing of the bark of kenaf plant and can yield dry weight of 6000–30,000 kg/ha per year. To produce 1 kg of kenaf, 15 MJ energy is required [26]. It has been used as nonwoven mat in automotive, textiles, fiberboard, and electronic industry. Recently its applications as reinforcement in thermoplastic polymers have been reported [44]. It is known as a cellulosic source with fibrous stalk and is capable to grow under extensive range of climatic conditions with negligible requirement of fertilizers, water, and pesticides [45]. It is extremely environmentally friendly as it collects CO_2 at significant high rate and absorbs nitrogen along with phosphorus from the environment [46]. It is also useful in biodiesel production due to its great cellulose content and little lignin contents. The 1 kg of kenaf fiber can produce 18,000 kJ of energy and it can produce 61.4 gal of biofuels/ton biomass [47]. Janoobi et al. have reported that the lignin content in kenaf is between 5.9 and 19% while that of hemicellulose (HC) is between 15.0 and 23.0% with cellulose content between 44.0 and 63.5% [48,49]. Asumani et al. studied comparison between the properties of (alkaline and silane treated) kenaf fiber reinforcement and glass reinforcement. They revealed that the specific tensile strength (TS), flexural strength (FS), and tensile modulus of 30 wt% kenaf fiber reinforced polypropylene (PP) were respectively 96, 89, and 82% those of glass reinforced PP [50]. Fairuz Fazillah Shuhimi et al. [51] reported that the wear resistance of epoxy can be improved in the presence of kenaf fiber (KF) as illustrated in Fig. 2.3.

2.2.3.3 Jute fibers (JFs)

Jute plants are huge source of biomass. It is assessed that during 100 days, 1 ha of jute plant can consume 15 tons of CO_2 and release about 11 tons of oxygen in atmosphere. A study has exposed that jute has negative greenhouse gas and has drawn the attention of the consumers to make more use of biodegradable jute products. Due to the natural advantages and development of new and diversified

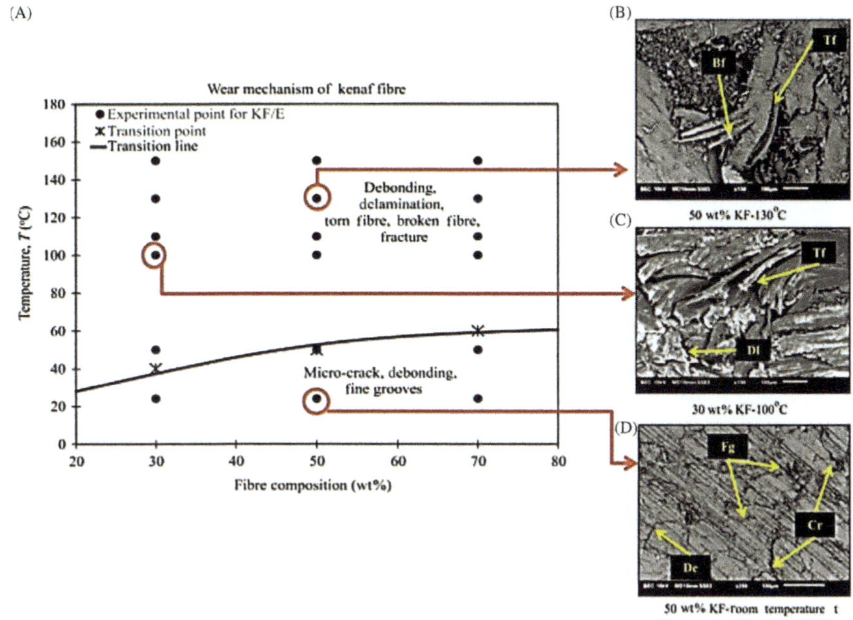

Figure 2.3 Wear mechanism on the Kenaf/Epoxy composite with SEM micrograph (*De*: debonding; *Dl*: delamination; *Cr*: Micro-crack; *Fg*: Fine grooves; *Bf*: Broken fiber; *Tf*: Torn fiber). (A) Wear mechanism for KF/epoxy composites, (B) Surface morphology of 50 wt% KF at 130°C, (C) Surface morphology of 30 wt% KF at 100°C, and (D) Surface morphology of 50 wt% KF at room temperature.
Reprinted from Fairuz Fazillah Shuhimi et al. 2016. Tribological characteristics comparison for oil palm fiber/epoxy and kenaf fiber/epoxy composites under dry sliding conditions, J. Tribol. Int. 101 (2016) 247–254. With permission from Elsevier.

jute products, the market has once again become receptive to jute for traditional products like sacking, clothes, and bags. Jute is a biodegradable plant fiber and is one of the cheapest and the strongest of all NFs [52]. JFs are mainly composed of cellulose and lignin. The chemical composition of jute fiber comprises of cellulose (61–71%), hemicellulose (13.6–20.4%), lignin (12–13%), ash (0.5–2%), pectin (~0.2%), wax (~0.5%), and moisture (~12.6%). JFs are degraded biologically by microorganisms resulting in weakening of fiber cell wall and lose in strength. Degradation also occurs through oxidation, hydrolysis, and dehydration reactions. Outdoor applications of JFs undergo photochemical degradation caused by ultraviolet light. Prolonged ultraviolet (UV) exposure may lead to decrease in mechanical integrity [52,53]. The jute yarns pretreated under UV conditions exhibited better tensile properties than those pretreated with gamma radiation.

2.2.3.4 Coir fibers (CFs)

The term coir is derived from kayaru meaning to be twisted. Coir is a NF which is extracted from the husk of coconut. It is present between the internal shell and the outer coat of a coconut. Coir fibers are found in tropical regions i.e., Srilanka, Thailand, Bangladesh, and India. CFs are important due to resilience, extensibility, fungi resistance, moth-proof, insulation against temperature and sound [54]. The individual CFs are hollow with thick wall made of cellulose with length of 10–30 cm. Initially they are pale and become hardened and yellowed as a layer of lignin is accumulated on their wall. The chemical composition of CFs includes lignin (45.84%), cellulose (43.44%), pectin (3%), hemicellulose (0.25%), ash (2.22%), and water soluble part (5.25%) [55]. CF contains 30–300 or more crystalline cellulose cells in cross section [56]. CFs have been reported to possess the highest elongation strength at break among typical NFs and they are capable of taking strain 4–6 times more than other NFs [57]. Cross section of CFs indicates that they comprises of numerous elementary fibers with large empty space located in the center of each fibers. This empty space is called lacuna and is shown in Fig. 2.4 [58]. The volume fraction of lacuna in CF is 2–3%. CFs have more lignin content but little cellulose content, this make them resilient, strong, tough, and durable. They are resistant to fungal and bacterial attacks and have ability of sound absorbing. They can stand water immersions for months without collapsing. They are used in number of products like brushes, ropes, bags, and mats. The individual CFs have uneven lumen and short length.

In the detailed images of CFs as shown in Figs. 2.5 and 2.6, it can be noticed that each basic CF consists of two cell wall layers containing number of microfibrils, stacked together with lamella glue. The structure of CF cell wall is similar to wood and plant fiber. The secondary wall is thicker with microfibrils oriented in 45

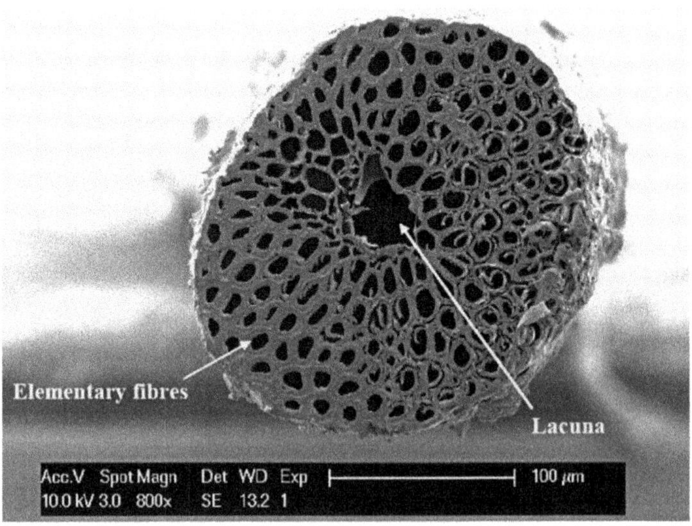

Figure 2.4 SEM images of cross section of a typical coir fiber with presence of lacuna and elementary fibers.
Reprinted from L.Q.N. Tran et al. 2015. Investigation of microstructure and tensile properties of porous natural coir fiber for use in composite materials, J. Industrial Crops Prod. 65, 437–445. With permission from Elsevier.

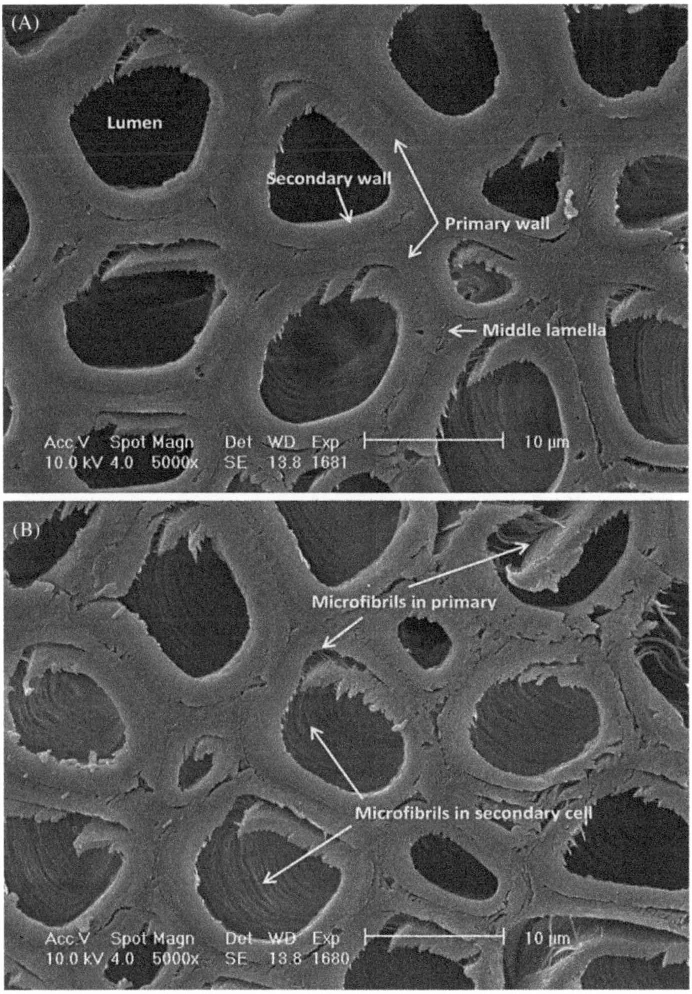

Figure 2.5 SEM images of elementary fibers which show (A) different lumen and cell walls and (B) some microfibrils the primary and secondary cell walls.
Reprinted from L.Q.N. Tran et al. 2015. Investigation of microstructure and tensile properties of porous natural coir fiber for use in composite materials, J. Industrial Crops Prod., 65, 437–445. With permission from Elsevier.

degree as compared to primary wall with micofibrils in 90 degree. Some negative characters of CFs include their hydrophilic nature, strong dependence of mechanical and physical properties on climate, high moisture absorption due presence of hydroxyl and polar groups in several constituents, all these factors lead to poor interfacial strength of CFs composites with polymer matrix [58]. CFs show less tensile strength as compared to other NFs listed in Table 2.3. This can be attributed to low cellulose contents (36–43%) and high microfibrillar angle (30–45 degree).

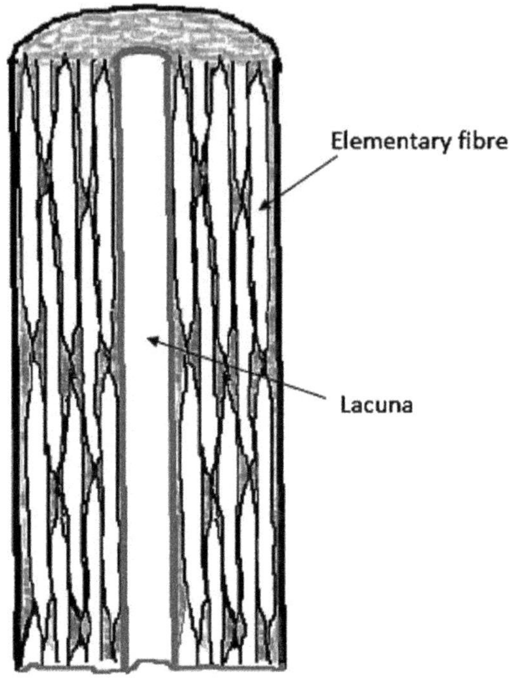

Figure 2.6 Schematic presentation of the orthogonal slice of technical coir fiber which shows the organization of elementary fibers (with lumen) inside the technical fiber. Reprinted from L.Q.N. Tran et al. 2015. Investigation of microstructure and tensile properties of porous natural coir fiber for use in composite materials, J. Industrial Crops Prod., 65, 437–445. With permission from Elsevier.

The moisture absorption of CFs can be decreased by suitable surface chemical treatments. Karthikeyan et al. studied alkali treated CF/polyester composites and found that alkali treated CF composites show less flexural strength even less than the flexural strength of bare polyester [55]. Yan et al. studied effect of 5 wt% NaOH solution at 20°C on mechanical properties of CF reinforced composites. It was found that TS and FS of alkali treated CF composites are 17.8 and 16.7% more than untreated CF composites [59]. Nam et al. studied the effect of NaOH treatment on mechanical properties of CF/poly butylene succinate (PBS) biodegradable composites using 10–30% fiber mass content. The TS and flexural properties of alkali treated CF/PBS composites showed increase of 54.5 and 45.7% as compared to untreated CF/PBS composites [60].

2.2.3.5 Sisal fibers

(SFs) originated from Mexico and Central USA, but are now extensively grown in tropical countries like Africa, the West Indies, and the Far East. SFs are extracted traditionally from fresh leaves of sisal plant (*Agava sisalana*) as shown in Fig. 2.7A

Figure 2.7 (A) Sisal plant, (B) Optical microscopy image of a longitudinal sisal fiber. Reprinted from Ahmed Belaadi et al. 2013. Tensile static and fatigue behavior of sisal fibers, J. Mater. Des. 46 (2013) 76–83. With permission from Elsevier.

by decorticators followed by washing and then drying in sun. The average sisal fiber diameter is 240 μm as illustrated in Fig. 2.7B [61].

The length of SFs varies from 0.6 to 1.5 m. The fiber comprises of bundles of subfibers with cell walls reinforced with cellulose within hemicellulose and lignin matrix. The chemical composition of SFs comprises of cellulose (65.8%), hemicellulose (12%) and lignin (9.9%), pectin (0.8%), wax (0.3%), and water soluble compounds [62]. Their hydrophobicity leads to meager load transfer between fibers and matrix and low mechanical properties [63]. This problem can be overcome by fiber surface treatment. SFs are mostly used in yarns, ropes, carpets, mattresses, ropes, handcraft articles, and twines [63]. Around 4.5 M tons of SFs are produced per year in all over the world especially in Tanazania and Brazil. Generally, the strength of NFs depend upon cellulose content and angle of microfibrils in inner secondary cell wall. The TS and modulus of single sisal bundles are 344–577 MPa and 9–19 GPa respectively [64]. Silva et al. studied tensile fatigue properties of single fiber bundles by sinusoidal loading with maximum forces between 0.7 and 3.5 N corresponding to five different loading ratios upto 10^6 cycles. An increase of the equivalent Young's modulus with the increase of the maximum fatigue stress was observed, no apparent hysteresis dependence over the number of cycles was recorded [65]. The SFs have a complex structure. Each fiber contains several sub fiber cells within size of 6–30 μm in dia. The cells in the fibers are joined using middle lamellae, which contains hemicellulose and lignin. The fibrillae are of thickness about 20 nm and contains cellulose chains with a thickness of around 0.7 nm. Structure and conducting tissues of SFs are shown in Fig. 2.8.

H. Zou et al. considered mercerization and silane treatment effect on SFs and revealed that increase in interfacial properties between SFs and polylactic acid (PLA) matrix leading to improved mechanical properties [66]. Yan Li et al. have studied tensile properties of SFs after treating them with different methods i.e., alkali treatment, acidic treatment, bonzol/alcohol dewax treatment, and thermal-alkali treatment [67]. It was found that thermal treatment at 150°C for 4 h is utmost desirable in terms of tensile strength and modulus properties as crystalinity increases from 62.4 to 66.2% at 150°C. At temperature above 200°C the degradation of SFs occurs and tensile properties decrease.

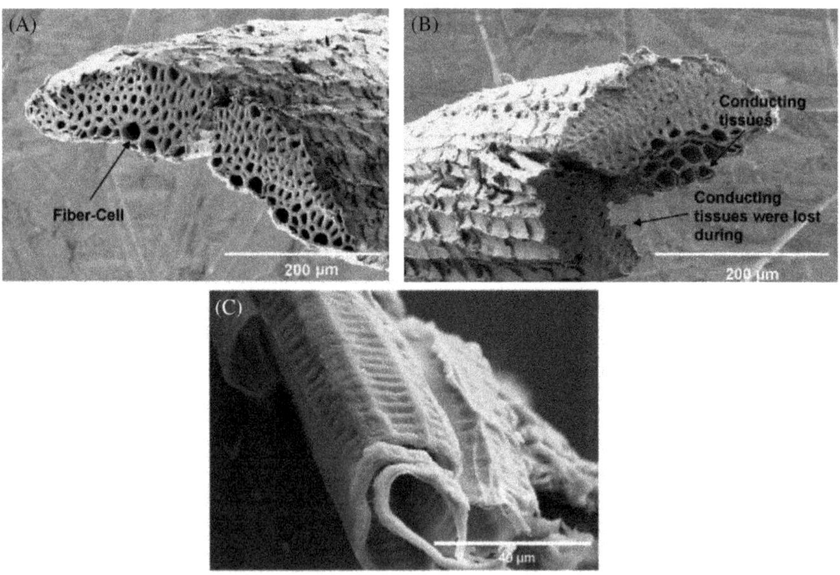

Figure 2.8 Various types of sisal fiber: (A) structural fiber, (B) arch fiber with conducting tissues, and (C) conducting tissue.
Reprinted from Flavio de Andrade Silva et al. 2009. An experimental investigation of the fatigue behavior of sisal fibers, Mater. Sci. Eng. A 516 (2009) 90–95. With permission from Elsevier.

2.2.3.6 Ramie fibers (RFs)

Ramie plant is a semitropical plant cultivated in warm and humid regions with an annual rainfall upto 1000 mm. It is cultivated primarily for its fibers. It is planted from roots instead of seed, root cuttings are planted three inches in the loose soil. The ancient Chinese are reported to extract fiber by stripping off the outside skin and peeling off the ribbon layer. The ribbon layer contains 20–30 wt% of gum. After the removal of gum, soft fibers were made ready for spinning. The chemical composition of RFs comprises of cellulose (68.6–76.2%), lignin (0.6–0.7%), hemicellulose (13.1–16.7%), pectin (1.9%), and wax (0.3%) [68]. The fibers extracted from ramie are white, long, and of silk luster. The fiber cells are longs ranging from 120 to 150 mm which is six times more than cotton, 10 times of flax and 8 times of silk fiber cells. The fibers have higher resistance to chemicals than other NFs and growth of bacteria and fungus including mildew. The global production of ramie in 2011 was 128,000 tones with most of the production from China. A sketch of Ramie plant is given in Fig. 2.9.

Typical industrial applications of ramie fibers includes industrial sewing thread, packing materials, fishing nets, handkerchiefs, woven fiber hoses, filter clothes, draperies, upholstery, bedspreads, dish towels. Ramie fibers are superior of steel wires in specific Young's modulus and close to E-glass fibers. The mechanical properties of ramie fibers vary in literature due to different production methods, fiber diameters, and difference in testing methods. Tensile strength was reported between 333 and 1000 MPa. Young's modulus values of rami fibers are in the range between 15.5 and 65 GPa [68]. The mechanical strength of ramie fibers depends upon fiber diameter and guage length. Bevitori et al. found that tensile strength of ramie fibers show hyperbolic type of inverse relationship with fiber diameter [69]. Angelini et al. found that mechanical properties of ramie fibers increases with decrease in guage length [70]. Nam et al. studied effect of thermal heating on ramie fibers by heating them at 100, 120, 140, 160, and 200°C in air circulating oven for two time periods of 0.5 and 2 h. It was found that heat treatment does not degrade fibers upto 160°C and fibers lose their modulus and tensile strength after two hours of heating at 200°C [71]. Xinqi et al. evaluated the effect of epoxy–silicon oil on

Figure 2.9 Sketch of rami plant.

structural and thermal properties of ramie polypropylene composites. The results showed increase in impact strength, elongation at break and tensile strength by 170, 196, and 39% respectively. The thermal degradation temperature was also found to be increased from 324°C to 370°C [68]. Like most NFs, the hydrophilic nature of ramie fibers causes poor interfacial strength with hydrophobic polymer matrices. This problem can be resolved either by chemical treatment or by addition of coupling agent. Goda et al. studied the effect of alkali treatment on mechanical properties of ramie fibers and found that 5% NaOH treatment of ramie fibers can increase Young's modulus (YM) by 29% as compared to untreated ramie fibers [72]. Yu et al. also found that NaOH treatment of ramie fibers significantly improves the tensile and flexural strength along with storage modulus in ramie fiber reinforced PLA composites [73]. Suizu et al. also studied the effect of 15% mercerization on ramie yarn reinforced polycarprolactone and corn-starch composites [74]. It was found that alkali treatment enhanced elongation and impact resistance but decreased tensile strength and modulus. Silane treatment is also extensively used as coupling agent to enhance interfacial strength between ramie fiber and hydrophobic polymer matrices. Yu et al. found increase in tensile and flexural strength as well as interfacial strength of ramie/ PLA composites [73]. M.J.A. van den Oever et al. found that the flexural stiffness of PLA increases directly with ramie fiber content [75]. Ramie dry fibers (at 105°C) showed slight improving in mechanical properties of PLA comparing with undried fiber. Ramie fibers can be well embedded and dispersed in PLA matrix as shown in Fig. 2.10. Xu et al. studied silane treatment effect on ramie fiber reinforced PLA and polycaprolactone composites. The tensile strength, impact strength and elastic modulus were found to be increased 93, 196, and 84% respectively [76].

2.2.3.7 Palm fibers (PFs)

Palm trees are grown along coastal areas of tropical zones. Palm leave consist of 20 different species depending on tree structure and tropical zone in which they grow.

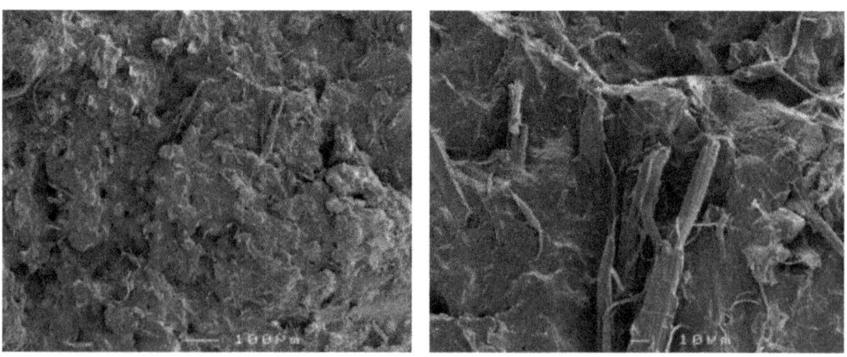

Figure 2.10 SEM of fracture surface of 30 mass% dried ramie–PLA composite. Reprinted from M.J.A. van den Oever et al. 2010. Agrofibre reinforced poly(lactic acid) composites: effect of moisture on degradation and mechanical properties, Compos. Part A 41 (2010) 1628–1635. With permission from Elsevier.

Borassus flabellifer (palmyra palm), *Corypha umbraculifera* (talipot palm), *Corypha taliera* (Roxb.) and *Raphia farinifera* (raffi a palm) leaf fibers are the most commonly used types. The palm plant grows best in swampy soils and high rainfall areas. Palm leaves are mainly used in textile products and construction, they are also used in ropes, beams, sticks, and roof coverings. The long thin fibers are produces by removing membrane on the underside of each individual frond leaf. The fiber pulp obtained from its trunk is used in paper manufacturing.

N. Saba, et al., (2016) stated that the addition of nano filler such as nano oil palm empty fruit bunch, montmorillonite (MMT) or organically modified nanoclays (OMMT) to the kenaf/epoxy composites can increase the storage modulus (E′) values significantly as shown in Fig. 2.11 [77].

2.3 Biodegradable/bio-based polymers as matrices

Matrix selection is a vital part of fiber reinforced composites. Matrix transfers load to fibers and protect them from mechanical abrasion and adverse environment. Natural matrices from renewable resources are being engineered to address environment concerns. Natural matrices are more biodegradable than synthetic polymers, like polyethylene (PE), polyvinylchloride (PVC), and polystyrene (PS). The biodegradability makes them suitable material to substitute synthetic plastics, therefor, several natural polymers like cellulose, starch, polylactate, polyester-amide, and polyhydroxyalkanoates (PHAs) are abundantly used. Some synthetic polymers like polyurethane polyethers (i.e., polyethylene glycols "PEGs", polypropylene glycols "PPGs", and polytetramethlene glycols) can be degraded by specific microorganisms to some extent. However, most of synthetic polymers i.e., polyamides, polyflorocarbons, polyethylene, polypropylene, polycarbonate are resistant to microbial degradation. Polymers from renewable resources can be classified into three groups:

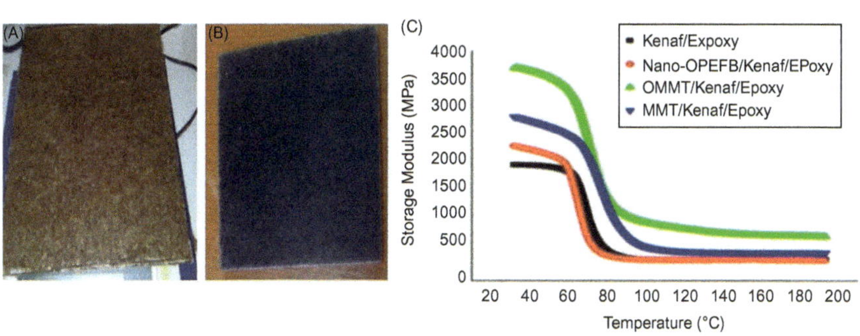

Figure 2.11 (A) Kenaf / epoxy composites, (B) nano oil empty fruit bunch (OPEFB/kenaf/ epoxy hybrid nanocomposite, and (C) Dynamic mechanical analysis (DMA) results. Reprinted from N. Saba et al. 2016. Dynamic mechanical properties of oil palm nano filler/ kenaf/epoxy hybrid nanocomposites, Constr. Build. Mater. 124 (2016) 133–138. With permission from Elsevier.

(1) natural polymers, such as starch, protein, and cellulose; (2) synthetic polymers from natural monomers, such as polylactic acid "polylactide" (PLA); (3) polymers from microbial fermentation, such as polyhydroxyalkanoates (PHAs). Poly lactic acid (PLA) as well as bio-based conventional polymers like polyethylene (PE), polyethylene terephtalate (PET), PHAs, thermoplastic starch polyhydroxybutyrate (PHB), and polyamide (PA) are natural-based or bio-based synthetic polymers [78]. The conventional polymers like PE or PET are not biodegradable. According to the standard (CEN/TR 15932:2010), biodegradation of NFs results in the reduction of molecular weight through chain scission in the backbone and end product should be low molecular weight compounds and biomass regardless of the mechanism of chain scission [78]. When biodegradable material is obtained solely from renewable resources, they are called green polymeric materials. Starch is a carbohydrate composed of many glucose units joined together through glycoside bonds as illustrated in Fig. 2.12A. It is a versatile biopolymer produced by all green vegetables. It is present in large quantities in maize (corn), rice, wheat (Fig. 2.12B) or potatoes, and cassava [79]. It is a mixture of linear amylose and branched amylopectin. It is converted into thermoplastics by destructurization in the presence of plasticizers. The pure starch is insoluble in cold water or alcohol. It is dissolved in hot water to be used as adhesive or stiffener.

Tensile strength, density, and modulus of elasticity properties of biodegradable polymers are illustrated in Fig. 2.13 and 2.14 using Cambridge Engineering Selector (CES) program. It can be noticed that the polyglycolic acid "polyglycolide" (PGA) has high tensile strength (70–117 MPa) and Young's modulus (6.1–7.2 GPa) which make it a good candidate for bioresorbable implantable medical applications notably sutures. Thermoplastic starch (plasticized) which composed of blend of modified corn-starch (polysaccharide), synthetic biodegradable polyester, plasticizer as listed in Table 2.8. Thermoplastic starch (plasticized) can be used for different applications according to the polymer processing. Injection molded: pencil sharpeners, rulers, cartridges, toys, plant pots, plastic bones and other toys for pets, plastic cutlery, hair combs. Thermo-formed: trays for fresh food packaging, especially fruit and vegetables. Film extrusion: shopping bags, bubble film for wrapping, plastic laminates for

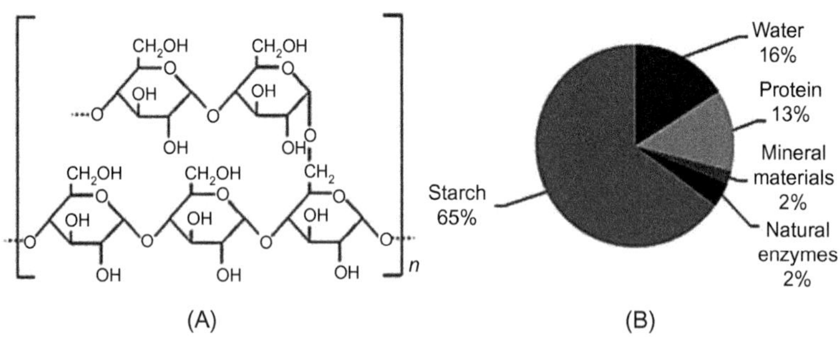

Figure 2.12 (A) Starch chemical formula and (B) Composition of a grain of wheat. Reprinted from A.T. Le et al. 2015. Influence of various starch/hemp mixtures on mechanical and acoustical behavior of starch-hemp composite materials, Compos. Part B 75 (2015) 201–211. With permission from Elsevier.

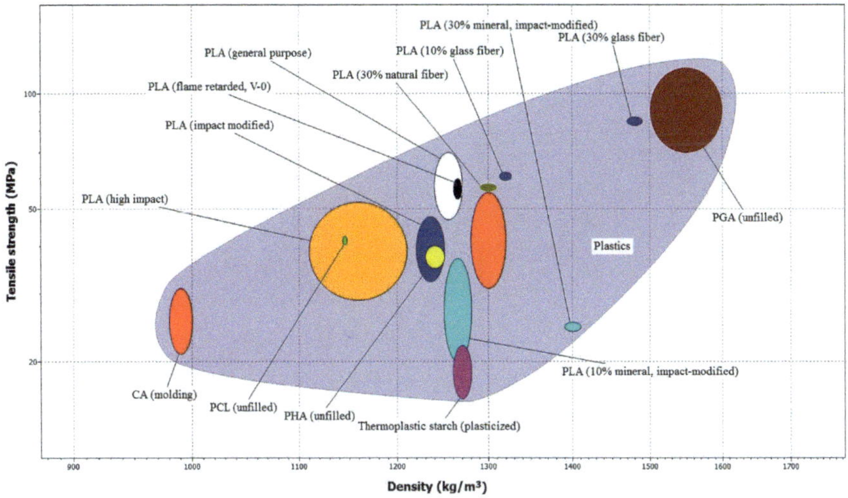

Figure 2.13 Density versus tensile strength of biodegradable polymers (Cambridge Engineering Selector) program.

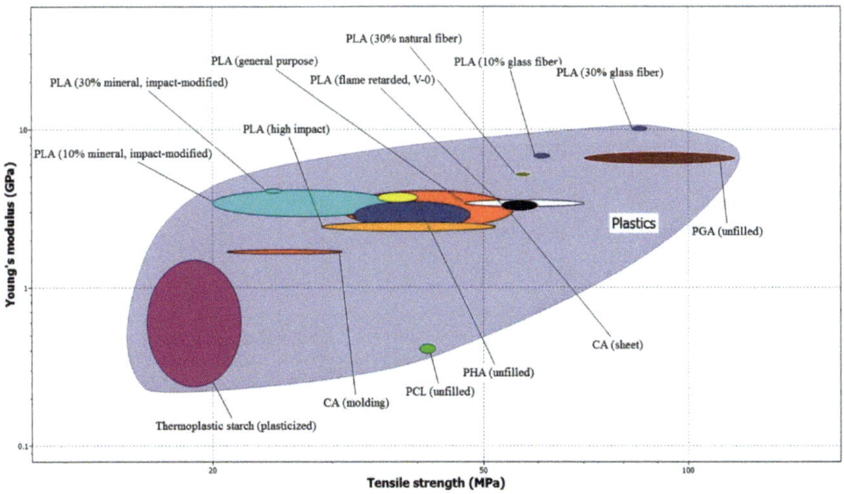

Figure 2.14 Tensile strength versus Modulus of elasticity of biodegradable polymers (Cambridge Engineering Selector) program.

paper cups and plates, bags for rubbish disposal, lining for baby nappies, mulching films for horticulture, wrapping for fruit, vegetables, and sanitary products.

Physical and mechanical properties of biodegradable polymers are listed in Table 2.9. Thermal, electrical, and absorption and permeability properties are given Tables 2.10, 2.11, and 2.12 respectively. Possible processing methods and operating

Table 2.8 **Chemical compositions of biodegradable polymers (CES program)**

	CA (molding)	CA (sheet)	PCL (unfilled)	PGA (unfilled)	PHA (unfilled)	PLA (general purpose)	Thermo plastic starch (plasticized)	PLA (10% glass fiber)	PLA (10% mineral, impact-modified)	PLA (30% glass fiber)	PLA (30% mineral, impact-modified)	PLA (30% natural fiber, (wood flour)	PLA (flame retarded, V-0)
Composition overview													
Base material	CA (Cellulose acetate)	CA (Cellulose acetate)	PCL (Polycap rolactone)	PGA (Poly glycolic acid)	PHA (Poly hydroxy alkano ates)	PLA (Polylactic acid/ polylactide)	Starch (Starch-based thermo plastic)	PLA (Poly lactic acid/ poly lactide)	PLA (Polylactic acid/ polylactide)	PLA (Polylactic acid/ polylactide)	PLA (Polylactic acid/ polylactide)	PLA (Polylactic acid/ polylactide)	PLA (Polylactic acid/ polylactide)
Filler/reinforcement form								Short fiber (<5 mm)	Particulate	Short fiber (<5mm)	Particulate	Short fiber (<5mm)	
Renewable content (%)	100	100	0	0	100	100	49	90	84.9	70	64.8	100	100
Polymer code	CA-P	CA-P	PCL	PGA	PHA	PLA	Starch	PLA-GF10	PLA-I-MD10	PLA-GF30	PLA-I-MD30	PLA-I-NX30	PLA-FR
Composition detail (polymers and natural materials)													
Polymer (%)	69.8	79.8	100	100	100	100	100	90	74.8	70	60	64.8	100
Flame retardant (%)													8.66
Impact modifier (%)	0	0	0	0	0	0	0	0	14.1	0	10	5	0
Plasticizer/oil (%)	29.6	19.4	0	0	0	0	0	0	0	0	0	0	0
Calcium carbonate (powder) (%)	0	0	0	0	0	0	0	0	0	0	30	0	0
Glass (fiber) (%)	0	0	0	0	0	0	0	10	0	30	0	0	0
Mineral (unspecified) (%)	0	0	0	0	0	0	0	0	10	0	0	0	0
Wood flour/ cellulose (%)	0	0	0	0	0	0	0	0	0	0	0	30	0

Table 2.9 **Physical and mechanical properties of biodegradable polymers (CES program)**

	CA (molding)	CA (sheet)	PCL (unfilled)	PGA (unfilled)	PHA (unfilled)	PLA (general purpose)	Thermoplastic starch (plasticized)	PLA (10% glass fiber)	PLA (10% mineral, impact-modified)	PLA (30% glass fiber)	PLA (30% mineral, impact-modified)	PLA (30% natural fiber, (wood flour)	PLA (flame retarded, V-0)
Physical properties													
Density (kg/m^3)	990	1300	1140	1550	1240	1250	1270	1320	1260	1480	1400	1300	1260
Mechanical properties													
Young's modulus (GPa)	1.7	3.14	0.414	6.63	3.74	3.45	0.6	6.89	3.43	10.2	4.14	5.25	3.34
Yield strength (elastic limit) (MPa)	26.5	33.1	28.5	66.3	37.4	62.9	18.8	79	35.3	110	32	74	63.4
Tensile strength (MPa)	25.5	41.4	41.4	90.5	37.4	57.4	18.8	60.8	27.2	84.7	24.6	57	56.4
Elongation (% strain)	33.9	31.6	802	7.21	12.2	3.87	28.3	2.94	13.9	2.94	5.88	1.5	22.8
Elongation at yield (% strain)						2.65		1.5	7.07	1.5	3	0.765	11.6
Compressive modulus (GPa)	1.7	3.14	0.414		3.83	3.45	0.6	6.89	3.43	10.2	4.14	5.25	3.34
Compressive strength (MPa)	40.6	39.7	28.4		42.4	75.5	23.7	94.8	42.4	132	38.4	88.9	76.1
Flexural modulus (GPa)	1.4	3.14	0.553	6.16	4.01	3.34	0.644	6.55	3.43	11.2	4.14	4.75	3.34

(Continued)

Table 2.9 **(Continued)**

	CA (molding)	CA (sheet)	PCL (unfilled)	PGA (unfilled)	PHA (unfilled)	PLA (general purpose)	Thermoplastic starch (plasticized)	PLA (10% glass fiber)	PLA (10% mineral, impact-modified)	PLA (30% glass fiber)	PLA (30% mineral, impact-modified)	PLA (30% natural fiber, (wood flour)	PLA (flame retarded, V-0)
Flexural strength (modulus of rupture) (MPa)	35.7	57.9	58.2	156	53.4	94.7	26.5	93.1	57.8	145	45	58	88.5
Shear modulus (GPa)	0.604	1.13		2.49	2.35	1.24	0.367	2.48	1.23	3.67	1.48	1.89	1.2
Bulk modulus (GPa)	3.07	4.79		8.24		5.99	0.897	11.4	5.19	18	6.26		5.05
Poisson's ratio	0.408	0.391	0.49	0.37	0.39	0.39	0.42	0.39	0.39	0.39	0.39	0.39	0.39
Hardness–Vickers (HV)	7.94	9.88	8.53		12	19.3	5.63	26.2	10.5	33.6	9.49		19
Hardness–Rockwell R	84.9	101	62.2	127	87.3	33.5	41.1	36	27.5	39.5	26	35.5	33.5
Fatigue strength at 10^7 cycles (MPa)	10.2	16.5	13.6	33	14.3	24.8	6.57	25.5	17.5	31.3	16.7	24.5	24.4
Mechanical loss coefficient (tan delta)	0.0276	0.0128		0.0104		0.077		0.124	0.0768	0.162	0.0872	0.103	0.0753
Impact and fracture properties													
Fracture toughness ($MPa.m^{0.5}$)	1.65	1.82	2.02	1.26	0.917	4	1.02	3.98	1.69	5.37	1.32	4.13	4.11
Ductility index	1.71	2.01	5.43		0.0894	0.0911	1.44	0.163	1.36	0.144	0.749	0.081	0.354

Impact strength, notched 23°C (kJ/m^2)	15.1	22.2	51.8	2.16	1.17	1.91	9.06	4.3	16	5.3	7.99	2	7.48
Impact strength, notched −30°C (kJ/m^2)	5.93	5.93		1.41									
Impact strength, unnotched 23°C (kJ/m^2)	595	595		35.4		20		32	93.8	32	79.9	8.5	82.3
Impact strength, Unnotched −30°C (kJ/m^2)	595	595		22.9									

Table 2.10 **Thermal properties of biodegradable polymers (CES program)**

	CA (molding)	CA (sheet)	PCL (unfilled)	PGA (unfilled)	PHA (unfilled)	PLA (general purpose)	Thermoplastic starch (plasticized)	PLA (10% glass fiber)	PLA (10% mineral, impact-modified)	PLA (30% glass fiber)	PLA (30% mineral, impact-modified)	PLA (30% natural fiber, (wood flour)	PLA (flame retarded, V-0)
Thermal properties													
Melting point (°C)			57.3	225	176	159		159	159	159	159	159	159
Glass temperature (°C)	107	82.6	− 65.2	38.2	13.4	55.9	14.1	146	77.2	163	75.5	53	82
Heat deflection temperature 0.45 MPa (°C)	120	81.5	34.6	175	91.3	53.4	50	143	75.7	160	74	52	80.4
Heat deflection temperature 1.8 MPa (°C)	89.4	68		165	92.3	50.8	20	106	72	149	70	54	78.4
Vicat softening point (°C)						53.4		143	75.7	160	74	64.5	80.4
Maximum service temperature (°C)	59.6	59.6	44.7	161	69.3	49.7	69.3	114	60.6	128	59.2	49.7	63.9
Minimum service temperature (°C)	− 63.2	− 44.9	− 54.8	− 56.6	− 64.8	− 15.5	− 54.8	− 15.5	− 15.5	− 15.5	− 15.5	− 15.5	− 15.5

Thermal conductivity (W/m · °C)	0.14	0.237	0.175	0.346	0.173	0.144	0.173	0.944	0.339	2.01	0.928	0.149	0.144
Specific heat capacity (J/kg · °C)	1400	1480	1460	1120	1500	1190	1600	1140	1160	1080	1090	1290	1190
Thermal expansion coefficient (μstrain/°C)	120	220	165	54	208	135	208	115	122	89.2	96.8	109	135
Latent heat of fusion (kJ/kg)				183									

Table 2.11 Electrical properties of biodegradable polymers (CES program)

	CA (molding)	CA (sheet)	PCL (unfilled)	PGA (unfilled)	PHA (unfilled)	PLA (general purpose)	Thermoplastic starch (plasticized)	PLA (10% glass fiber)	PLA (10% mineral, impact-modified)	PLA (30% glass fiber)	PLA (30% mineral, impact-modified)	PLA (30% natural fiber, (wood flour)	PLA (flame retarded, V-0)
Electrical properties													
Electrical resistivity (μohm · cm)	9.95E + 18	9.95E + 18	6.33E + 19	1.00E + 20	1.00E + 17	4.31E + 17	1.00E + 17	1.52E + 22	9.97E + 20	3.50E + 22	3.04E + 17	3.46E + 17	4.31E + 17
Dielectric constant (relative permittivity)	5.12	5.12	4.29	4.18	3.87	3.1	4.47	3.87	3.43	4.9	4.64	4.47	3.1
Dissipation factor (dielectric loss tangent)	0.0245	0.0245	0.0021	0.0106	0.0866	0.01	0.0866	0.0156	0.00909	0.0232	0.00729	0.106	0.01
Dielectric strength (dielectric breakdown) (MV/m)	17.9	15.2	16.6	14.5	13.9	16.7	13.9	16.1	16.1	15.4	14.1	13.4	16.7

Table 2.12 **Absorption and permeability of biodegradable polymers (CES program)**

	CA (molding)	CA (sheet)	PCL (unfilled)	PGA (unfilled)	PHA (unfilled)	PLA (general purpose)	Thermoplastic starch (plasticized)	PLA (10% glass fiber)	PLA (10% mineral, impact-modified)	PLA (30% glass fiber)	PLA (30% mineral, impact-modified)	PLA (30% natural fiber, (wood flour)	PLA (flame retarded, V-0)
Absorption and permeability													
Water absorption @ 24 h (%)	2.51	2.83	0.34	0.474	0.548	0.25	7.07	0.125	0.125	0.125	0.125	0.877	0.125
Water absorption @ sat (%)				1.41	1.58	1.02	36.6	0.504	0.504	0.504	0.504	1.34	0.504
Humidity absorption @ sat (%)				0.424		0.305	11	0.15	0.15	0.15	0.15	0.406	0.15
Water vapor transmission ($g \cdot mm/m^2 \cdot day$)	0.394	0.394		0.199		8.04		8.04	8.04	8.04	8.04	8.04	8.04
Permeability (O_2) ($cm^3 \cdot mm/m^2 \cdot day \cdot atm$)	5.87	5.87		0.0767		24.1		24.1	24.1	24.1	24.1	24.1	24.1
Permeability (CO_2) ($cm^3 \cdot mm/m^2 \cdot day \cdot atm$)	149	149		0.255		74.8		74.8	74.8	74.8	74.8	74.8	74.8
Permeability (N_2) ($cm^3 \cdot mm/m^2 \cdot day \cdot atm$)	1.86	1.86		0.0309		5.1		5.1	5.1	5.1	5.1	5.1	5.1

conditions of biodegradable polymers are tabulated in Table 2.13. Acids, alkalis, and oil resistance of biodegradable polymers are pointed out in Table 2.14. Processing energy and footprint (CO_2) values are listed in Table 2.15.

Anderson et al. have evaluated interfacial shear strength (ISS) of flax fiber/ starch acetate composites with using stress strain curve of composite in tension. The ISS was found to be between 5.0 and 20.5 MPa depending upon composition of material which is within range of good matrix adhesion. The interfacial shear strength (IFSS) trend was found to be increasing with increase fiber loading and decreasing with increase content of plasticizers [80]. M. Iman et al. prepared nano-composites using starch, jute, glutaraldehyde, nanoclay, and glycerol. It was found that by adding 50% glutaraldehyde (GA) (crosslinking agent) increased mechanical, thermal, flame retardency, and dimensional stability via interaction of the aldehydes group of GA with the hydroxyl group of starch and jute [81]. A. Le et al. fabricated starch-hemp composites using various hemp/starch ratios [79]. It was found that hemp reinforced starch composite materials are good sound absorbing materials and can be used in the manufacturing of soundproof wall. Zainduddin et al. prepared biodegradable composites of starch and chemically treated kenaf fibers. They used glycerol/sorbitol as filler with different compositions (0–10 wt%) and found 6% filler content optimum with tensile strength (TS) of 8.2 MPa [82]. Cellulose is semicrysalline natural polymer and one of the main components of NFs. It is composed of repeating d-gucopyranose units. The structure of cellulose esters including cellulose acetate (CA), cellulose acetate propionate (CAP), and cellulose acetate butyrate (CAB). CAP and CAB are used in variety of plastic applications. CA represents an important class of cellulose based thermoplastic with different degrees of acetylation. It is used in various applications as filters, membranes, films, food packaging, and capsules for drug delivery. Gutiérrez et al. prepared bio-composites using cellulose acetate and chemically treated short curauá fibers using large scale extrusion and injection molding [83]. They found that 50% increase in Young's modulus and 100% increase in thermal stability in chemically treated fiber composites as compared to plasticized cellulose acetate/ curauá fibers composites.

2.3.1 Polylactic acid (PLA)

PLA is a biodegradable, highly versatile polyester produced from renewable sources. It is mainly based on agricultural bases such as corn, beet, wheat, and other starchy products. It has high TS and YM than PS, PP, and PE. The elastic modulus of PLA is comparable to PET but, PLA is very brittle and has low toughness and less than 10% elongation at break. These factors limit its use in applications involving plastic deformation at high stress value. Rasal et al have reported rise in the impact resistance of PLA by introducing ethylene-co-vinyl acetate polymer. PLA is obtained through the polymerization of lactic acid by a fermentation process using 100% annually renewable resources [84,85]. In a study by Oksman et al. 15% tiacetin (glycerol triacetate) was added to reduce brittle nature of PLA in PLA-flax composites. It was found that composites with 30–40 wt% flax have tensile strength 50% better than PP-flax composites [86]. Several drawbacks limit PLA

Table 2.13 **Processing of biodegradable polymers (CES program)**

	CA (molding)	**CA (sheet)**	**PCL (unfilled)**	**PGA (unfilled)**	**PHA (unfilled)**	**PLA (general purpose)**	**Thermoplastic starch (plasticized)**	**PLA (10% glass fiber)**	**PLA (10% mineral, impact-modified)**	**PLA (30% glass fiber)**	**PLA (30% mineral, impact-modified)**	**PLA (30% natural fiber, (wood flour)**	**PLA (flame retarded, V-0)**
Processing properties													
Polymer injection molding	Excellent	Unsuitable	Acceptable	Acceptable	Acceptable	Acceptable	Acceptable	Acceptable	Acceptable	Acceptable	Acceptable	Acceptable	Acceptable
Polymer extrusion	Limited use	Unsuitable	Acceptable	Acceptable	Limited use	Acceptable	Acceptable	Limited use	Limited use	Limited use	Limited use	Limited use	Limited use
Polymer thermo forming	Limited use	Limited use	Unsuitable	Acceptable	Unsuitable	Acceptable	Unsuitable	Limited use	Limited use	Limited use	Limited use	Limited use	Limited use
Linear mold shrinkage (%)	0.52	0.52	0.324		0.274	0.346	0.274	0.346	0.346	0.346	0.346	0.253	0.49
Melt temperature (°C)	205	205	119	244	184	189	185	184	184	184	184	177	180
Mold temperature (°C)	44.7	44.7	21.7	110	64.3	15.8	32.4	91.7	91.7	91.7	91.7	13.2	49
Molding pressure range (MPa)	110	110	51	70.7	62.4	74.2	62	75.3	75.3	75.3	75.3	74.2	74.2

Table 2.14 **Durability of biodegradable polymers (CES program)**

	CA (molding)	CA (sheet)	PCL (unfilled)	PGA (unfilled)	PHA (unfilled)	PLA (general purpose)	Thermoplastic starch (plasticized)	PLA (10% glass fiber)	PLA (10% mineral, impact-modified)	PLA (30% glass fiber)	PLA (30% mineral, impact-modified)	PLA (30% natural fiber, (wood flour)	PLA (flame retarded, V-0)
Durability													
Water (fresh)	Excellent	Excellent	Limited use	Limited use	Acceptable	Acceptable	Limited use	Acceptable	Acceptable	Acceptable	Acceptable	Acceptable	Acceptable
Water (salt)	Excellent	Excellent	Limited use	Limited use	Acceptable	Acceptable	Limited use	Acceptable	Acceptable	Acceptable	Acceptable	Acceptable	Acceptable
Weak acids	Unacceptable	Unacceptable	Limited use	Limited use	Unacceptable	Acceptable	Limited use	Acceptable	Acceptable	Acceptable	Acceptable	Acceptable	Acceptable
Strong acids	Unacceptable	Unacceptable	Unacceptable	Unacceptable	Unacceptable	Unacceptable	Unacceptable	Unacceptable	Unacceptable	Unacceptable	Unacceptable	Unacceptable	Unacceptable
Weak alkalis	Unacceptable	Unacceptable	Limited use	Unacceptable	Acceptable	Acceptable	Limited use	Acceptable	Acceptable	Acceptable	Acceptable	Acceptable	Acceptable
Strong alkalis	Unacceptable	Unacceptable	Unacceptable	Unacceptable	Unacceptable	Unacceptable	Unacceptable	Unacceptable	Unacceptable	Unacceptable	Unacceptable	Unacceptable	Unacceptable
Organic solvents	Unacceptable	Unacceptable	Limited use	Limited use	Unacceptable	Limited use	Unacceptable	Limited use	Limited use	Limited use	Limited use	Limited use	Limited use
Oils and fuels			Excellent	Excellent									

Table 2.15 Processing energy and CO_2 footprint of biodegradable polymers (CES program)

	CA (molding)	CA (sheet)	PCL (unfilled)	PGA (unfilled)	PHA (unfilled)	PLA (general purpose)	Thermoplastic starch (plasticized)	PLA (10% glass fiber)	PLA (10% mineral, impact-modified)	PLA (30% glass fiber)	PLA (30% mineral, impact-modified)	PLA (30% natural fiber, (wood flour)	PLA (flame retarded, V-0)
Processing energy and CO_2 footprint													
Polymer extrusion energy (MJ/kg)	6.06	6.08	5.9	6.24	6.04	5.94	6.07	5.93	5.93	5.91	5.92	5.96	5.94
Polymer extrusion CO_2 (kg/kg)	0.455	0.456	0.442	0.464	0.453	0.445	0.455	0.445	0.445	0.444	0.444	0.447	0.445
Polymer molding energy (MJ/kg)	18	18.6	13.4	22.8	17.5	14.3	18.2	14	14.1	13.7	13.7	14.8	14.3
Polymer molding CO_2 (kg/kg)	1.35	1.39	1.01	1.71	1.31	1.07	1.36	1.05	1.06	1.02	1.03	1.11	1.07
Coarse machining energy (per unit wt removed) (MJ/kg)	0.909	0.805	0.747		0.841	0.545	0.685	0.546	0.522	0.557	0.518	0.544	0.544
Coarse machining CO_2 (per unit wt removed) (kg/kg)	0.0681	0.0604	0.0561		0.0631	0.0409	0.0514	0.041	0.0391	0.0418	0.0388	0.0408	0.0408
Fine machining energy (per unit wt removed) (MJ/kg)	4.59	3.55	2.98		3.92	0.954	2.36	0.96	0.714	1.07	0.676	0.938	0.945
Fine machining CO_2 (per unit wt removed) (kg/kg)	0.344	0.266	0.223		0.294	0.0716	0.177	0.072	0.0535	0.0803	0.0506	0.0703	0.0708

(Continued)

Table 2.15 (Continued)

	CA (molding)	CA (sheet)	PCL (unfilled)	PGA (unfilled)	PHA (unfilled)	PLA (general purpose)	Thermoplastic starch (plasticized)	PLA (10% glass fiber)	PLA (10% mineral, impact-modified)	PLA (30% glass fiber)	PLA (30% mineral, impact-modified)	PLA (30% natural fiber, (wood flour)	PLA (flame retarded, V-0)
Grinding energy (per unit wt removed) (MJ/kg)	8.68	6.61	5.46		7.33	1.41	4.22	1.2	0.821	1.64	0.85	1.37	1.39
Grinding CO_2 (per unit wt removed) (kg/kg)	0.651	0.495	0.409		0.55	0.106	0.317	0.0895	0.0615	0.123	0.0635	0.103	0.104

performance like low heat resistance and its high cost. Its rapid physical aging makes it brittle material with low impact strength [87]. Ding et al. investigated molecular and crystallization behavior of PLA/northern bleached softwood kraft (NBSK)/Poly ethlylene glycol composite foams which were fabricated using injection molding process. The results showed that the incorporation of cellulolic fibers of NBSK increases the crystallization temperature and degree of crystallinity and decrease the crystallization half time [88].

2.3.2 Polyhydroxyalkanoates (PHAs)

Polyhydroxyalkanoates (PHAs) are naturally occurring biodegradable polyesters, which are produced by bacterial fermentation. They are important because they are easily biodegradable by a multiplicity of microorganisms within one year [89]. On the basis of quantity of carbon atoms in PHA monomers, they can be parted in two groups, the short chain length PHAs (scl-PHAs) with C_3-C_5 atoms and medium chain length PHAs (mcl-PHAs) with C_6-C_{14} atoms. Poly(3-hydroxy burate) (P(3HB)) is the example of scl-PHAs. PHB has an elasticity modulus of 3 GPa and ultimate TS of 25 MPa. Some drawbacks are associated with PHB like its poor mechanical properties, high fragility and it has low thermal stability and is difficult to process in molten state as it starts to degrade at temperature close to 200°C. The toughness and processability of PHB can be increased by addition of 3-hydroxyvalerate (3HV) in the bacterial fermentation process. Increasing 3HV content increased the yield strength and YM of PHB, meanwhile, increases the cost of materials. The melting temperature of PHBV declines with increasing component of 3HV, with a minimum value 75°C at approximately 40 mol% 3HV. Barkoula et al. studied addition of flax fiber in PHB matrix promotes toughness and stiffness of matrix. However, drop in TS was found in the initial stage of degradation [90]. Hossain et al. prepared nanoclay infused Jute/PHB-HV composites using 2%, 3%, and 4% by weight nanoclay and found moisture absorption can be reduced by infusing 4% nanoclay in matrix [91].

Matrix selection for specific fiber depends upon its degradation temperature; usually NFs are thermally stable till 200°C. The thermoplastics like polyolefin, polystyrene, polyethylene, and polypropylene can be processed below average degradation temperature of NFs.

2.3.3 Aliphatic polyesters

Aliphatic polyesters are obtained synthetically by polycondensation reaction of glycols and aliphatic dicarboxylic acid. Generally the monomers used are petroleum based. However, some of the monomers e.g., 1,3-propanediol and succinic acids are prepared from natural resources [92]. This methods is limited by low degree of polymerization, yielding low molecular weight polymers, failing to produce block polymers. This method produces oligomers having molecular weight of few thousands. Oligomers possess poor thermomechanical properties, unsuitable for commercial production. However, chain extension can be performed on these oligomers to get high molecular weight polymers. Therefore, oligomer produced by the reaction of octadecanedicarboxylic acid with caprolactone diol that can be subjected to a sequential chain extension reaction with sebacoyl chloride. Another synthetic route for the synthesis of aliphatic polyesters is through ring opening polymerization of lectones and cyclic diesters which yields high molecular weight polymers under favorable conditions. Aliphatic polyesters also originate from nature in the form of polyhyroxyalkanoates by bacterial fermentation of sugars or lipids.

The physical properties of the polyester depends upon multiple factors. Monomer composition, chain flexibility, molecular mass, polarity of the groups attached, crystallinity, extent of branching, and orientation are some of the factors affecting its physical properties. These properties can be tailored by the process of blending, copolymerization or by altering the macromolecular architecture.

2.3.4 *Aliphatic aromatic copolyesters*

The biodegradability of aliphatic polyesters can be combined with the mechanical strength of aromatic polyesters by copolyesterification of both monomers. Aliphatic acids and aromatic dicarboxylic acids react with aliphatic glycols in a melt polycondensation process to form aliphatic aromatic copolymers. The best available aliphatic aromatic copolymers is Ecoflex; developed by BASF. Ecoflex originated as a result of a study undertaken by BASF on the request of German state government to make biodegradable plastics for packaging. It is made of adipic acid, terephthalic acid, 1,4-butanediol and modular units. Ecoflex is designed to have a specific set of properties in processability, biodegradability, and utilization, which is achieved by producing a suitable molecular structure obtained with the help of modular units [93]. The modular units connects to the hydrophilic part of the copolyesters monomer in a branched fashion which leads to lengthening of the molecular chain and thus the molecular weight of the tailor-made product. The concept of modular system is shown in the Fig. 2.15.

Ecoflex possesses comparable thermal and mechanical properties to low density polyethylene (LDPE). A comparison of mechanical properties of blown film Ecoflex and LDPE (Table 2.16) reveals that Ecoflex bears high strength and failure energy, superseding LDPE values. Moreover, Ecoflex films are slightly permeable to water vapors making the film breathable and this property can be manipulated in different batches of Ecoflex.

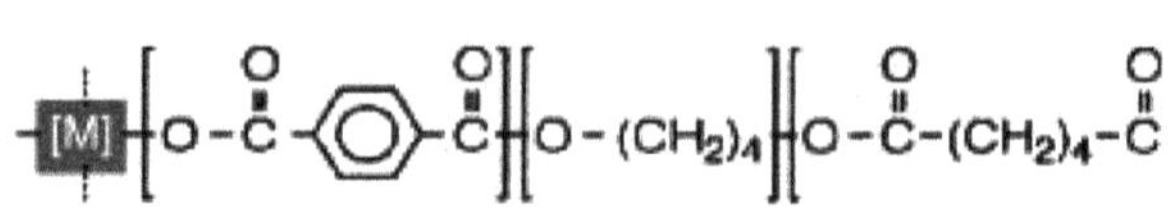

Figure 2.15 The connection of modular structure with the copolyesters.

Table 2.16 **Physical, mechanical and permeability of Ecoflex and PE-LD**

Property	Unit	Test method	Ecoflex	Lupolen ® 2420 F (LDPE)
Transparency	%	ASTM D 1003	82	89
Tensile strength	N mm^{-2}	ISO 527	32/36	26/20
Ultimate strength	N mm^{-2}	ISO 527	32/36	–
Ultimate elongation	%	ISO 527	580/ 820	300/600
Failure energy (Dyna-test)	J mm^{-1}	DIN 53373	14.3	5.5
Tear propagation resistance	N mm^{-2}	DIN 53363	236/ 124	–
Permeation rates				
Oxygen	Cm^3 (m^2 d · bar)	DIN 53380	1600	2900
Water vapor	g (m^{-2} · d)	DIN 53122	1400	1.7

2.3.5 *Polyester amides*

Polyester amides (PEAs) are biodegradable polymers containing ester and amide linkage in the chain and exhibit properties from the parent polyester and polyamide families. In physiological ambiance, polyesters degrade from the ester linkage cleavage. Moreover, they exhibit a better solubility in organic solvents, and better mechanical properties. Polyamides, on the other hand, possess better thermomechanical properties due to the presence of hydrogen bond among the individual chains. Polyamides have a very low degradation rate in the human body making it essentially nondegradable. A combination of ester and amide inherits biodegradation and better mechanical properties and opens new horizons for material design for commercial and biomedical use.

2.3.6 *Polybutylene succinates*

Polybutylene succinate (PBS) is a well-known and stable biodegradable polymer in the family of polyesters, which is produced by succinic acid, butanediol, or other carboxylates and alkyldiols. Its is a white crystalline solid with density, melting point and glass transition temperature in the range of 1.25 g/cm^3, 90–120°C and

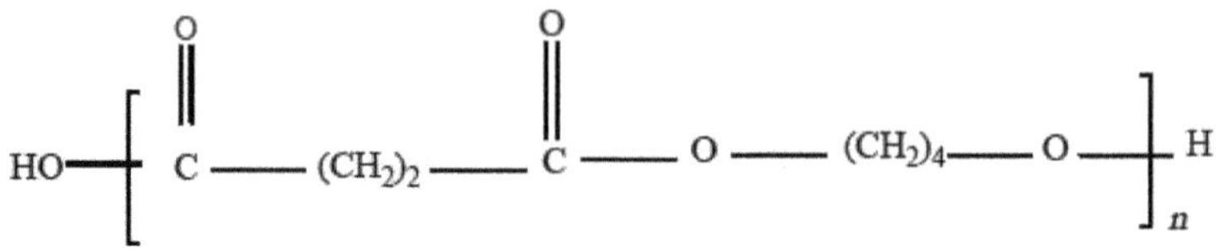

Figure 2.16 Poly butylene succinate recurring unit.

$-45°C$ to $-10°C$ respectively. The properties and processibilty of PBS is comparable to the polyethylene. It possesses good mechanical properties, thermally stable upto 200°C, and is biodegradable in nature. The recurring unit in PBS is shown in the Fig. 2.16.

PBS is produced industrially by condensation polymerization of succinic acid (derived from petroleum resources) and butanediolm, both of the reagents obtained from maleic anhydride. Due to excellent mechanical properties, PBS can be processed through conventional melt processing equipment of extrusion, injection, and blow molding. General applications of PBS include packaging, mulch films, bags, and disposable hygiene products. In view of the rapidly increasing market for biodegradable plastics, its demand is expected to increase in future.

2.3.7 Polyvinyl alcohol

Polyvinyl alcohol (PVA) is a translucent water soluble resin mainly used in fabric sizing and paper coating processes. PVA is unique in its production as it does not result as a polymerization of monomers. Instead, it is produced by reacting poly vinyl acetate with methanol in the presence of an alkaline catalyst like sodium hydroxide [85]. This hydrolysis reaction results in replacement of acetate group with the hydroxyl group, with any modification in the long vinyl chain, thus making polyvinyl alcohol with the repeating unit:

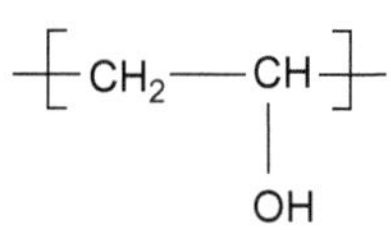

The PVA is sold as completely and partially hydrolyzed. When completely hydrolyzed, PVA becomes strongly soluble in water but very little soluble in organic solvents. Partially hydrolyzed PVA is less soluble in water and more soluble in organic solvent. Polyvinyl alcohol is among the few vinyl polymers that can have higher biodegradation rate due to OH group. PVA is used in sizing agents to control the absorbency of the yarn, giving it greater strength to textile yarns. It is also used in paper industry to for coating paper to make it resilient to oils and greases [85].

References

[1] U. Javaid, Z.M. Khan, M. Khan, M. Bassyouni, S.-S. Abdel-Hamid, M. Abdel-Aziz, et al., Compos. Part B Eng. 91 (2016) 257–266.

[2] M. Bassyouni, N. Iqbal, S.S. Iqbal, S.-S. Abdel-Hamid, M. Abdel-Aziz, U. Javaid, et al., Polym. Degrad. Stab. 110 (2014) 195–202.

[3] O. Al-Qabandi, A. De Silva, S. Al-Enezi, M. Bassyouni, J. Reinf. Plast. Compos., (2014) 0731684414561538.

[4] M. Bassyouni, S.A. Gutub, Polym. Polym. Compos. 21 (2013) 463.

[5] S.A. Sherif, M. Sadek, F. Ashour, M. Bassyouni, Polym. Polym. Compos. 17 (2009) 481.

[6] M. Bassyouni, S.W. Ul Hasan, The use of rice straw and husk fibers as reinforcements in composites, Biofiber Reinf. Compos. Mater. 51 (2015) 385–422.

[7] M. Bassyouni, I. Taha, M.-S. Shereen, L. Steuernagel, J. Reinf. Plast. Compos. 31 (2012) 303–312.

[8] N.G. Jústiz-Smith, G.J. Virgo, V.E. Buchanan, Mater. Characterization 59 (2008) 1273–1278.

[9] B. Hachemane, R. Zitoune, B. Bezzazi, C. Bouvet, Compos. Part B Eng. 51 (2013) 1–10.

[10] S.V. Joshi, L. Drzal, A. Mohanty, S. Arora, Compos. Part A Appl. Sci. Manuf. 35 (2004) 371–376.

[11] J. Holbery, D. Houston, Jom 58 (2006) 80–86.

[12] S. Mishra, A. Mohanty, L. Drzal, M. Misra, S. Parija, S. Nayak, et al., Compos. Sci. Technol. 63 (2003) 1377–1385.

[13] C. Hong, I. Hwang, N. Kim, D. Park, B. Hwang, C. Nah, J. Industrial Eng. Chem. 14 (2008) 71–76.

[14] C. Scarponi, C. Pizzinelli, Int. J. Mater. Prod. Technol. 36 (2009) 278–303.

[15] J.I. Morán, V.A. Alvarez, V.P. Cyras, A. Vázquez, Cellulose 15 (2008) 149–159.

[16] B. Xiao, X. Sun, R. Sun, Polym. Degrad. Stab. 74 (2001) 307–319.

[17] M.A. Fuqua, S. Huo, C.A. Ulven, Polym. Rev. 52 (2012) 259–320.

[18] A. Aguilar-Rios, P. Herrera-Franco, A. de J. Martínez-Gómez, A. Valadez-Gonzalez, Express Polym. Lett. 8 (2014).

[19] A. Bledzki, J. Gassan, Progr. Polym. Sci. 24 (1999) 221–274.

[20] R. Lundmark, Scand. J. For. Res. 18 (2003) 184–192.

[21] M. Bassyouni, S.W. ul Hasan, M.H. Abdel-Aziz, S.M.-S. Abdel-hamid, S. Naveed, A. Hussain, et al., Date palm waste gasification in downdraft gasifier and simulation using ASPEN HYSYS, Energy Convers. Manage. 88 (2014) 693–699.

[22] K. Li, X. Lei, L. Lu, C. Camm, Pulp Paper Can. 111 (2010) 28–33.

[23] D. Tranchida, S. Piccarolo, J. Loos, A. Alexeev, Macromolecules. 40 (2007) 1259–1267.

[24] V. Mediavilla, M. Leupin, A. Keller, Industrial Crops Prod. 13 (2001) 49–56.

[25] V.K. Thakur, M.K. Thakur, Carbohydr. Polym. 109 (2014) 102–117.

[26] H. Akil, M. Omar, A. Mazuki, S. Safiee, Z. Ishak, A.A. Bakar, Mater. Des. 32 (2011) 4107–4121.

[27] P. Chen, C. Lu, Q. Yu, Y. Gao, J. Li, X. Li, J. Appl. Polym. Sci. 102 (2006) 2544–2551.

[28] F.L. Matthews, R.D. Rawlings, Composite Materials: Engineering and Science, Elsevier, New York, NY, 1999.

[29] K. Pickering, Properties and Performance of Natural-fibre Composites, Elsevier, New York, NY, 2008.
[30] M.P. Gashti, M.P. Gashti, J. Dispers. Sci. Technol. 34 (2013) 853–858.
[31] S. Kalia, B. Kaith, I. Kaur, Polym. Eng. Sci. 49 (2009) 1253–1272.
[32] X. Li, L.G. Tabil, S. Panigrahi, J. Polym. Environ. 15 (2007) 25–33.
[33] A. Alawar, A.M. Hamed, K. Al-Kaabi, Compos. Part B Eng. 40 (2009) 601–606.
[34] D. Ray, B. Sarkar, J. Appl. Polym. Sci. 80 (2001) 1013–1020.
[35] L.Y. Mwaikambo, M.P. Ansell, J. Appl. Polym. Sci. 84 (2002) 2222–2234.
[36] D.B. Dittenber, H.V. GangaRao, Compos. Part A Appl. Sci. Manuf. 43 (2012) 1419–1429.
[37] R. Mahjoub, J.M. Yatim, A.R.M. Sam, S.H. Hashemi, Constr. Build. Mater. 55 (2014) 103–113.
[38] V. Fiore, G. Di Bella, A. Valenza, The effect of alkaline treatment on mechanical properties of kenaf fibers and their epoxy composites, Compos. Part B Eng. 68 (2015) 14–21.
[39] O. Faruk, A.K. Bledzki, H.P. Fink, M. Sain, Macromol. Mater. Eng. 299 (2014) 9–26.
[40] B. Barari, E. Omrani, A.D. Moghadam, P.L. Menezes, K.M. Pillai, P.K. Rohatgi, Mechanical, physical and tribological characterization of nano-cellulose fibers reinforced bio-epoxy composites: an attempt to fabricate and scale the 'Green'composite, Carbohydr. Polym. 147 (2016) 282–293.
[41] C. Pang, R.A. Shanks, F. Daver, Compos. Part A Appl. Sci. Manuf. 70 (2015) 52–58.
[42] J.A. Fernández, N. Le Moigne, A. Caro-Bretelle, R. El Hage, A. Le Duc, M. Lozachmeur, et al., Industrial Crops Prod. 85 (2016) 93–108.
[43] G. Marques, J.C. del Río Andrade, A. Gutiérrez Suárez, Fate of lipophilic extractives from several non-wood species during alkaline pulping and TCF/ECF bleaching, 2009.
[44] B. Tajeddin, R. Abdul Rahman, L. Chuah, N.A. Ibrahim, Y.A. Yusof, Eur. J. Sci. Res. 32 (2009) 223–230.
[45] S.H. Aziz, M.P. Ansell, S.J. Clarke, S.R. Panteny, Modified polyester resins for natural fibre composites, Compos. Sci. Technol. 65 (2005) 525–535.
[46] K. Abe, Y. Ozaki, Soil Sci. Plant Nutr. 44 (1998) 599–607.
[47] N. Saba, M. Jawaid, K. Hakeem, M. Paridah, A. Khalina, O. Alothman, Renew. Sustain. Energy Rev. 42 (2015) 446–459.
[48] M. Jonoobi, J. Harun, P. Tahir, L. Zaini, A.S. Saiful, M. Makinejad, Characteristics of nanofibers extracted from kenaf core, BioResources 5 (2010) 2556–2566.
[49] P.A. Ling, H. Ismail, BioResources 8 (2012) 806–817.
[50] O.M.L. Asumani, R.G. Reid, R. Paskaramoorthy, The effects of alkali–silane treatment on the tensile and flexural properties of short fibre non-woven kenaf reinforced polypropylene composites, Compos. Part A Appl. Sci. Manuf. 43 (9) (2012) 1431–1440.
[51] F.F. Shuhimi, M.F.B. Abdollah, M.A. Kalam, M. Hassan, A. Mustafa, H. Amiruddin, Tribological characteristics comparison for oil palm fibre/epoxy and kenaf fibre/epoxy composites under dry sliding conditions, J. Tribol. Int. 101 (2016) 247–254.
[52] R.M. Rowell, J.S. Han, J.S. Rowell, Natural Polymers and Agrofibers Bases Composites, Embrapa Instrumentacao Agropecuaria, Sao Carlos, SP, Brazil, 2000, pp. 115–134.
[53] Y. Cai, S. David, M. Pailthorpe, Dyeing of jute and jute/cotton blend fabrics with 2: 1 pre-metallised dyes, Dyes Pigments 45 (2000) 161–168.
[54] M.N. Islam, M.R. Rahman, M.M. Haque, M.M. Huque, Compos. Part A Appl. Sci. Manuf. 41 (2010) 192–198.
[55] A. Karthikeyan, K. Balamurugan, J. Sci. Industrial Res. 71 (2012) 627–631.

[56] H.U. Zaman, M. Beg, J. Compos. Mater. 48 (2014) 3293–3301.
[57] S.S. Munawar, K. Umemura, S. Kawai, J. Wood Sci. 53 (2007) 108–113.
[58] L.Q.N. Tran, T. Nguyen Minh, C.A. Fuentes, T. Truong Chi, A.W. Van Vuure, I. Verpoest, Investigation of microstructure and tensile properties of porous natural coir fibre for use in composite materials, Ind, Crops Prod. 65 (2015) 437–445.
[59] L. Yan, N. Chouw, X. Yuan, J. Reinf. Plast. Compos., 2012 0731684412439494.
[60] T.H. Nam, S. Ogihara, H.T. Nguyen, S. Kobayashi, Mechanical and thermal properties of short coir fibre reinforced poly (butylene succinate) biodegradable composites, J. Solid Mech. Mater. Eng. 5 (6) (2011) 251–262.
[61] A. Belaadi, A. Bezaz, M. Bourchak, F. Scarpa, Tensile static and fatigue behaviour of sisal fibres, Mater. Design 46 (2013) 76–83.
[62] A. Barreto, M. Esmeraldo, D. Rosa, P. Fechine, S. Mazzetto, Polym. Compos. 31 (2010) 1928–1937.
[63] F. de Andrade Silva, N. Chawla, R.D. de Toledo Filho, Compos. Sci. Technol. 68 (2008) 3438–3443.
[64] S.G. Bergström, H.-E. Gram, Durability of alkali-sensitive fibres in concrete, Int. J. Cem. Compos. Lightweight Concr. 6 (1984) 75–80.
[65] F. de Andrade Silva, N. Chawla, R.D. de Toledo Filho, An experimental investigation of the fatigue behavior of sisal fibers, Mater. Sci. Eng. A 516 (2009) 90–95.
[66] H. Zou, L. Wang, H. Gan, C. Yi, Effect of fiber surface treatments on the properties of short sisal fiber/poly (lactic acid) biocomposites, Polym. Compos. 33 (2012) 1659–1666.
[67] Y. Li, Y.-W. Mai, L. Ye, Compos. Sci. Technol. 60 (2000) 2037–2055.
[68] X. Li, L. He, H. Zhou, W. Li, W. Zha, Carbohydr. Polym. 87 (2012) 2000–2004.
[69] A.B. Bevitori, I.L.A. da Silva, R.S. Carreiro, F.M. Margem, S.N. Monteiro, Elastic modulus variation with diameter for ramiefibers, Charact. Miner. Metals Mater. (2012) 395–402.
[70] L.G. Angelini, M. Scalabrelli, S. Tavarini, P. Cinelli, I. Anguillesi, A. Lazzeri, Ramie fibers in a comparison between chemical and microbiological retting proposed for application in biocomposites, Ind. Crops Prod. 75 (2015) 178–184.
[71] S. Nam, A.N. Netravali, Green composites. I. Physical properties of ramie fibers for environment-friendly green composites, Fibers Polym. 7 (4) (2006) 372–379.
[72] K. Goda, T. Asai, T. Yamane, J. Soc. Mater. Sci. Japan 52 (2003) 1245–1252.
[73] T. Yu, J. Ren, S. Li, H. Yuan, Y. Li, Effect of fiber surface-treatments on the properties of poly (lactic acid)/ramie composites, Compos. Part A Appl. Sci. Manuf. 41 (2010) 499–505.
[74] N. Suizu, T. Uno, K. Goda, J. Ohgi, J. Mater. Sci. 44 (2009) 2477–2482.
[75] M.J.A. Van den Oever, B. Beck, J. Müssig, Agrofibre reinforced poly (lactic acid) composites: Effect of moisture on degradation and mechanical properties, Compos. Part A Appl. Sci. Manuf. 41 (11) (2010) 1628–1635.
[76] H. Xu, L. Wang, C. Teng, M. Yu, Biodegradable composites: Ramie fibre reinforced PLLA-PCL composite prepared by in situ polymerization process, Polym. Bull. 61 (5) (2008) 663–670.
[77] N. Saba, M.T. Paridah, K. Abdan, N.A. Ibrahim, Dynamic mechanical properties of oil palm nano filler/kenaf/epoxy hybrid nanocomposites, Constr. Build. Mater. 124 (2016) 133–138.
[78] B. Imre, B. Pukánszky, Compatibilization in bio-based and biodegradable polymer blends, Eur. Polym J 49 (2013) 1215–1233.

[79] A. Le, A. Gacoin, A. Li, T. Mai, N. El Wakil, Influence of various starch/hemp mixtures on mechanical and acoustical behavior of starch-hemp composite materials, Compos. Part B Eng. 75 (2015) 201–211.
[80] J. Andersons, J. Modniks, R. Joffe, B. Madsen, K. Nättinen, Apparent interfacial shear strength of short-flax-fiber/starch acetate composites, Int. J. Adhes. Adhes. 64 (2016) 78–85.
[81] M. Iman, T.K. Maji, Effect of crosslinker and nanoclay on starch and jute fabric based green nanocomposites, Carbohydr. Polym. 89 (2012) 290–297.
[82] S.Y.Z. Zainuddin, I. Ahmad, H. Kargarzadeh, I. Abdullah, A. Dufresne, Potential of using multiscale kenaf fibers as reinforcing filler in cassava starch-kenaf biocomposites, Carbohydr. Polym. 92 (2013) 2299–2305.
[83] M.C. Gutierrez, M.-A. De Paoli, M.I. Felisberti, Biocomposites based on cellulose acetate and short curauá fibers: effect of plasticizers and chemical treatments of the fibers, Compos. Part A Appl. Sci. Manuf. 43 (8) (2012) 1338–1346.
[84] R.M. Rasal, A.V. Janorkar, D.E. Hirt, Poly (lactic acid) modifications, Progr. Polym. Sci. 35 (2010) 338–356.
[85] J. Lu, T. Wang, L.T. Drzal, Preparation and properties of microfibrillated cellulose polyvinyl alcohol composite materials. Compos, Part A Appl. Sci. Manuf. 39 (2008) 738–746.
[86] K. Oksman, M. Skrifvars, J.-F. Selin, Natural fibres as reinforcement in polylactic acid (PLA) composites, Compos. Sci. Technol. 63 (2003) 1317–1324.
[87] H. Cai, V. Dave, R.A. Gross, S.P. McCarthy, Effects of physical aging, crystallinity, and orientation on the enzymatic degradation of poly (lactic acid), J. Polym. Sci. Part B Polym. Phys. 34 (1996) 2701–2708.
[88] W. Ding, D. Jahani, E. Chang, A. Alemdar, C.B. Park, M. Sain, Development of PLA/cellulosic fiber composite foams using injection molding: crystallization and foaming behaviors, Compos. Part A Appl. Sci. Manuf. 83 (2016) 130–139.
[89] A.J. Anderson, E.A. Dawes, Occurrence, metabolism, metabolic role, and industrial uses of bacterial polyhydroxyalkanoates, Microbiol. Rev. 54 (4) (1990) 450–472.
[90] N. Barkoula, S. Garkhail, T. Peijs, Biodegradable composites based on flax/polyhydroxybutyrate and its copolymer with hydroxyvalerate, Ind. Crops Prod. 31 (2010) 34–42.
[91] M.K. Hossain, M.W. Dewan, M. Hosur, S. Jeelani, Mechanical performances of surface modified jute fiber reinforced biopol nanophased green composites, Compos. Part B Eng. 42 (2011) 1701–1707.
[92] A. Díaz, R. Katsarava, J. Puiggalí, Synthesis, properties and applications of biodegradable polymers derived from diols and dicarboxylic acids: from polyesters to poly (ester amide) s, Int. J. Mol. Sci. 15 (5) (2014) 7064–7123.
[93] U. Witt, T. Einig, M. Yamamoto, I. Kleeberg, W.-D. Deckwer, R.-J. Müller, Biodegradation of aliphatic–aromatic copolyesters: evaluation of the final biodegradability and ecotoxicological impact of degradation intermediates, Chemosphere 44 (2) (2001) 289–299.

Water soluble polymer based hybrid nanocomposites

3

Johnsy George[1], S.N. Sabapathi[1] and Siddaramaiah[2]
[1]Defence Food Research Laboratory, Mysore, India, [2]Sri Jayachamarajendra College of Engineering, Mysore, India

Chapter Outline

3.1 Hybrid polymer nanocomposites

Hybrid materials are a creative alternative to design new materials and compounds; their improved properties helps to develop materials for innovative applications [1]. Hybrid nanocomposites can be prepared by combining more than one type of nanomaterial such as organic and inorganic nanomaterials together in a polymer matrix. Generally, these hybrid nanocomposites are not simply physical mixtures of different nanomaterials, but they are intimately mixed, more homogenous, and exhibit synergistic properties [1]. They are different from the normal polymer nanocomposites where only one type of nanomaterial is incorporated in the polymer. Here the

Hybrid Polymer Composite Materials: Processing. DOI: http://dx.doi.org/10.1016/B978-0-08-100789-1.00003-4

negative attributes of one type of nanomaterial addition can be compensated by adding one or more different types of nanomaterials. The properties of these hybrids are not only due to the individual nanoparticles contribution but also due to the characteristics of interfaces. Mixing at the nanoscopic level results in the formation of different interfaces but gives rise to a homogeneous material where different phases will have interactions between them. Different varieties of well-defined hybrid polymers with a variety of architectures have also been reported. These hybrid polymers can also be considered as a type of hybrid nanocomposite, as they are either constructed by polymerization in the presence of inorganic compositions, or by the coupling reaction of functional polymers with inorganic nanoparticles [2].

On the basis of the nature of the interface, these types of nanocomposites are classified into class I and class II hybrids [3]. In the class I hybrid, the organic and inorganic phases exchange weak interactions such as hydrogen bonds, electrostatic forces, van der Waal forces, etc., while in class II hybrids the two phases are linked by strong chemical bonds such as covalent or ionic bonds. In class II hybrids, often an organic phase containing a reactive group acts as a network modifier, which covalently bonds with the inorganic phase. The improvement in properties that can be achieved by these multiphase structures together in a single polymer matrix helps in designing new multifunctional materials with superior properties. These hybrid nanocomposites provide a versatile platform for bringing new functionalities and helps in the development of high end polymeric materials suitable for a wide range of emerging applications in the field of biosensors, electronics, automobiles, photovoltaic cells, etc. [4]. Many tailor-made hybrid nanocomposites are increasingly utilized in numerous fields of applications such as catalysis, energy, environment, medicine, optics, etc. [5].

3.2 Gelatin-based hybrid polymer nanocomposites

Water soluble polymers have a wide range of industrial applications including food and pharmaceuticals. Polysaccharide and protein based polymers were the two dominant groups among water soluble polymers. Among these polymers, protein based films gained more interest especially in the field of food packaging due to their nutritive values [6]. Gelatin is a polymer that belongs to this category, having better film-forming ability; hence, it is widely studied for its applications in food packaging [7]. Gelatin is obtained from collagen, which exists as a triple helix comprising three discrete and three dimensional α chains interconnected by hydrogen bonds [8]. The collagen which is insoluble in water can be converted into water soluble gelatin by acid or alkali treatments. This process involves the cleavage of large number of covalent cross links. These chemical treatments result in the formation of a three dimensional gel-type network composed of crystallites interlaced with amorphous regions [9]. Gelatin is a transparent and slightly brittle protein, which finds wide applications in the field of foods, drugs, and photographic film applications. It melts when heated and solidifies when cooled again. Together with water,

Table 3.1 **Some physical properties of gelatin**

Physical properties	
True density ($g\ mL^{-1}$)	0.98
Bulk density ($g\ mL^{-1}$)	0.15–0.61
Viscosity at 40°C (cps)	44.0–55.0
Ash (% max)	1.5
Specific gravity	1.3–1.4
Refractive index	1.35

it forms a semi-solid colloid gel. Gelatin forms a solution of high viscosity in water, which sets to a gel on cooling, and its chemical composition is closely similar to that of its parent, collagen. Gelatin is also soluble in most polar solvents. The physical properties of gelatin are given in Table 3.1.

Gelatin has been extensively applied in the food industry as an ingredient to improve the consistency and stability of foods. It is also used as a gelling agent in cooking and for the clarification of juices. Gelatin, which can form thermoreversible gels with a melting point close to body temperature, has substantially increased its applications as a protective colloid and stabilizer in various frozen foods and dairy products [10]. Another important application of gelatin is in poultry coatings and sausage casings [11]. However, the large scale use of gelatin is limited because of their poor mechanical properties and water sensitivity. Incorporation of nanomaterials having different size and shapes into these water soluble polymers enhances the mechanical properties significantly. Other physical properties can also be improved by the introduction of nanomaterials.

3.3 Nanomaterials suitable for fabricating gelatin-based hybrid polymer nanocomposites

Nanomaterials are well known for reinforcing polymers, when they are infused into different polymer matrices. The reinforcing efficiency of these nanomaterials depends on the physical and chemical properties such as size, shape, aspect ratio, hydrogen bonds, and chemical interactions [12]. By combining different nanomaterials together, hybrid nanocomposites with several advantages were emerged. In this investigation, three different nanomaterials such as cellulose nanocrystals, amine functionalized nanoclay, and silver nanoparticles were used for the fabrication of hybrid nanocomposites.

Cellulose nanocrystals have attracted great interest for the fabrication of nanocomposites due to their renewability, biodegradability, spectacular mechanical properties, and the ability to reinforce polymers [13]. Cellulose nanocrystals are rod-like nanomaterials possessing a high aspect ratio (3–5 nm diameter, 50–500 nm in length), which makes them suitable for reinforcing different polymers. They are produced by the hydrolysis of large cellulose fibers by selectively hydrolyzing the amorphous fractions. In this investigation, bacterial cellulose fibers, the main raw material used for preparing cellulose nanocrystals, was synthesized using the bacterial strain, *Gluconacetobacter xylinum*. The preparation procedures of these pellicles were reported in our previous studies. [14,15]. The bacterial cellulose obtained in the form of pellicles were mechanically disintegrated to obtain microfibrils and subsequently converted into nanocrystals by acid hydrolysis. A detailed synthesis protocols for the preparation of nanocrystals and their applications in reinforcing polymers are reported elsewhere [16,17]. During hydrolysis, para-crystalline domains that are regularly distributed along the microfibrils undergoes hydrolysis easily, leaving behind highly crystalline rod-shaped nanocrystals. Due to their highly crystalline nature and peculiar size and shape, they can be used for the preparation of high-performance nanocomposites. They are attractive nanomaterials because of their affordability, renewability, high crystallinity, and environmentally friendly attributes. Cellulose nanocrystals possess a greater elastic modulus than Kevlar, and its mechanical properties are comparable to other nanomaterials; their properties are inferior only to carbon nanotubes [18].

Cellulose nanocrystals have the potential to make low cost, lightweight, and high-strength polymer nanocomposites [19]. Nakayama et al., reported the use of bacterial cellulose reinforced gelatin for the fabrication of double network hydrogels having excellent mechanical properties [20]. Gelatin is very brittle and will break down easily under modest compression. In contrast, by the addition of bacterial cellulose and subsequent formation of double network hydrogel, the fracture strength and elastic modulus increased several orders of magnitude higher than that of pure gelatin gel. This hydrogel composed of two kinds of hydrophilic polymer networks exhibited excellent elastic modulus and compressive fracture strength, despite of containing 90% water. In another report, a cellulose nanocrystal and its enzymatically modified form was incorporated to gelatin to prepare nanocomposites with enhanced rehydration ability [21]. The polar functional groups of gelatin together with the porous network formed due to the presence of nanocellulose effectively improved the rehydration properties. In a recent study, microfibrillated cellulose isolated from bagasse was used to prepare novel nanocomposites using cross-linked gelatin [22]. The addition of microfibrillated cellulose improved the wet and dry tensile strength and modulus of cross-linked gelatin but reduced percentage elongation. Microfibrillated cellulose did not affect the water absorption of cross-linked gelatin but significantly improved its moisture barrier property. There are still significant scientific and technical challenges to be addressed in the field of biopolymers reinforced with nanomaterials, and research in this area is progressing steadily. Despite few reports about gelatin films reinforced with cellulose nanomaterials, data available on the hybrid polymer nanocomposites based on gelatin is scarce.

Nanoclay is another type of nanomaterial that received considerable attention in the fabrication of polymer nanocomposites. Nanoclay is a term generally used to describe a group of layered silicates like hydrous aluminum phyllosilicate minerals that are made up of individual platelets typically in the order of nanometers [23]. Nanoclays are made up of several phyllosilicate minerals rich in silicon and aluminum oxides/hydroxides which include variable amounts of water [24]. Clays are generally classified into two types depending on their origin such as natural and synthetic. Natural clays are formed by the chemical weathering of silicate-bearing rocks by carbonic acid; they can also be formed by hydrothermal activities. One of the main applications of this clay is in the fabrication of high-barrier polymer nanocomposites. However, the main limitation of these naturally occurring clays is that they are not compatible with most of the polymers in their pristine form [25–26]. Synthetic clays on the other hand are made by the self-assembly of hydrolyzed silane precursors to form phyllosilicate like inorganic framework [27]. The condensation between silanols and aqueous metal species leads to the formation of silane group sandwiched brucite sheets [28]. Layered magnesium organosilicates are also having a structure analogous to 2:1 trioctahedral smectites. Synthetic clays contain covalently linked organosilicates, with a composition resembling that of brucite, $R_8Si_8Mg_6O_{16}(OH)_4$, where R can be alkyl or alkylamine groups [29]. In these synthetic nanoclays, various types of organic groups are covalently attached to inorganic clay sheets during the clay synthesis itself, which can be tailor-made to make it either hydrophobic or hydrophilic and to make it more compatible with a particular polymer [30]. The incorporation of amino groups in the silicate network makes the clay completely water-dispersible and hence very much suitable for making water soluble polymer based nanocomposites [31].

ANC was prepared by using 3-aminopropyltriethoxysilane, an organosilane precursor and magnesium chloride in ethanol under constant stirring at room temperature. White gel-type clay sheets obtained after the reaction was centrifuged, washed, and dried. This clay can be easily exfoliated in water and hence, they are used to impart better properties to gelatin. Gelatin–aminoclay (AC) nanocomposites were prepared using solution intercalation method and found to have improved mechanical strength, which varied with the AC content. The use of AC enables the complete exfoliation of clay sheets at higher concentrations, due to the protonation of amino groups within the layers upon addition of water that creates repulsive forces. This type of easy exfoliation is very much desirable to get excellent properties when used in the preparation of polymer nanocomposites. Secondly, the presence of covalently attached groups within the clay platelets gives more functionality and therefore improves the properties of the resulting nanocomposites [32]. Only few studies have been conducted so far to explore the effect of these clays on the properties of polymer nanocomposites.

One of the most commonly used nanomaterials for the fabrication of antimicrobial food packaging films is silver nanoparticles (AgNPs), which is well known for its strong activity towards a wide range of microorganisms [33,34]. AgNPs can be prepared by employing either a physical or a chemical method [35,36]. Among these two techniques, chemical reduction is the most preferred method, as the synthesized AgNPs will be obtained as stable colloidal dispersions. The reduction of silver cations in an aqueous solution produces colloidal silver with varying particle diameters in the order of several nanometers [37]. The chemical reduction initially favors the formation of silver atoms (Ag^0), which subsequently aggregates into oligomeric clusters, which further grows to form AgNPs [38]. The synthesis is often performed by reducing metal salts in the presence of capping agents and stabilizers to avoid reaggregation of the nanoparticles [39]. The chemical reduction process offers great control over the particle size and shape by changing the experimental parameters like concentration of reactants, temperature, pH, type of reducing agents and stabilizers. AgNPs synthesized in this way were found to have antimicrobial activity against several microorganisms [40–42]. AgNPs were incorporated in various polymers, and nanocomposite materials with antimicrobial activity were fabricated. Several water soluble polymers were used to make AgNPs reinforced nanocomposites with better properties [43,44,45]. AgNPs used in this study were synthesized by reducing a silver salt such as silver nitrate ($AgNO_3$) in the presence of gelatin, which acts as a capping agent. Sodium borohydride ($NaBH_4$) was used as a reducing agent. Stotiometric quantities of silver salt and reducing agents were selected to obtain 1 wt% of AgNPs.

3.4 Hybrid gelatin nanocomposites containing a combination of BCNC and AgNPs

Most reports currently available about polymer nanocomposites were focused on the synthesis and characterization of nanocomposites containing only one type of nanomaterial and reports about hybrid nanocomposites containing a combination of various nanomaterials like cellulose nanocrystals, nanoclay, and AgNPs together in a single polymer matrix is scarce. There is a great demand for the development of these multifunctional nanocomposites, as they provide some unique properties. But there are several difficulties and challenges to overcome in their fabrication, such as optimizing the concentration of nano-additives. Hybrid multifunctional gelatin-based nanocomposites were prepared by incorporating 2 and 4 wt% of BCNC suspension to a fixed quantity of gelatin solution (10%, 100 mL) by sonication. AgNPs were synthesized in situ by reducing silver nitrate in the presence of both gelatin and BCNC using a reducing agent such as $NaBH_4$. The presence of AgNPs was confirmed by the formation of yellow color as shown in the figure (Fig. 3.1). The quantities of AgNPs were fixed at 1 wt% in all formulations. The polymer solution containing two different nanomaterials were casted in PP dishes and dried at room temperature for 2 days and these films were used for further characterization. The physico-mechanical, barrier, thermal, morphological, and optical behavior of these films were evaluated. The mechanical, thermal, and barrier properties were evaluated. The optical and antimicrobial properties of these multifunctional nanocomposites were also investigated.

3.4.1 Morphology

The morphology of these hybrid nanocomposites were analyzed using Transmission electron microscopy (TEM). TEM images of nanocomposites depicted in Fig. 3.2 shows that these nanoparticles are dispersed well within the polymer-BCNC matrix.

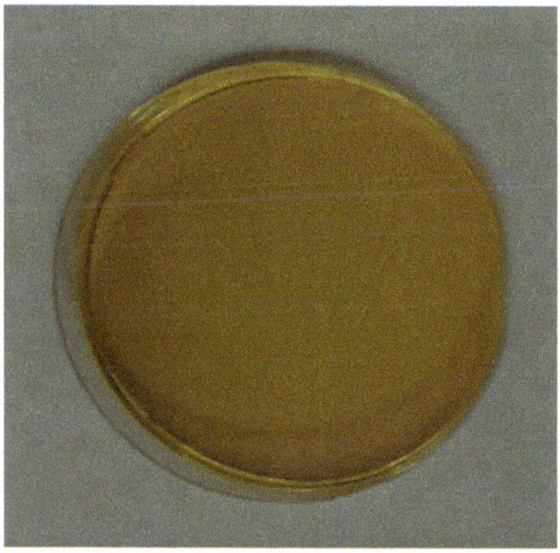

Figure 3.1 Optical images of hybrid gelatin nanocomposites containing 1 wt% AgNPs.

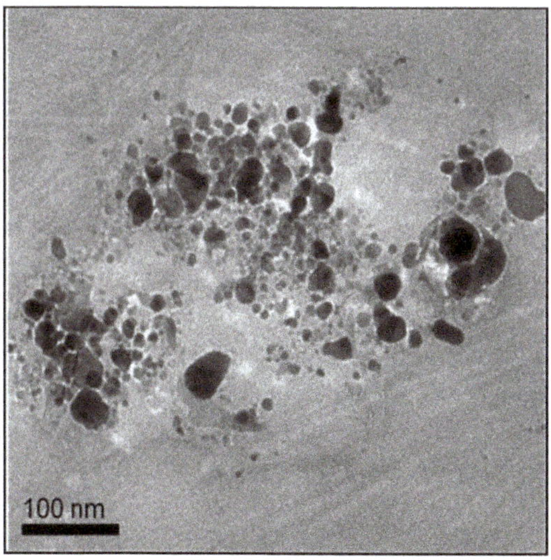

Figure 3.2 TEM images of Gelatin nanocomposites containing BCNC and AgNPs.

From the TEM image, it is very clear that AgNPs are distributed throughout the nanocomposite matrix. The polymer matrix was able to stabilize AgNPs due to the presence of hydroxyl groups, which tightly hold the silver cations and further reduces to nanoparticles. These hydroxyl groups have strong interactions with AgNPs once it is formed, and such interactions prevent the agglomeration of nanoparticles, which normally have a tendency to form large particles. The average size of the nanoparticle formed is less than 30 nm, as noticed from the TEM image. The presence of rod-like cellulose nanocrystals and AgNPs distributed within the polymer matrix can be noticed in the image.

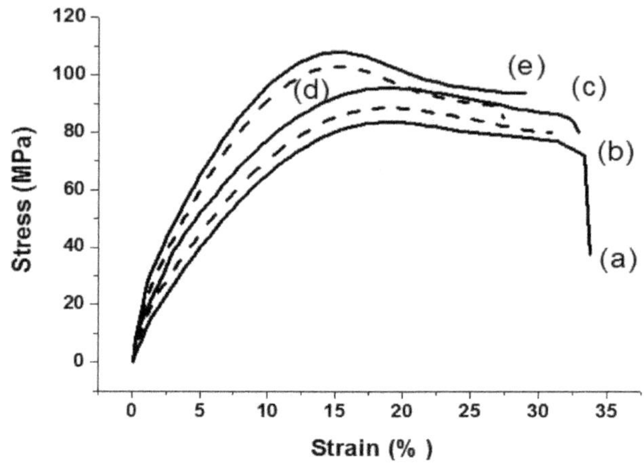

Figure 3.3 Stress versus strain curves of (a) gelatin, (b) gelatin—2% BCNC, (c) gelatin—2% BCNC—1% AgNPs, (d) gelatin—4% BCNC and (e) gelatin—4% BCNC—1% AgNPs nanocomposites.

3.4.2 Mechanical properties

The tensile behavior of these hybrid nanocomposites was studied and the deformation patterns of these nanocomposites were depicted in Fig. 3.3 and a consolidated result of the mechanical properties of these nanocomposites is given in Table 3.2. From the results it is very clear that AgNPs containing gelatin nanocomposites exhibited improved strength and ductility. Tensile strength improved from 83 to 88 MPa and 103 MPa with an increase in BCNC from 2 to 4 wt%, respectively, while the addition of AgNPs resulted in an increase to 95 and 108 MPa respectively for the same weight percentage of BCNC. The addition of 2 and 4 wt% of BCNC resulted in reducing the percentage of elongation from 33.7 to 29.8 and 27.1% respectively, whereas the addition of AgNPs helped to regain it slightly to 32.1 and 29.2% respectively. Similar types of improvement in the mechanical properties of PVA and HPMC nanocomposites were reported [46,47].

3.4.3 Moisture sorption properties

Moisture sorption analysis of gelatin nanocomposite films were carried out using a dynamic vapor-sorption analyzer, which measures how quickly moisture is absorbed or desorbed from a sample and also to quantify the same. This technique shows the equilibrium amount of water vapor sorbed as a function of steady state vapor pressure at a constant temperature. In this experiment, the sample is exposed to a series of step changes in RH and the resultant change in weight is measured as a function of time. The equilibrium mass values at each RH step are used to generate the sorption isotherm. This technique has a lot of potential to measure moisture and flavor-diffusion properties of different water-sensitive polymers when they are

Table 3.2 Mechanical properties of gelatin nanocomposites containing BCNC and AgNPs as bi-functional fillers

Sample	Thickness[a] (μm)	Tensile strength[a] (MPa)	Tensile modulus[a] (MPa)	Elongation at break[a] (%)	Stiffness[a] (KN m^{-1})
Gelatin	200.1 ± 11	83 ± 8.3	2190 ± 240	33.7 ± 2.7	268.5 ± 24
Gelatin + 2% BCNC	203.2 ± 09	88 ± 9.1	2260 ± 190	29.8 ± 1.7	279.2 ± 20
Gelatin + 2% BCNC +1% AgNPs	206.8 ± 14	95 ± 9.4	2290 ± 240	32.1 ± 2.2	285.3 ± 24
Gelatin + 4% BCNC	204.6 ± 12	103 ± 8.9	2350 ± 305	27.1 ± 2.8	304.8 ± 21
Gelatin + 4% BCNC +1% AgNPs	207.2 ± 15	108 ± 9.7	2400 ± 255	29.2 ± 2.9	314.2 ± 30

[a]Mean ± SE, $n = 10$.

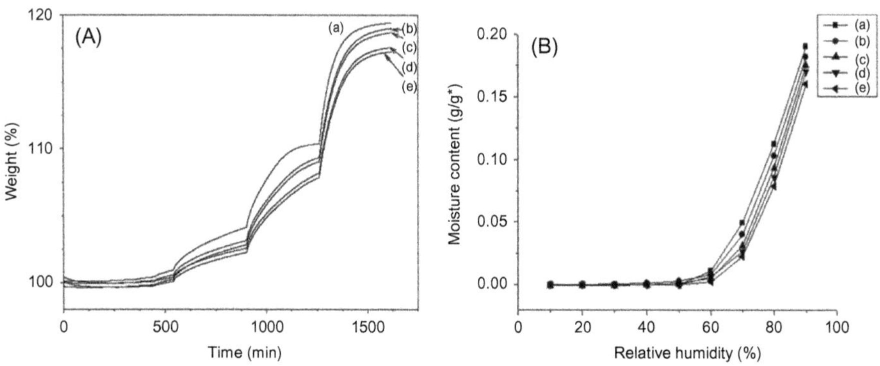

Figure 3.4 The plots of (A) weight gain as a function of time and (B) moisture sorption isotherm curves of (a) gelatin, (b) gelatin having 2% BCNC, (c) gelatin having 4% BCNC, (d) gelatin—2% BCNC—1% AgNPs and (e) gelatin—4% BCNC—1% AgNPs nanocomposites.

used in packaging applications. The weight gain as a function of time as well as the moisture sorption isotherm curves of these nanocomposites are given in Fig. 3.4.

All the gelatin nanocomposite samples demonstrated a similar tendency to adsorb large amount of moisture especially at higher RH values. Pure gelatin tends to adsorb more moisture, but the addition of BCNC and AgNPs reduced this tendency. This may be due to the fact that some of the −OH groups of gelatin and BCNC are strongly coupled with AgNPs, which are not available to interact with water molecules and eventually reduce the total moisture sorption. Moisture sorption isotherms obtained from these experiments describe the relationship between moisture content and water activity. A large number of isotherm equations have been developed for food and related products, while only a few isotherm models are reported for nanocomposite films. In order to examine the interaction of moisture on these nanocomposite films, Guggenheim−Anderson−DeBoer (GAB) model was employed as it can adequately describe moisture sorption parameters for water activities up to 0.9 [48].

$$\mathrm{EMC} = \frac{W_\mathrm{m} C_\mathrm{g} k a_\mathrm{w}}{(1 - k a_\mathrm{w})(1 - k a_\mathrm{w} + C_\mathrm{g} k a_\mathrm{w})}$$

where, equilibrium moisture content is represented as EMC, monolayer moisture content as W_m, Guggeheim constant as C_g, k as a factor related to the properties of multilayer molecules, and aw is the water activity. The calculated GAB model constants and root mean square values (r^2) values of gelatin and its nanocomposites were given in Table 3.3.

The W_m value calculated for gelatin is 10.3, which is similar to the monolayer values obtained for bovine gelatin [49]. The addition of BCNC reduced the W_m, while the addition of AgNPs along with BCNC considerably reduced it further. This implies that the addition of AgNPs decreased hydrophilicity, which is a very encouraging result, because the moisture absorption into these films during storing can be reduced. The values of C_g, decreased with the addition of AgNPs, while K value increased. The R^2 values obtained is more than 0.985, implying a good fit with the experimental data.

Table 3.3 GAB model constants of gelatin and its nanocomposites (*Mean ± SE, $n = 5$)

	GAB model constants			
Samples	W_m (% w w^{-1})	C_g	k	r^2
Gelatin	10.3 ± 0.03	18.6 ± 0.6	0.78 ± 0.04	0.985
Gelatin + 2% BCNC	10.28 ± 0.06	18.3 ± 0.7	0.80 ± 0.02	0.987
Gelatin + 2% BCNC + 1% AgNPs	10.19 ± 0.04	18.1 ± 0.7	0.82 ± 0.01	0.994
Gelatin + 4% BCNC	10.22 ± 0.02	18.2 ± 0.4	0.81 ± 0.03	0.986
Gelatin + 4% BCNC + 1% AgNPs	10.02 ± 0.07	17.9 ± 0.5	0.85 ± 0.02	0.992

3.4.4 Thermal properties

Some of the thermal properties of these hybrid nanocomposite was analyzed with the help of Differential Scanning Calorimetry (DSC). DSC analysis (Fig. 3.5) was helpful in studying the effect of BCNC and AgNPs on the thermal properties of gelatin. The results showed that the addition of nanoparticles affected most of the thermal properties like glass transition temperature (T_g), melting temperature (T_m) and enthalpy of melting (ΔH_m) of gelatin. The addition of 2 and 4 wt% of BCNC has resulted in an increase in T_g, T_m, and ΔH_m, as evident for the Table 3.4. Here the addition of AgNPs resulted in increasing the peak melting temperature and its associated enthalpy of gelatin nanocomposites. However, the T_g was increased after the addition of BCNC, but the use of AgNPs helped to bring down the T_g. This is an indication that the restriction on the segmental mobility was reduced to acertain extent by the addition of AgNPs. Since the AgNPs are dispersed properly within the polymer matrix, they can interfere with the radius of gyration of a polymer chain and increase the free volume, thus enhancing the segmental mobility. The overall trend observed with these nanocomposites is that the addition of AgNPs is affecting the T_g as well as the T_m of polymer. AgNPs interacts strongly with the hydroxyl groups of gelatin and reduces hydrophilicity.

3.5 Gelatin nanocomposites containing a combination of amine functionalized clay and AgNPs

The use of synthetic nanoclays having tailor-made pendant groups such as ANC enables the complete exfoliation of clay sheets even at higher concentrations, due

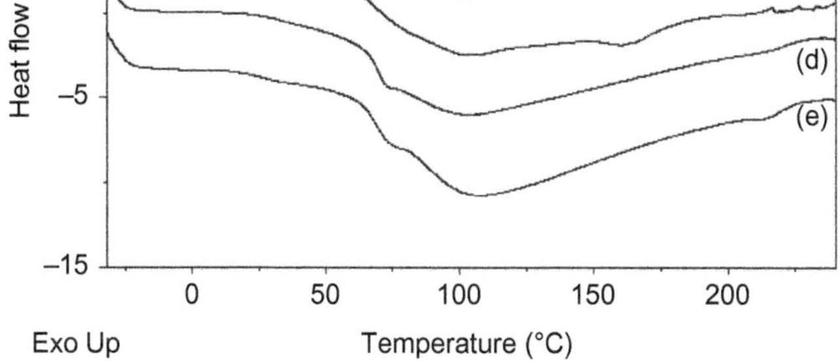

Figure 3.5 DSC thermograms of (a) gelatin, (b) gelatin—2% BCNC, (c) gelatin—2% BCNC—1% AgNPs, (d) gelatin—4% BCNC and (e) gelatin—4% BCNC—1% AgNPs nanocomposites.

Table 3.4 **DSC results of gelatin and its nanocomposites**

Sample	Glass transition temperature[a] T_g (°C)	Peak melting temperature[a] (°C)	Enthalpy of melting[a] (ΔH_m) (J g^{-1})
Gelatin	66.9 ± 2.2	101.5 ± 1.5	63.7 ± 5.6
Gelatin + 2% BCNC	68.1 ± 2.9	103.7 ± 2.4	82.1 ± 5.2
Gelatin + 2% BCNC + 1% AgNPs	67.8 ± 2.4	104.3 ± 2.7	86.3 ± 6.7
Gelatin + 4% BCNC	69.8 ± 2.1	107.3 ± 2.6	96.8 ± 5.8
Gelatin + 4% BCNC + 1% AgNPs	68.2 ± 2.7	108.1 ± 1.8	97.3 ± 6.2

[a]Mean ± SE, $n = 3$.

to the protonation of amino groups within the layers upon addition of water that creates repulsive forces. This type of easy exfoliation is very much desirable to get excellent properties when used in the preparation of polymer nanocomposites. Secondly, the presence of chemically modified clay platelets improves the properties of the resulting nanocomposites. This type of improvement in the properties of water soluble polymers is reported in our previous study [31]. The presence of another nanomaterial such as AgNPs can result in further improvement in the properties.

Hybrid multifunctional nanocomposites of gelatin containing both ANC and AgNPs were fabricated by solution casting method. ANC solutions at different concentrations in water (2 and 4 wt%) were exfoliated separately by sonicating for 5 min by keeping the ratio of clay to water constant at 1 wt% and later added to the polymer solution. AgNPs were synthesized in situ by reducing silver nitrate in the presence of polymer and ANC using $NaBH_4$ as reducing agent. The quantities of AgNPs were fixed at 1 wt%. The polymer solution containing both ANC and AgNPs were casted in PP petri dishes and dried at room temperature for 2 days and the films were used for further characterization.

3.5.1 Mechanical properties

The tensile behavior of gelatin nanocomposites containing both ANC and AgNPs were shown in Fig. 3.6. The data obtained from the experiments are tabulated in Table 3.5. AgNPs containing gelatin nanocomposites also exhibited improved strength and ductility. Tensile strength improved from 83 to 86 and 93 MPa with an

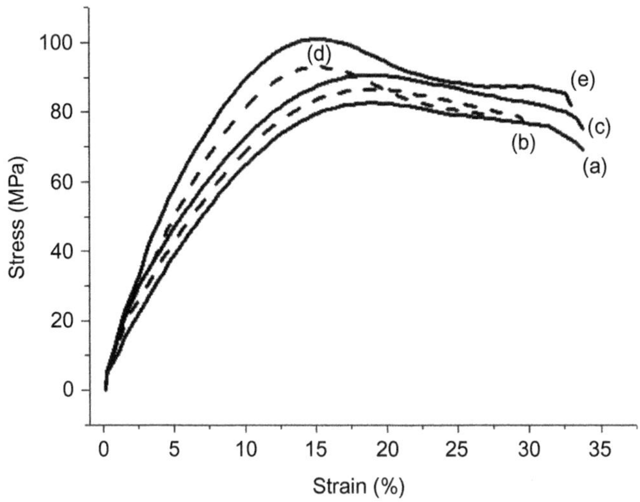

Figure 3.6 Stress versus strain curves of (a) gelatin, (b) gelatin—2% ANC, (c) gelatin—2% ANC—1% AgNPs, (d) gelatin—4% ANC and (e) gelatin—4% ANC – 2% AgNPs nanocomposites.

Table 3.5 Mechanical properties of gelatin nanocomposites containing ANC and AgNPs as bi-functional fillers

Sample	Tensile strength[a] (MPa)	Tensile modulus[a] (MPa)	Elongation at break[a] (%)	Stiffness[a] (KN m^{-1})
Gelatin	82.7 ± 7.5	2190 ± 240	33.7 ± 3.9	268.5 ± 24
Gelatin + 2% ANC	86.3 ± 9.1	2260 ± 190	29.8 ± 2.7	279.2 ± 24
Gelatin + 2% ANC + 1% AgNPs	91 ± 9.4	2290 ± 240	32.1 ± 2.6	285.3 ± 27
Gelatin + 4% ANC	93.1 ± 8.9	2350 ± 305	27.1 ± 2.8	288.8 ± 25
Gelatin + 4% ANC + 1% AgNPs	101 ± 9.7	2400 ± 255	31.2 ± 2.9	298.2 ± 29

[a]Mean ± SE, n = 10.

increase in ANC content, while the addition of silver nanoparticles resulted in further increase to 91MPa and 101 MPa respectively for the same weight percentage of ANC (Table 3.4). The addition of 2 and 4 wt% of ANC resulted in reducing the percentage of elongation from 33 to 29 and 27% respectively; however, the addition of AgNPs brought it back to 32 and 31% respectively.

3.5.2 Thermal properties

DSC analysis of gelatin nanocomposites containing ANC and AgNPs have been carried out to study the effect of AgNPs on the thermal characteristics of these nanocomposites. The DSC thermograms obtained for these nanocomposites are shown in Fig. 3.7. The results showed that the addition of clay increased T_g and T_m of gelatin (Table 3.6). These results also suggest that the AgNPs are able to increase the T_g values further. Glass transition temperature is affected by the interaction of polymer chains with nanoparticles and dispersion of nanoparticles. The presence of small amount of AgNPs resulted in enhancing T_g to 67.8°C and 69.1°C for 2 and 4 wt% ANC loading, respectively.

3.5.3 Barrier properties

The barrier properties—oxygen transmission rate and CO_2 transmission rate of gelatin-ANC nanocomposite films with and without AgNPs—were shown in

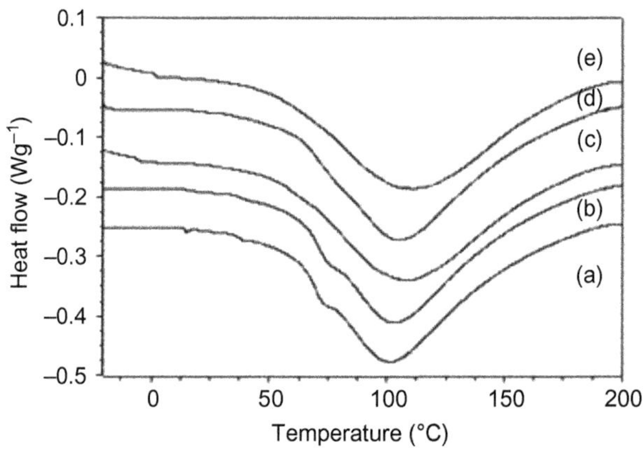

Figure 3.7 DSC thermograms of (a) gelatin, (b) gelatin—2% ANC, (c) gelatin—2% ANC—1% AgNPs, (d) gelatin—4% ANC and (e) gelatin—4% ANC—1% AgNPs.

Table 3.6 DSC results of PVA, HPMC, and gelatin nanocomposites containing ANC and AgNPs as bi-functional fillers

Sample	T_g[a] (°C)	T_m[a] (°C)	ΔH_m[a] (J g^{-1})
Gelatin	66.9 ± 2.2	105.5 ± 1.5	63.7 ± 5.6
Gelatin + 2% ANC	67.3 ± 3.2	106.2 ± 2.7	63.4 ± 6.5
Gelatin + 2% ANC + 1% AgNPs	67.8 ± 3.8	106.7 ± 3.6	63.6 ± 5.9
Gelatin + 4% ANC	68.2 ± 2.8	107.1 ± 3.2	63.6 ± 5.9
Gelatin + 4% ANC + 1% AgNPs	69.1 ± 2.6	107.6 ± 3.1	63.8 ± 5.8

[a]Mean ± SE, $n = 3$.

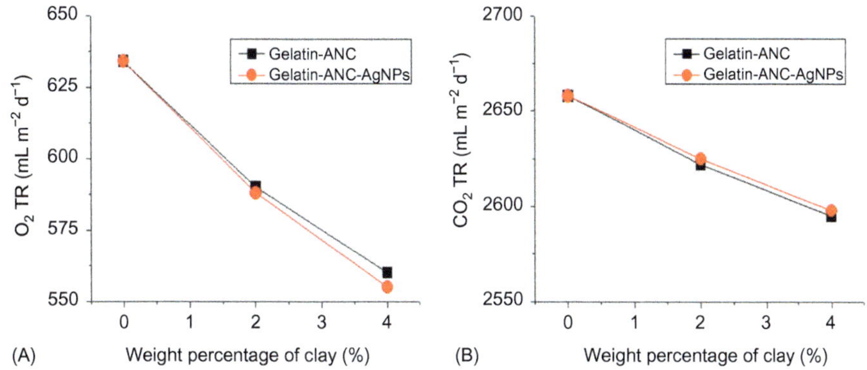

Figure 3.8 (A) Oxygen transmission rate and (B) CO_2 transmission rate of gelatin-ANC nanocomposites with and without AgNPs.

Fig. 3.6. The incorporation of ANC reduced OTR and CO_2TR of polymer films considerably. ANCs are capable of completely exfoliating in water and were able to disperse in a polymer matrix without re-aggregating. These exfoliated platelets act as impermeable obstacles in the path of gas diffusion process. The addition of AgNPs along with ANC marginally reduces the barrier properties as evident from the Fig. 3.8. The AgNPs are comparatively smaller particles than clay layers and hence, they were not able to create additional obstacles to the incoming gas molecules. Even though the size and shape of AgNPs don't favor the formation of a tortuous path to the incoming gas molecules, they influence the barrier properties by causing changes to the polymer matrix itself at the interfacial regions. Since the hydroxyl groups of gelatin is having favorable interactions with AgNPs, polymer chains are partially immobilized. When the gas molecules travel through these zones, there will be attenuated hopping rates between free volume holes, which indirectly reduces the gas transmission rate [30,31].

3.6 Conclusions

Hybrid nanocomposites with improved properties were developed using an innovative combination of different nanomaterials to form nanocomposites with gelatin. Each nanomaterial is well known for improving different properties of polymers such as BCNC, which can improve the mechanical properties by forming a percolating network within the polymer matrix, while ANC can improve the barrier properties by creating tortuous path to the incoming gas molecules. AgNPs can improve the mechanical properties and also can impart antimicrobial properties. Using different nanomaterials together in a polymer matrix helped in synergistically bringing together the advantages of each individual nanomaterial. In the first combination, BCNC was used along with AgNPs in fabricating a polymer nanocomposite system, while in the second system ANC was used along with AgNPs. Combining AgNPs along with BCNC and ANC improved the mechanical, barrier, and thermal properties of gelatin. Synthesizing nanocomposites having different nanoparticles existing together within a polymer matrix is a simple and efficient method to integrate and improve desired multifunctional properties without compromising much of the inherent properties of the polymer.

References

[1] C. Sanchez, B. Julian, P. Belleville, M. Popall, Applications of hybrid organic–inorganic nanocomposites, J. Mater. Chem. 15 (35–36) (2005) 3559–3592.
[2] Z. Zhang, P. Zhang, Y. Wang, W. Zhang, Recent advances in organic–inorganic well-defined hybrid polymers using controlled living radical polymerization techniques, Polym. Chem. 7 (2016) 3950–3976.
[3] T. Krasia-Christoforou, Organic–inorganic polymer hybrids: synthetic strategies and applications, Hybrid and Hierarchical Composite Materials, Springer International Publishing, 2015, pp. 11–63.
[4] C. Sanchez, P. Belleville, M. Popall, L. Nicole, Applications of advanced hybrid organic–inorganic nanomaterials: from laboratory to market, Chem. Soc. Rev. 40 (2) (2011) 696–753.
[5] L. Nicole, C. Laberty-Robert, L. Rozes, C. Sanchez, Hybrid materials science: a promised land for the integrative design of multifunctional materials, Nanoscale 6 (12) (2014) 6267–6292.
[6] A. Gennadios, R.Y. Yada, Edible films and coatings from proteins, Proteins in Food Processing, Woodhead Publishing, Cambridge, UK, 2004, pp. 442–467.
[7] P.J.A. Sobral, F.C. Menegalli, M.D. Hubinger, M.A. Roques, Mechanical, water vapor barrier and thermal properties of gelatin based edible films, Food Hydrocolloids 15 (4–6) (2001) 423–432.
[8] E.S. Gil, D.J. Frankowski, M.K. Bowman, A.O. Gozen, S.M. Hudson, R.J. Spontak, Mixed protein blends composed of gelatin and Bombyx m ori silk Fibroin: effects of solvent-induced crystallization and composition, Biomacromolecules. 7 (3) (2006) 728–735.
[9] L. Slade, H. Levine, Polymer-chemical properties of gelatin in foods, in: A.M. Pearson, T.R. Dutson, A.Q.J. Bailey (Eds.), Advances in Meat Research (USA), 4, Van Nostrand Reinhold, New York, USA, 1987, pp. 251–266.
[10] D. Achet, X.W. He, Determination of the renaturation level in gelatin films, Polymer. (Guildf). 36 (4) (1995) 787–791.
[11] A. Gennadios, M.A. Hanna, L.B. Kurth, Application of edible coatings on meats, poultry and seafoods: a review, LWT-Food Sci. Technol. 30 (4) (1997) 337–350.
[12] J. Kalfus, J. Jancar, Reinforcing mechanisms in amorphous polymer nano-composites, Compos. Sci. Technol. 68 (15–16) (2008) 3444–3447.
[13] Y. Habibi, L.A. Lucia, O.J. Rojas, Cellulose nanocrystals: chemistry, self-assembly, and applications, Chem. Rev. 110 (6) (2010) 3479–3500.
[14] J. George, K.V. Ramana, S.N. Sabapathy, A.S. Bawa, Physico-mechanical properties of chemically treated bacterial (*Acetobacter xylinum*) cellulose membrane, World J. Microbiol. Biotechnol. 21 (8–9) (2005) 1323–1327.

[15] J. George, K.V. Ramana, S.N. Sabapathy, J.H. Jagannath, A.S. Bawa, Characterization of chemically treated bacterial (*Acetobacter xylinum*) biopolymer: some thermo-mechanical properties, Int. J. Biol. Macromol. 37 (4) (2005) 189–194.
[16] J. George, A.S. Bawa, Siddaramaiah, Synthesis and characterization of bacterial cellulose nanocrystals and their PVA nanocomposites, Adv. Mater. Res. 123 (2010) 383–386.
[17] J. George, K.V. Ramana, A.S. Bawa, Siddaramaiah, Bacterial cellulose nanocrystals exhibiting high thermal stability and their polymer nanocomposites, Int. J. Biol. Macromol. 48 (1) (2011) 50–57.
[18] R.J. Moon, A. Martini, J. Nairn, J. Simonsen, J. Youngblood, Cellulose nanomaterials review: structure, properties and nanocomposites, Chem. Soc. Rev. 40 (7) (2011) 3941–3994.
[19] A. Dufresne, Comparing the mechanical properties of high performances polymer nanocomposites from biological sources, J. Nanosci. Nanotechnol. 6 (2) (2006) 322–330.
[20] A. Nakayama, A. Kakugo, J.P. Gong, Y. Osada, M. Takai, T. Erata, et al., High mechanical strength double-network hydrogel with bacterial cellulose, Adv. Funct. Mater. 14 (11) (2004) 1124–1128.
[21] S.-B. Lin, C.-P. Hsu, L.-C. Chen, H.-H. Chen, Adding enzymatically modified gelatin to enhance the rehydration abilities and mechanical properties of bacterial cellulose, Food Hydrocolloids 23 (8) (2009) 2195–2203.
[22] S.M. Fadel, M.L. Hassan, K. Oksman, Improving tensile strength and moisture barrier properties of gelatin using microfibrillated cellulose, J. Compos. Mater. 47 (16) (2012) 1977–1985.
[23] M. Okamoto, Recent advances in polymer/layered silicate nanocomposites: an overview from science to technology, Mater. Sci. Technol. 22 (7) (2006) 756–779.
[24] S. Sinha Ray, M. Okamoto, Polymer/layered silicate nanocomposites: a review from preparation to processing, Progr. Polym. Sci. 28 (11) (2003) 1539–1641.
[25] R.A. Vaia, E.P. Giannelis, Lattice model of polymer melt intercalation in organically-modified layered silicates, Macromolecules. 30 (25) (1997) 7990–7999.
[26] R.A. Vaia, E.P. Giannelis, Polymer melt Intercalation in organically-modified layered silicates: model predictions and experiment, Macromolecules. 30 (25) (1997) 8000–8009.
[27] L. Ukrainczyk, R.A. Bellman, A.B. Anderson, Template synthesis and characterization of layered Al – and Mg – Silsesquioxanes, J. Phys. Chem. B. 101 (4) (1997) 531–539.
[28] S.L. Burkett, A. Press, S. Mann, Synthesis, characterization, and reactivity of layered inorganic – organic nanocomposites based on 2: 1 trioctahedral phyllosilicates, Chem. Mater. 9 (5) (1997) 1071–1073.
[29] E. Muthusamy, D. Walsh, S. Mann, Morphosynthesis of organoclay microspheres with sponge-like or hollow interiors, Adv. Mater. 14 (13–14) (2002) 969–972.
[30] B. Lebeau, J. Brendlé, C. Marichal, A.J. Patil, E. Muthusamy, S. Mann, One-step synthesis and solvent-induced exfoliation of hybrid organic–inorganic phyllosilicate-like materials, J. Nanosci. Nanotechnol. 6 (2) (2006) 352–359.
[31] G. Johnsy, K.K.R. Datta, V.A. Sajeevkumar, S.N. Sabapathy, A.S. Bawa, M. Eswaramoorthy, Aminoclay: a designer filler for the synthesis of highly ductile polymer-nanocomposite film, ACS Appl. Mater. Interfaces 1 (12) (2009) 2796–2803.
[32] F. Wypych, K.G. Satyanarayana, Functionalization of single layers and nanofibers: a new strategy to produce polymer nanocomposites with optimized properties, J. Colloid. Interface. Sci. 285 (2) (2005) 532–543.

[33] S.Y. Liau, D.C. Read, W.J. Pugh, J.R. Furr, A.D. Russell, Interaction of silver nitrate with readily identifiable groups: relationship to the antibacterial action of silver ions, Lett. Appl. Microbiol. 25 (4) (1997) 279–283.

[34] R. Kumar, H. Munstedt, Silver ion release from antimicrobial polyamide/silver composites, Biomaterials. 26 (14) (2005) 2081–2088.

[35] M.M. Oliveira, D. Ugarte, D. Zanchet, A.J.G. Zarbin, Influence of synthetic parameters on the size, structure, and stability of dodecanethiol-stabilized silver nanoparticles, J. Colloid. Interface. Sci. 292 (2) (2005) 429–435.

[36] E.M. Egorova, A.A. Revina, Synthesis of metallic nanoparticles in reverse micelles in the presence of quercetin, Colloids Surf. A Physicochem. Eng. Aspects 168 (1) (2000) 87–96.

[37] B. Wiley, Y. Sun, B. Mayers, Y. Xia, Shape-controlled synthesis of metal nanostructures: the case of silver, Chem. Eur. J. 11 (2) (2005) 454–463.

[38] S. Kapoor, D. Lawless, P. Kennepohl, D. Meisel, N. Serpone, Reduction and aggregation of silver ions in aqueous gelatin solutions, Langmuir. 10 (9) (1994) 3018–3022.

[39] V.K. Sharma, R.A. Yngard, Y. Lin, Silver nanoparticles: green synthesis and their antimicrobial activities, Adv. Colloid. Interface. Sci. 145 (1–2) (2009) 83–96.

[40] K.-H. Cho, J.-E. Park, T. Osaka, S.-G. Park, The study of antimicrobial activity and preservative effects of nanosilver ingredient, Electrochim. Acta. 51 (5) (2005) 956–960.

[41] Q.L. Feng, J. Wu, G.Q. Chen, F.Z. Cui, T.N. Kim, J.O. Kim, A mechanistic study of the antibacterial effect of silver ions on *Escherichia coli* and *Staphylococcus aureus*, J. Biomed. Mater. Res. 52 (4) (2000) 662–668.

[42] I. Sondi, B. Salopek-Sondi, Silver nanoparticles as antimicrobial agent: a case study on *E. coli* as a model for gram-negative bacteria, J. Colloid. Interface. Sci. 275 (1) (2004) 177–182.

[43] P.K. Khanna, N. Singh, S. Charan, V.V.V.S. Subbarao, R. Gokhale, U.P. Mulik, Synthesis and characterization of Ag/PVA nanocomposite by chemical reduction method, Mater. Chem. Phys. 93 (1) (2005) 117–121.

[44] M.R. de Moura, L.H.C. Mattoso, V. Zucolotto, Development of cellulose-based bactericidal nanocomposites containing silver nanoparticles and their use as active food packaging, J. Food Eng. 109 (3) (2012) 520–524.

[45] C. Aimé, I.B. Rietveld, T. Coradin, Hydrazine-induced thermo-reversible optical shifts in silver–gelatin bionanocomposites, Chem. Phys. Lett. 505 (1–3) (2011) 37–41.

[46] J. George, V.A. Sajeevkumar, K.V. Ramana, S.N. Sabapathy, Siddaramaiah, Augmented properties of PVA hybrid nanocomposites containing cellulose nanocrystals and silver nanoparticles, J. Mater. Chem. 22 (42) (2012) 22433–22439.

[47] J. George, R. Kumar, V.A. Sajeevkumar, K.V. Ramana, R. Rajamanickam, V. Abhishek, et al., Hybrid HPMC nanocomposites containing bacterial cellulose nanocrystals and silver nanoparticles, Carbohydr. Polym. 105 (2014) 285–292.

[48] J.-F. Su, Z. Huang, Y.-H. Zhao, X.-Y. Yuan, X.-Y. Wang, M. Li, Moisture sorption and water vapor permeability of soy protein isolate/poly(vinyl alcohol)/glycerol blend films, Industrial Crops Prod. 31 (2) (2010) 266–276.

[49] E.O. Timmermann, J. Chirife, H.A. Iglesias, Water sorption isotherms of foods and foodstuffs: BET or GAB parameters? J. Food Eng. 48 (1) (2001) 19–31.

Dynamic fabrication of amylosic supramolecular composites in an enzymatic polymerization field

4

Jun-ichi Kadokawa
Kagoshima University, Kagoshima, Japan

Chapter Outline

4.1 Introduction

Biological macromolecules such as polysaccharides, proteins, and nucleic acids are widely distributed in biological systems and play important in vivo functions as vital materials [1–3]. The highly biological functions of such macromolecules are exhibited not only by their primary chemical structures, but also by controlling their higher-order structures. Biological macromolecules are solely present in nature; hybrid structures from one or multiple kinds of biological macromolecules often appear in important vital systems, such as glycoproteins and proteoglycans/peptideglycans, which are hybrid macromolecular systems from glycans and peptides [4–7]. Accordingly, the artificial hybridization of biological and synthetic macromolecules has been expected to provide high performance functional composite materials. The hybridization by covalent linkages has widely been performed in polymer science to obtain a variety of composite materials such as block, graft, and branched polymers (Fig. 4.1) [8]. Hierarchically hybridization from several kinds of macromolecules by controlled noncovalent linkages has also been achieved to fabricate supramolecular composite materials (Fig. 4.1) [9–11]. Because of their controlled higher-order structures, natural polysaccharides are one of

Hybrid Polymer Composite Materials: Processing. DOI: http://dx.doi.org/10.1016/B978-0-08-100789-1.00004-6

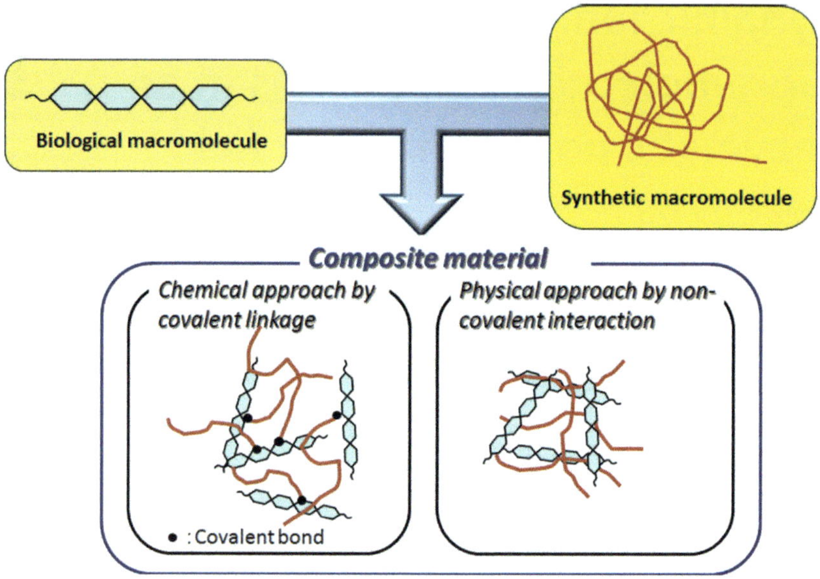

Figure 4.1 Composite formation from biological and synthetic macromolecules via chemical and physical approaches.

the most useful candidates to construct supramolecular composite materials with synthetic polymers.

Amylose is a representative natural polysaccharide as a component in starch, which is composed of $\alpha(1\rightarrow4)$-linked glucose repeating units [3]. Besides its important role as the energy provider in nature, it also acts as functional polymeric material, which forms supramolecular inclusion complexes, owing to its left-handed helical conformation [12]. Because of the presence of hydroxy groups of glucose units at outer of the helix, the cavity inside is hydrophobic. Amylose, therefore, acts as a host molecule to form supramolecular inclusion complexes with hydrophobic guest molecules, typically with low molecular weight, by hydrophobic interaction (Fig. 4.2) [13]. In addition to traditional natures of the amylosic supramolecular complexes, in order to provide supramolecular composite materials with further functions and properties suitably for the practical applications, such as good mechanical property, stimuli responsive behavior, and processability to multifashions, polymeric guest molecules with high molecular weight are better candidates for complexation with amylose, instead of low molecular weight guests. However, there haven't been many research examples of efficient methods for the direct formation of amylose-polymer supramolecular composites (Fig. 4.2). Because driving force for the complexation in the cavity of amylose is hydrophobic interaction with the guest molecules, it has been well identified that amylose does not have sufficient ability to directly include the long chain of polymeric guests only by such weak interaction. For the direct incorporation of polymeric guests into the cavity of

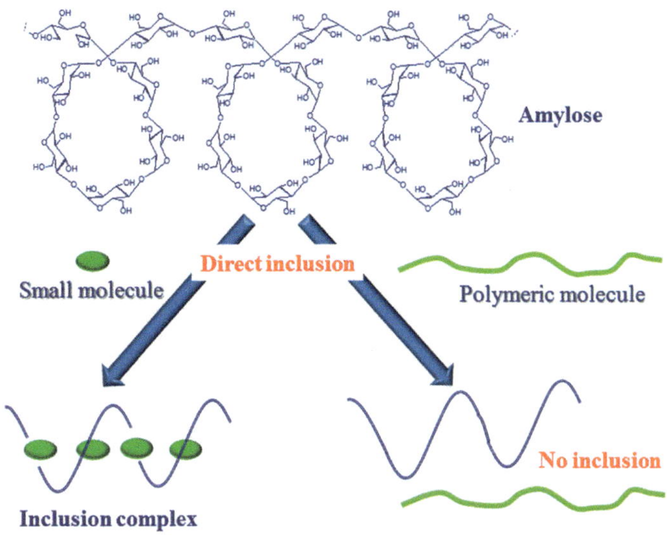

Figure 4.2 Amylose forms inclusion complex with relatively low molecular weight (small) hydrophobic molecule but mostly does not form it with polymeric molecule.

amylose, suitable hydrophilic groups have been incorporated at the polymer chain ends, which imply enhancement of the complexation in aqueous media [14,15]. The guest exchange approach from small to large (polymeric) guest molecules was also achieved [16]. The other way to directly form amylose-polymer inclusion composites has been achieved by an inclusion polymerization approach [17,18]. In this approach, polymerizable monomers with the appropriate chemical structure as the guest molecule are complexed in the cavity of the amylose helix. Then, polymerization of the monomers in the cavity is conducted to form amylose-polymer inclusion composites. Even the direct mixing method under the selected conditions has also achieved to obtain the inclusion composites from amylose and a guest polymer, poly(tetrahydrofuran) (PTHF) [19–21]. A partially-methylated amylose, on the other hand, has play an effective host molecule, which complexes with PTHF to form an inclusion composite [22]. This is because the double helix, which is a well-known higher-ordered structure of the amylose molecules, is slightly loosened, because the intra- and intermolecular hydrogen bonds are weakened by derivatization, leading to allowance for the incorporation of guest molecules compared with amylose.

It has been well-known that amylose with a well-defined structure is synthesized by phosphorylase-catalyzed enzymatic polymerization (Fig. 4.3) [23–26], as it has been well accepted that an enzymatic method is a powerful approach to synthesize well-defined polysaccharides [27–31]. Phosphorylase is the enzyme that catalyzes glucosylation of $\alpha(1\rightarrow4)$-glucan such as maltooligosaccharide as a glycosyl acceptor with α-D-glucose 1-phosphate (G-1-P) as a glycosyl donor to produce a

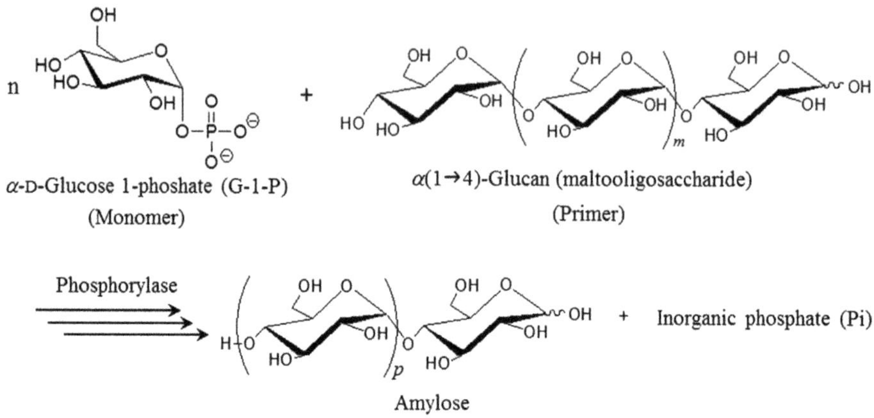

Figure 4.3 Phosphorylase-catalyzed enzymatic polymerization to produce amylose.

α(1→4)-glucosidic linkage [32,33]. Under the suitable conditions for the way to chain elongation, typically in the presence of large excess molar ratio of G-1-P to the acceptor, successive enzymatic glucosylations take place as the manner of polymerization by phosphorylase catalysis. The polymerization is initiated by transfer reaction of a G residue from G-1-P to the nonreducing end of maltooligosaccharide, accordingly, which is often called a primer of the polymerization. Then, the propagation occurs by the successive transfer reactions of G residues from G-1-P to the elongating nonreducing end to produce polysaccharide, that is, amylose. The polymerization is progressed analogously to living polymerization because of no occurrence of significant termination and chain-transfer reactions. Accordingly, molecular weights of the produced amyloses are controlled by monomer/primer feed ratios and their distributions are typically narrow [34].

The author has previously found that when the phosphorylase-catalyzed enzymatic polymerization is conducted in the presence of appropriate synthetic polymers, the propagating amylose chain gradually complexes with the polymers in the cavity to produce supramolecular inclusion composites [35–40]. The elongation from the shorter α(1→4)-glucan (maltooligosaccharide) to the longer α(1→4)-glucan (amylose) can be considered to provide the dynamic field for the more facile complexation with polymeric guests than the direct complexation between the polymeric host (amylose) and guest. Because the process of the present method for the dynamic fabrication of supramolecular composites is similar to the way that vines of plants grow with twining around a rod, this polymerization approach has been named "vine-twining polymerization" (Fig. 4.4). Furthermore, the vine-twining approach has been employed for the dynamic fabrication of supramolecular network/higher-ordered composites. The present chapter reviews dynamic fabrication of amylosic supramolecular inclusion composites in the phosphorylase-catalyzed enzymatic polymerization field by means of the vine-twining polymerization manner.

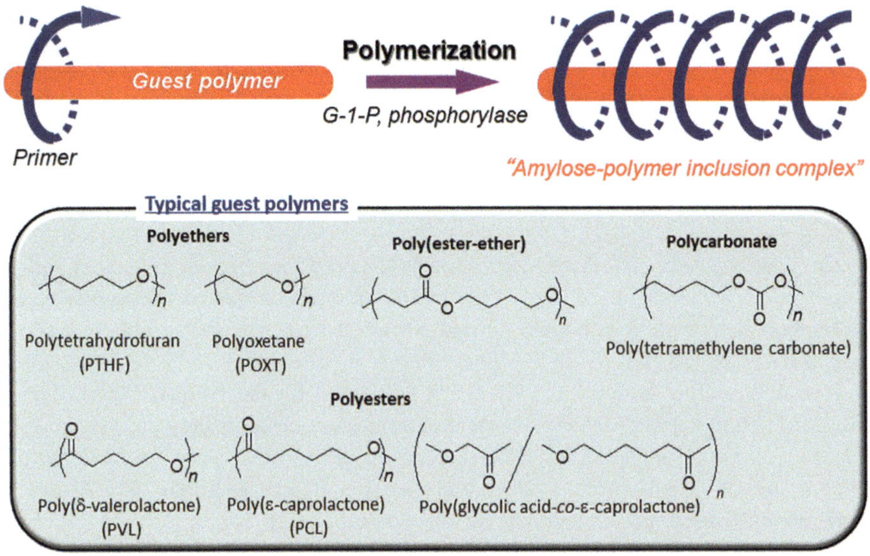

Figure 4.4 Image of vine-twining polymerization and typical guest polymer structures.

4.2 Dynamic formation of amylosic supramolecular inclusion composites by vine-twining polymerization and related system

The followings are the characteristics typically for the guest polymers that have dynamically formed supramolecular inclusion composites with amylose in the vine-twining polymerization. As mentioned earlier, hydrophobicity is required and is included in the cavity of amylose by hydrophobic interaction. Because the vine-twining polymerization is conducted in aqueous buffer solvent as the reaction field of the phosphorylase-catalyzed enzymatic polymerization, however, guest polymers have to be dispersed in such aqueous media. Accordingly, relatively polar groups should be present in the main-chain of guest polymers. A third one is suitability of slender polymeric fashion without bulky side groups as the guest polymer structure because of the fact that the cavity size of the amylose helix is not sufficiently large to include bulky molecules. On the basis of these characteristics, in addition to the previous studies on the aforementioned direct formation of inclusion composites with amylose, PTHF has been considered as one of the most appropriate candidates as the guest polymer for the vine-twining polymerization because it is the hydrophobic polymer composed of the relatively polar ether groups, but without any side groups. Indeed, the first example of the vine-twining polymerization was achieved using PTHF as a hydrophobic guest polymer [41,42]. When the phosphorylase-catalyzed enzymatic polymerization of G-1-P from maltoheptaose (G_7) as a primer was carried out in the presence of PTHF dispersed in citrate buffer as a polymerization solvent, the product was gradually precipitated from the reaction media. The analytical results of the isolated product fully supported the structure of an amylose-PTHF supramolecular inclusion composite. Complexation between amylose and PTHF did not occur by mixing the two polymers in the buffer solvent, which suggested the formation of the composite during the progress of the enzymatic polymerization according to the vine-twining approach. Poly(oxetane) (POXT) of another hydrophobic polyether was also found to form a supramolecular inclusion composite with amylose in the vine-twining polymerization [41]. On the contrary, the vine-twining polymerization system in the presence of poly(ethylene glycol) (PEG) of a hydrophilic polyether did not induce the composite formation. These results strongly indicated the importance of hydrophobicity of guest polymers on complexation with amylose to obtain supramolecular inclusion composites in the vine-twining polymerization.

Hydrophobic polyesters have also been employed for the vine-twining polymerization to obtain supramolecular inclusion composites because they are composed of relatively polar ester linkages. Poly(ε-caprolactone) (PCL), poly(δ-valerolactone) (PVL), and a copolyester, poly(glycolic acid-*co*-ε-caprolactone) (P(GA-*co*-CL)) were used as guest polymers to form the corresponding supramolecular inclusion composites in the vine-twining polymerization [43–45]. The homopolyester, PGA, was not acted as a guest polymer for complexation with amylose owing to high crystallinity and low dispersibility in aqueous media.

A hydrophobic poly(ester-ether) (PEE, $-CH_2CH_2C(C=O)OCH_2CH_2CH_2CH_2O-$) consisting of alternating ester and ether linkages also formed a supramolecular inclusion composite with amylose in the vine-twining polymerization [44]. On the other hand, a hydrophilic analog of the above PEE, that is, $-CH_2CH_2C(C=O)OCH_2CH_2O-$, did not behave as a guest polymer for complexation in the vine-twining polymerization, supporting the significance in hydrophobicity of guest polymers whether they are suitable for complexation with amylose in the vine-twining polymerization. A hydrophobic polycarbonate, poly(tetramethylene carbonate) (PTMC), composed of relatively polar carbonate linkages, also formed a supramolecular inclusion composite with amylose in the vine-twining polymerization [46].

In addition to difficulty in the formation of supramolecular inclusion composites from hydrophilic polymers, strongly hydrophobic polymers did not afford complexation with amylose in the vine-twining polymerization, due to hard dispersibility in aqueous media. For example, poly(oxapane), poly(octamethylene carbonate), poly (decamethylene carbonate), and poly(dodecamethylene carbonate), which are a polyether and polycarbonates with the longer methylenes than PTHF and PTMC, have not been mostly complexed with amylose in the vine-twining polymerization. These results indicated the requirement of moderate hydrophobicity in guest polymers to form supramolecular inclusion complexes.

To obtain a supramolecular inclusion composite from a strongly hydrophobic polyester, parallel enzymatic polymerization system has been achieved, where two enzymatic polymerizations, that is, the phosphorylase-catalyzed polymerization and lipase-catalyzed polycondensation, which individually produce the host amylose and guest polyester, have been simultaneously conducted in one reaction media (Fig. 4.5) [47]. As the monomers for the polycondensation, sebacic acid and 1,8-octanediol were used, which were converted into the strongly hydrophobic polyester by the

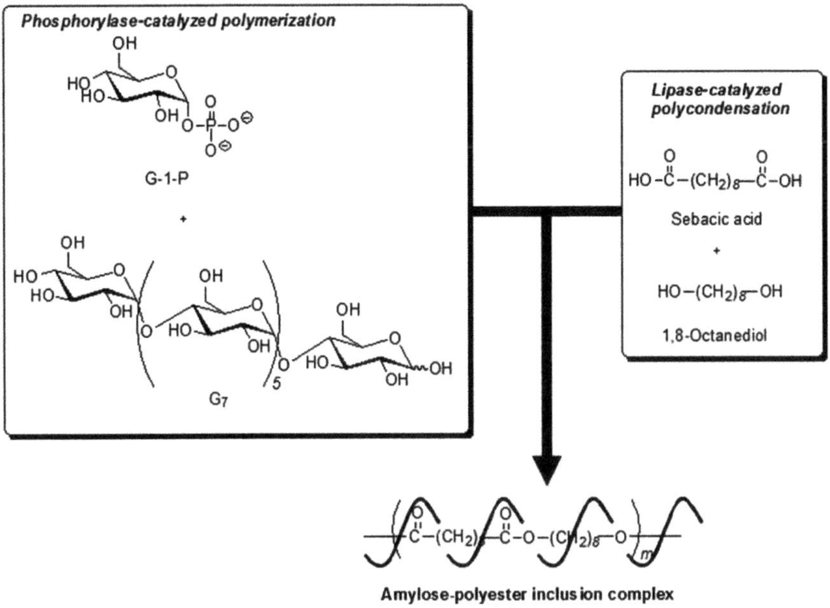

Figure 4.5 Formation of supramolecular inclusion composite composed of amylose and strongly hydrophobic polyester in parallel enzymatic polymerization system.

lipase-catalyzed polycondensation under aqueous conditions [48,49]. The analytical results of the product by the parallel enzymatic polymerization system supported the formation of such composite from amylose and the polyester. To confirm the fact that the supramolecular inclusion composite was obtained only by the parallel enzymatic polymerization system, the following two experiments were additionally conducted. When in the presence of the strongly hydrophobic polyester in aqueous media, the phosphorylase-catalyzed enzymatic polymerization was conducted, which accorded to the vine-twining polymerization manner, amylose was enzymatically produced, but it did not induce complexation with the polyester. As the other experiment, the lipase-catalyzed polycondensation of sebacic acid and 1,8-octanediol was conducted in the presence of amylose in aqueous media. Consequently, amylose complexed with some monomers, but did not complex with the polyester although the enzymatic polycondensation was progressed. These results concluded that only the parallel enzymatic polymerization system suitably afforded complexation between amylose and the strongly hydrophobic polyester.

4.3 Selective complexation of amylose in vine-twining polymerization

The above findings in fundamental investigations of the vine-twining polymerization have suggested the demand for moderate hydrophobicity of guest polymers in

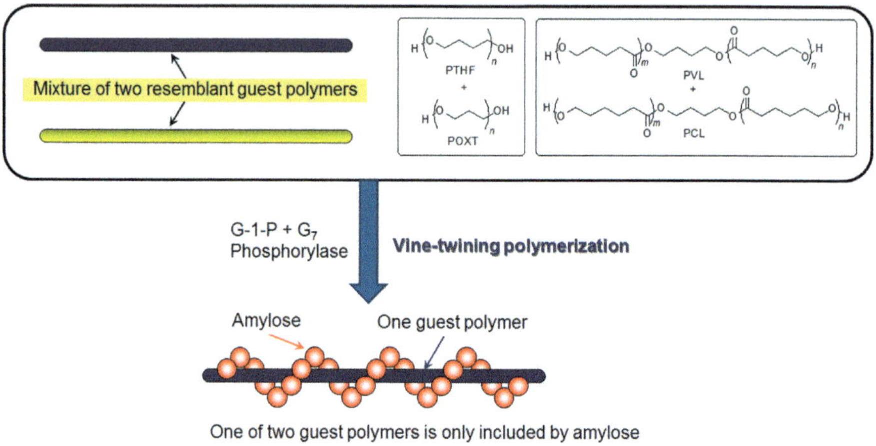

Figure 4.6 Amylose selectively includes one of two resemblant polyethers or polyesters.

deciding whether amylose complexes with them or not. Amylose, therefore, exhibits different complexation behaviors depending on subtle changes in the structures of hydrophobic guest polymers, which implies selective inclusion toward guest polymers with resemblant chemical structures. For example, amylose selectively included PTHF to form the supramolecular inclusion composite when a mixture of PTHF/POXT was present in the vine-twining polymerization system (Fig. 4.6) [50]. The selectivity was attributed to the slight difference in their hydrophobicities. As another example, from a mixture of PCL/PVL, amylose selectively formed a supramolecular inclusion composite with PVL in the vine-twining polymerization (Fig. 4.6) [51].

Amylose also showed selective complexation toward a specific range in molecular weights of guest polymers in the vine-twining polymerization. Because synthetic polymers generally have molecular weight distribution, they can be considered as mixtures of analogous molecules with the different numbers of repeating units. The polymer properties are somewhat changed in accordance with the numbers of repeating units. For example, molecular weights of PTHF affect its hydrophobicity and water-solubility, in which PHTF with certain molecular weight is hydrophobic and insoluble in water, whereas PTHF with considerably low molecular weight shows water-solubility. Accordingly, when the several vine-twining polymerization systems were conducted using PTHFs with different average molecular weights, amylose selectively recognized the specific range in molecular weights of all PTHFs to form the supramolecular inclusion composites [52].

In addition to chemical structure and molecular weight, amylose exhibited selectivity toward chirality in guest polymers in the vine-twining polymerization. Such stereoselective inclusion by amylose was observed in the vine-twining polymerization using chiral polyesters, poly(lactide)s (PLAs) as guest polymers (Fig. 4.7) [53]; there are three kinds of the stereoisomers, i.e., poly(L-lactide) (PLLA), poly(D-lactide) (PDLA), and racemic poly(DL-lactide) (PDLLA). When the vine-twining

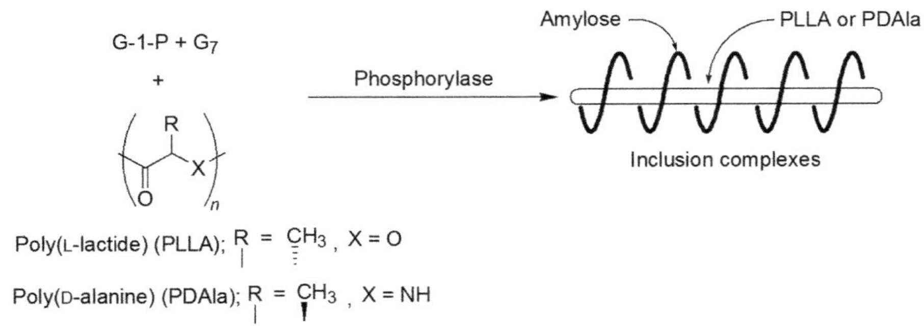

Figure 4.7 Stereoselective inclusion complexation by amylose in vine-twining polymerization using poly(L-lactide) (PLLA) and poly(D-alanine) (PDAla).

polymerization was conducted using PLLA, complexation occurred to produce an amylose-PLLA supramolecular inclusion composite. On the other hand, the vine-twining polymerization systems using PDLA and PDLLA did not induce the formation of inclusion composites. The similar stereoselective inclusion by amylose was observed when polyalanine (PAlas) stereoisomers, which were chiral polypeptides, were used in the vine-twining polymerization (Fig. 4.7) [54]. A supramolecular inclusion composite was obtained from poly(D-alanine) (PDAla), whereas the other stereoisomers, poly(L-alanine) (PLAla), and poly(DL-alanine) (PDLAla) did not form composites with amylose. The stereoselective inclusion behavior of amylose toward PLLA and PDAla in the vine-twining polymerization has been explained by the helical directions of host and guest polymers. Because these guest polymers construct left-handed helical conformation, which is the same direction as that of the host amylose, they are selectively included by amylose. The opposite and irregular helical conformations in PDLA/PLAla and PDLLA/PDLAla, respectively, on the other hand, are not suitable for complexation with the amylose helix.

4.4 Dynamic fabrication of amylosic supramolecular inclusion composite materials by vine-twining polymerization

The vine-twining polymerization approach has been employed for dynamic fabrication of amylosic supramolecular inclusion composite materials such as gel and film (Fig. 4.8) [55]. To fabricate such supramolecular materials, polymeric networks composed of amylosic supramolecular inclusion complexes as crosslinking points have been designed as the vine-twining polymerization products using graft copolymers having hydrophobic graft chains. In such a system, the produced amylose chains potentially include the hydrophobic graft chains as guest polymers to produce inclusion complexes between the intermolecular graft copolymers, which act as crosslinking points to yield hydrogels. In spite of hydrophobicity of graft chains as guest polymers for amylose, a whole nature of the graft copolymer has to be water-soluble as a component of hydrogels. Taking these antagonistic properties in the graft copolymer into account, poly(acrylic acid sodium salt-*graft*-δ-valerolactone) (PAA-Na-*g*-PVL) was first used, in which PAA-Na is a strong hydrophilic polymer contributing to enhancement of water-solubility and PVL allows complexation by amylose in the vine-twining polymerization [56]. When the vine-twining polymerization was performed in the presence of PAA-Na-*g*-PVL in an acetate buffer, the reaction mixture gradually turned into hydrogel form with the progress of the enzymatic polymerization. The analytical results of a powdered sample lyophilized from the hydrogel supported the presence of inclusion complexes as crosslinking points, which were produced by amylose with the intermolecular (PAA-Na-*g*-PVL)s by the vine-twining polymerization manner.

Because amylose is enzymatically hydrolyzed and produced by the amylase- and phosphorylase-catalyzed reactions, respectively, the enzymatic disruption and

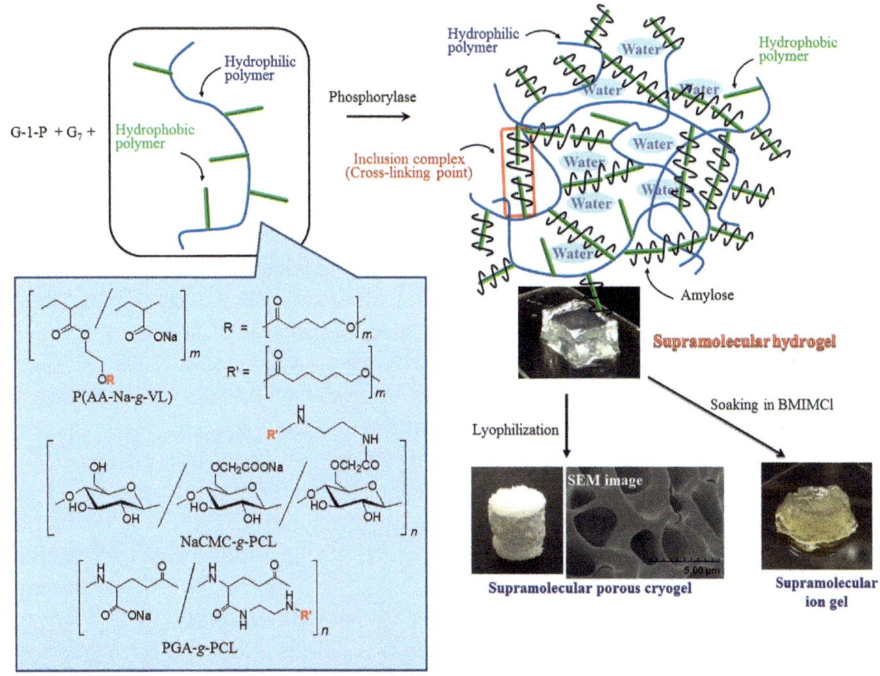

Figure 4.8 Fabrication of amylosic supramolecular inclusion composite materials by vine-twining polymerization using graft copolymers having hydrophilic main chains and hydrophobic guest graft chains and conversion of supramolecular hydrogel into cryo- and ion gels.

reproduction of the supramolecular hydrogel by the two enzymatic reactions were investigated. The hydrogel totally transferred into a solution in the presence of β-amylase under the enzymatic reaction conditions. This disruption was owing to the disappearance of crosslinking points due to the enzymatic hydrolysis of amylose components in the hydrogel. When the resulting solution was incubated in the presence of phosphorylase and G-1-P under the enzymatic reaction conditions, rehydrogelation was observed. Because maltooligosaccharides, which were produced by partial hydrolysis of amylose by the β-amylase-catalyzed reaction, acted as the primer for the enzymatic polymerization of G-1-P in the solution, the enzymatically produced amylose included the PVL graft chains to form supramolecular crosslinking points, leading to rehydrogelation.

A supramolecular film was also fabricated through hydrogelation by the vine-twining polymerization using another graft copolymer. In this case, carboxymethyl cellulose (CMC) was employed as the main-chain of the graft copolymer because of its film formability in addition to water-solubility. Accordingly, CMC-*g*-PCL was synthesized and used for the vine-twining polymerization [57]. The reaction mixture by the vine-twining polymerization using CMC-*g*-PCL totally turned into a hydrogel. The hydrogel was further converted into a supramolecular film by moisturizing the powdered sample, which was prepared by lyophilization of the hydrogel.

Because mechanical properties of the hydrogels formed by using PAA-Na-*g*-PVL and CMC-*g*-PCL were not sufficiently strengthened for further applications, poly (γ-glutamic acid) (PGA) was then employed as the main-chain of a graft copolymer to improve properties of the supramolecular hydrogel [58]. PGA is a well-known natural polypeptide with high water retention and moisturizing properties, which potentially contribute to the enhancement of the mechanical property. When the vine-twining polymerization was performed, using PGA-*g*-PCL, the supramolecular hydrogel with self-standing property was formed, indicating much better mechanical property of the present hydrogel compared with the above supramolecular hydrogels.

The macroscopic interfacial healing of the supramolecular hydrogel consisting of PGA-*g*-PCL was observed through the phosphorylase-catalyzed enzymatic polymerization. The hydrogel formed initially from the vine-twining polymerization was cut into two pieces, and a sodium acetate buffer solution containing G-1-P and phosphorylase was dropped on the surfaces of the hydrogel pieces. After the surfaces were placed in contact with one another, the materials were left standing under the conditions for the progress of the enzymatic polymerization. Consequently, the two hydrogel pieces fused at the point of contact. The healing of the gels on a macroscopic level was induced by complexation of the enzymatically produced amyloses with the PCL graft chains at the interface. Furthermore, porous cryogel was fabricated by lyophilization of the hydrogel, while ion gel was prepared by soaking the hydrogel in an ionic liquid of 1-butyl-3-methylimidazolium chloride (BMIMCl) (Fig. 4.8).

Supramolecular polymeric composites consisting of amylose-PTHF and amylose-PLLA inclusion complexes were dynamically fabricated by the vine-twining polymerizations using primer-guest conjugates, i.e., G_7-*block*-PTHF and G_7-*block*-PLLA (Fig. 4.9A) [59,60]. In these systems, a propagating amylose chain initiated by the phosphorylase catalysis from a G_7 segment in the conjugate potentially included a PTHF or PLLA segment of another conjugate, whereby the successive occurrence of such a propagating manner gave rise to linear supramolecular polymeric composites.

Similarly, a branched G_7-$PLLA_2$ conjugate was employed as the substrate for the vine-twining polymerization to produce a hyperbranched supramolecular polymeric composite (Fig. 4.9B) [61]. The resulting hyperbranched product then formed an ion gel with BMIMCl, which was further converted into a hydrogel upon exchange of the dispersion media by soaking in water. Lyophilization of the hydrogel resulted in a cryogel with a porous morphology. The gelation behavior of the hyperbranched composite was a result of its three-dimensionally extended structure, whereas the aforementioned linear supramolecular polymeric composites did not exhibit such gelation behavior.

The relative chain orientation of amylose and PLLA in the supramolecular inclusion composite formed in the vine-twining polymerization was investigated by using two G_7–PLLA conjugates, which were composed of a G_7 moiety

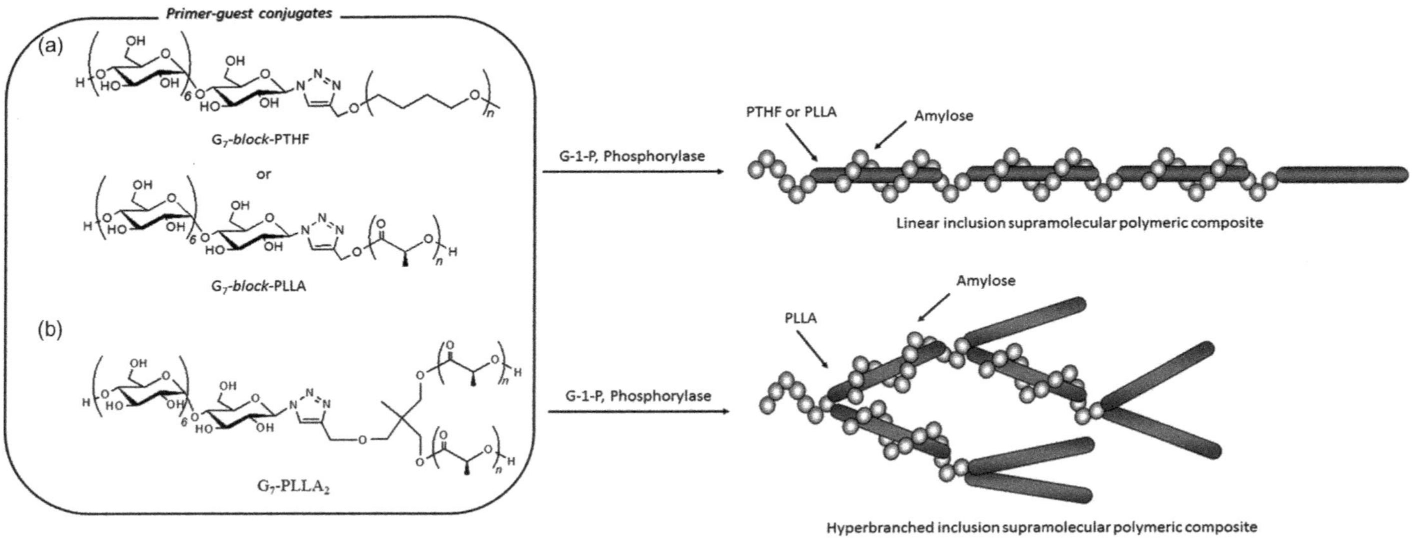

Figure 4.9 Fabrication of (A) linear and (B) hyperbranched supramolecular polymeric composites by vine-twining polymerization using primer-guest conjugates.

functionalized at the carboxylate or hydroxy termini of PLLA [62]. The amylose-PLLA supramolecular polymeric composites were obtained in the enzymatic polymerization in the presence of both of the two PLLA conjugates, suggesting that, regardless of the chain orientation of PLLA, the amylose cavity recognized the PLLA segment. On the other hand, the amylose–PDLA diblock copolymers, which did not contain supramolecular inclusion composite structures, were produced by the phosphorylase-catalyzed enzymatic polymerization in the presence of both of the two G_7–PDLA conjugates. This was because of the effect of chirality on the inclusion behavior of the amylose cavity, which was irrespective of the PDLA chain orientation. The left-handed helical directions from both of amylose and PLLA imply the formation of the supramolecular inclusion composite, whereas the directions of methyl substituents in PLA toward amylose, which are oppositely changed in accordance with the relative chain orientation, are not a significant factor for complexation.

4.5 Conclusions

This chapter reviewed the dynamic fabrication of amylosic supramolecular inclusion composites in the phosphorylase-catalyzed enzymatic polymerization field according to the vine-twining polymerization manner. During the course of the investigations, the importance in moderate hydrophobicity of guest polymers was revealed whether amylose took up them into the cavity. Amylose showed selectivity in chemical structure, molecular weight, and chirality of guest polymers on inclusion complexation. The supramolecular composite materials were dynamically fabricated by the vine-twining polymerization using the designed graft copolymers composed of hydrophilic main chains and hydrophobic guest graft chains. The vine-twining polymerization using guest-primer conjugates also afforded the dynamic fabrication of the supramolecular polymeric composites composed of continuum of inclusion complexes. As described in this chapter, the vine-twining polymerization approach efficiently produced supramolecular composites from amylose and other bio-related polymers, such as PCL, PLLA, PGA, and so on. Therefore, the produced materials have a potential to practically apply as functional substrates in environmentally benign and sustainable fields. Furthermore, this approach will provide additional useful eco-friendly materials with high performance functions by using newly designed bio-related substrates in the future.

References

[1] J.M. Berg, J.L. Tymoczko, L. Stryer, Biochemistry, W.H. Freeman, New York, 2012.
[2] S. Kasapis, I.T. Norton, J.B. Ubbink, Modern Biopolymer Science: Bridging the Divide Between Fundamental Treatise and Industrial Application, Academic Press, San Diego, 2009.

[3] C. Schuerch, Polysaccharides, in: H.F. Mark, N. Bilkales, C.G. Overberger (Eds.), Encyclopedia of Polymer Science and Engineering, Vol. 13, John Wiley & Sons, New York, 1986, pp. 87–162.
[4] J. Montreuil, J.F.G. Vliegenthart, H. Schachter, Glycoproteins, Elsevier, Amsterdam; New York, 1995.
[5] J. Montreuil, J.F.G. Vliegenthart, H. Schachter, Glycoproteins II, Elsevier, Amsterdam; New York, 1997.
[6] R.V. Iozzo, Proteoglycans: Structure, Biology, and Molecular Interactions, Marcel Dekker, New York, 2000.
[7] D.E.S. Stewarttull, The immunological activities of bacterial peptidoglycans, Annu. Rev. Microbiol. 34 (1980) 311–340.
[8] M.K. Mishra, S. Kobayashi, Star and Hyperbranched Polymers, Marcel Dekker, New York, 1999.
[9] J.M. Lehn, Supramolecular polymer chemistry- scope and perspectives, Polym. Int. 51 (10) (2002) 825–839.
[10] J.M. Lehn, Dynamers: dynamic molecular and supramolecular polymers, Progr. Polym. Sci. 30 (8–9) (2005) 814–831.
[11] J.W. Steed, P.A. Gale, Supramolecular Chemistry: From Molecules to Nanomaterials, Wiley, Hoboken, NJ, 2012.
[12] A. Sarko, P. Zugenmaier, Crystal structures of amylose and its derivatives, in: A.D. French, K.H. Gardner (Eds.), Fiber Diffraction Methods, American Chemical Society, Washington, DC, 1980, pp. 459–482. ACS Symposium Series 141.
[13] J.A. Putseys, L. Lamberts, J.A. Delcour, Amylose-inclusion complexes: formation, identity and physico-chemical properties, J. Cereal Sci. 51 (3) (2010) 238–247.
[14] R.L. Shogren, Complexes of starch with telechelic poly(epsilon-caprolactone) phosphate, Carbohydr. Polym. 22 (2) (1993) 93–98.
[15] R.L. Shogren, R.V. Greene, Y.V. Wu, Complexes of starch polysaccharides and poly (ethylene coacrylic acid) - structure and stability in solution, J. Appl. Polym. Sci. 42 (6) (1991) 1701–1709.
[16] A. Star, D.W. Steuerman, J.R. Heath, J.F. Stoddart, Starched carbon nanotubes, Angew. Chem. Int. Ed. 41 (14) (2002) 2508–2512.
[17] K. Kumar, A.J.J. Woortman, K. Loos, Synthesis of amylose-polystyrene inclusion complexes by a facile preparation route, Biomacromolecules 14 (6) (2013) 1955–1960.
[18] M. Ikeda, Y. Furusho, K. Okoshi, S. Tanahara, K. Maeda, S. Nishino, et al., A luminescent poly(phenylenevinylene)-amylose composite with supramolecular liquid crystallinity, Angew. Chem. Int. Ed. 45 (39) (2006) 6491–6495.
[19] R. Rachmawati, A.J.J. Woortman, K. Loos, Facile preparation method for inclusion complexes between amylose and polytetrahydrofurans, Biomacromolecules 14 (2) (2013) 575–583.
[20] R. Rachmawati, A.J.J. Woortman, K. Loos, Tunable properties of inclusion complexes between amylose and polytetrahydrofuran, Macromol. Biosci. 13 (6) (2013) 767–776.
[21] R. Rachmawati, A.J.J. Woortman, K. Loos, Solvent-responsive behavior of inclusion complexes between amylose and polytetrahydrofuran, Macromol. Biosci. 14 (1) (2014) 56–68.
[22] T. Kida, T. Minabe, S. Okabe, M. Akashi, Partially-methylated amyloses as effective hosts for inclusion complex formation with polymeric guests, Chem. Commun. 15 (2007) 1559–1561.
[23] K. Fujii, H. Takata, M. Yanase, Y. Terada, K. Ohdan, T. Takaha, et al., Bioengineering and application of novel glucose polymers, Biocatal. Biotransform. 21 (4–5) (2003) 167–172.

[24] J. Seibel, H.J. Jordening, K. Buchholz, Glycosylation with activated sugars using glycosyltransferases and transglycosidases, Biocatal. Biotransform. 24 (5) (2006) 311–342.
[25] M. Yanase, T. Takaha, T. Kuriki, α-Glucan phosphorylase and its use in carbohydrate engineering, J. Sci. Food Agric. 86 (11) (2006) 1631–1635.
[26] G. Ziegast, B. Pfannemüller, Linear and star-shaped hybrid polymers. 4. Phosphorolytic syntheses with di-functional, oligo-functional and multifunctional primers, Carbohydr. Res. 160 (1987) 185–204.
[27] S. Kobayashi, A. Makino, Enzymatic polymer synthesis: an opportunity for green polymer chemistry, Chem. Rev. 109 (11) (2009) 5288–5353.
[28] S. Kobayashi, H. Uyama, S. Kimura, Enzymatic polymerization, Chem. Rev. 101 (12) (2001) 3793–3818.
[29] S. Shoda, R. Izumi, M. Fujita, Green process in glycotechnology, Bull. Chem. Soc. Japan 76 (1) (2003) 1–13.
[30] S. Shoda, H. Uyama, J. Kadokawa, S. Kimura, S. Kobayashi, Enzymes as green catalysts for precision macromolecular synthesis, Chem. Rev. 116 (4) (2016) 2307–2413.
[31] J. Kadokawa, Precision polysaccharide synthesis catalyzed by enzymes, Chem. Rev. 111 (7) (2011) 4308–4345.
[32] M. Kitaoka, K. Hayashi, Carbohydrate-processing phosphorolytic enzymes, Trends Glycosci. Glycotechnol. 14 (75) (2002) 35–50.
[33] H. Nakai, M. Kitaoka, B. Svensson, K. Ohtsubo, Recent development of phosphorylases possessing large potential for oligosaccharide synthesis, Curr. Opin. Chem. Biol. 17 (2) (2013) 301–309.
[34] S. Kitamura, Starch polymers, natural and synthetic, in: C. Salamone (Ed.), The Polymeric Materials Encyclopedia, Synthesis, Properties and Applications, Vol. 10, CRC Press, New York, 1996, pp. 7915–7922.
[35] Y. Kaneko, J. Kadokawa, Vine-twining polymerization: a new preparation method for well-defined supramolecules composed of amylose and synthetic polymers, Chem. Rec. 5 (1) (2005) 36–46.
[36] Y. Kaneko, J. Kadokawa, Synthesis of nanostructured bio-related materials by hybridization of synthetic polymers with polysaccharides or saccharide residues, J. Biomater. Sci. Polym. Ed. 17 (11) (2006) 1269–1284.
[37] Y. Kaneko, J. Kadokawa, Preparation of polymers with well-defined nanostructure in the polymerization field, in: J.N. Lee (Ed.), Modern Trends in Macromolecular Chemistry, Nova Science Publishers, Inc, Hauppauge, NY, 2009, pp. 199–217.
[38] J. Kadokawa, Preparation and applications of amylose supramolecules by means of phosphorylase-catalyzed enzymatic polymerization, Polymers 4 (2012) 116–133.
[39] J. Kadokawa, Architecture of amylose supramolecules in form of inclusion complexes by phosphorylase-catalyzed enzymatic polymerization, Biomolecules 3 (2013) 369–385.
[40] J. Kadokawa, Chemoenzymatic synthesis of functional amylosic materials, Pure Appl. Chem. 86 (5) (2014) 701–709.
[41] J. Kadokawa, Y. Kaneko, S. Nagase, T. Takahashi, H. Tagaya, Vine-twining polymerization: amylose twines around polyethers to form amylose - polyether inclusion complexes, Chem. A Eur. J. 8 (15) (2002) 3321–3326.
[42] J. Kadokawa, Y. Kaneko, H. Tagaya, K. Chiba, Synthesis of an amylose-polymer inclusion complex by enzymatic polymerization of glucose 1-phosphate catalyzed by phosphorylase enzyme in the presence of polythf: a new method for synthesis of polymer-polymer inclusion complexes, Chem. Commun. 5 (2001) 449–450.
[43] J. Kadokawa, Y. Kaneko, A. Nakaya, H. Tagaya, Formation of an amylose-polyester inclusion complex by means of phosphorylase-catalyzed enzymatic polymerization of

α-D-glucose 1-phosphate monomer in the presence of poly(ε-caprolactone), Macromolecules 34 (19) (2001) 6536–6538.
[44] J. Kadokawa, A. Nakaya, Y. Kaneko, H. Tagaya, Preparation of inclusion complexes between amylose and ester-containing polymers by means of vine-twining polymerization, Macromol. Chem. Phys. 204 (11) (2003) 1451–1457.
[45] S. Nomura, T. Kyutoku, N. Shimomura, Y. Kaneko, J. Kadokawa, Preparation of inclusion complexes composed of amylose and biodegradable poly(glycolic acid-*co*-ε-caprolactone) by vine-twining polymerization and their lipase-catalyzed hydrolysis behavior, Polym. J. 43 (12) (2011) 971–977.
[46] Y. Kaneko, K. Beppu, J. Kadokawa, Preparation of amylose/polycarbonate inclusion complexes by means of vine-twining polymerization, Macromol. Chem. Phys. 209 (10) (2008) 1037–1042.
[47] Y. Kaneko, Y. Saito, A. Nakaya, J. Kadokawa, H. Tagaya, Preparation of inclusion complexes composed of amylose and strongly hydrophobic polyesters in parallel enzymatic polymerization system, Macromolecules 41 (15) (2008) 5665–5670.
[48] S. Kobayashi, H. Uyama, S. Suda, S. Namekawa, Dehydration polymerization in aqueous medium catalyzed by lipase, Chem. Lett. 1 (1997), 105–105.
[49] S. Suda, H. Uyama, S. Kobayashi, Dehydration polycondensation in water for synthesis of polyesters by lipase catalyst, Proc. Japan Acad. Ser. B Phys. Biol. Sci. 75 (7) (1999) 201–206.
[50] Y. Kaneko, K. Beppu, J. Kadokawa, Amylose selectively includes one from a mixture of two resemblant polyethers in vine-twining polymerization, Biomacromolecules 8 (10) (2007) 2983–2985.
[51] Y. Kaneko, K. Beppu, T. Kyutoku, J. Kadokawa, Selectivity and priority on inclusion of amylose toward guest polyethers and polyesters in vine-twining polymerization, Polym. J. 41 (4) (2009) 279–286.
[52] Y. Kaneko, K. Beppu, J. Kadokawa, Amylose selectively includes a specific range of molecular weights in poly(tetrahydrofuran)s in vine-twining polymerization, Polym. J. 41 (9) (2009) 792–796.
[53] Y. Kaneko, K. Ueno, T. Yui, K. Nakahara, J. Kadokawa, Amylose's recognition of chirality in polylactides on formation of inclusion complexes in vine-twining polymerization, Macromol. Biosci. 11 (10) (2011) 1407–1415.
[54] R. Gotanda, K. Yamamoto and J. Kadokawa, Amylose stereoselectively includes poly(D-alanine) to form inclusion complex in vine-twining polymerization: a novel saccharide-peptide supramolecular conjugate. *Macromol. Chem. Phys.* 217 (9) (2016) 1074–1080.
[55] J. Kadokawa, Hierarchically fabrication of amylosic supramolecular nanocomposites by means of inclusion complexation in phosphorylase-catalyzed enzymatic polymerization field, in: K.V. Thakur, K.M. Thakur (Eds.), Eco-friendly Polymer Nanocomposites: Processing and Properties, Springer India, New Delhi, 2015, pp. 513–525.
[56] Y. Kaneko, K. Fujisaki, T. Kyutoku, H. Furukawa, J. Kadokawa, Preparation of enzymatically recyclable hydrogels through the formation of inclusion complexes of amylose in a vine-twining polymerization, Chem. Asian J. 5 (7) (2010) 1627–1633.
[57] J. Kadokawa, S. Nomura, D. Hatanaka, K. Yamamoto, Preparation of polysaccharide supramolecular films by vine-twining polymerization approach, Carbohydr. Polym. 98 (1) (2013) 611–617.
[58] J. Kadokawa, K. Tanaka, D. Hatanaka, K. Yamamoto, Preparation of multiformable supramolecular gels through helical complexation by amylose in vine-twining polymerization, Polym. Chem. 6 (35) (2015) 6402–6408.
[59] T. Tanaka, S. Sasayama, S. Nomura, K. Yamamoto, Y. Kimura, J. Kadokawa, An amylose-poly(L-lactide) inclusion supramolecular polymer: enzymatic synthesis by means

of vine-twining polymerization using a primer-guest conjugate, Macromol. Chem. Phys. 214 (24) (2013) 2829–2834.
[60] T. Tanaka, A. Tsutsui, R. Gotanda, S. Sasayama, K. Yamamoto, J. Kadokawa, Synthesis of amylose-polyether inclusion supramolecular polymers by vine-twining polymerization using maltoheptaose-functionalized poly(tetrahydrofuran) as a primer-guest conjugate, J. Appl. Glycosci. 62 (2015) 135–141.
[61] T. Tanaka, R. Gotanda, A. Tsutsui, S. Sasayama, K. Yamamoto, Y. Kimura, et al., Synthesis and gel formation of hyperbranched supramolecular polymer by vine-twining polymerization using branched primer-guest conjugate, Polymer 73 (2015) 9–16.
[62] T. Tanaka, S. Sasayama, K. Yamamoto, Y. Kimura, J. Kadokawa, Evaluating relative chain orientation of amylose and poly(L-lactide) in inclusion complexes formed by vine-twining polymerization using primer–guest conjugates, Macromol. Chem. Phys. 216 (2015) 794–800.

Advanced composites with strengthened nanostructured interface

5

Mohit Sharma[1], *Himani Sharma*[2] *and Santiranjan Shannigrahi*[1]
[1]Institute of Materials Research and Engineering, Singapore, Singapore, [2]Doon University Dehradun, Dehradun, Uttarakhand, India

Chapter Outline

5.1 Introduction: necessity to strengthen the fiber–matrix interface

The fiber-reinforced composites are promptly replacing conventional metallic materials counterparts for structural applications. The reinforcement imparts essential performance properties to the composites system, like high specific strength, higher performance-to-weight ratio, high stiffness, and wear resistance. Tougher matrix supports the overall structural integrity of composites and provide inertness, corrosion resistance, fluid tightness, and durability. The polymer matrix eases the stress transfer among the reinforcing laminates as well as aligns and protects the reinforcements from critical flaws and breakage. The conventional steels and other similar monolithic structure elements have high density and are more prone to environmental degradations like corrosion, which shorten their life and cause early catastrophic failure. Contrarywise, reinforced composites have additional ability to be custom

Hybrid Polymer Composite Materials: Processing. DOI: http://dx.doi.org/10.1016/B978-0-08-100789-1.00005-8

made as per the intended applications, which makes them a potential contender to replace the conventional materials. The reinforced composites materials are predominantly categorized as particle-reinforced composites (large particulate, micro/nano powder additives), fiber-reinforced composites (continuous, unidirectional, bidirectional, randomly oriented reinforcements), and structural composites (laminates and sandwiched panel). The composites are extensively used in a number of areas for civil and automotive structure. The advanced composites structural utilities including aerospace structures, cryogenic tanks, compressed gas storage and transportation, fuel cells, deep sea drilling platforms, wind turbines, automotive energy systems, turbo machinery, antistatic, and electromagnetic shielding materials and lightweight space vehicle applications.

The composites interface/interphase is defined as a predetermined interlayer with discrete physical and chemical properties between the fiber and matrix. Strengthening of the fiber–matrix interface is crucial to enhance the composites performance and mechanical strength properties. The fiber–matrix interfaces/interphases act as vast assortment in composition, structure, and design. Strong fiber–matrix mechanical interlocking is required to ease the stress transfer and improved performance properties. However, a feeble interface can help to improve the impact properties of the composites by energy absorption [1].

The fiber–matrix adhesion attributes to the physical mechanical locking and chemical interaction between the matrix and reinforcements. The mechanical interlocking can be achieved by varying roughness and can lead to perforation, pits, and small grooves on the reinforcement surface. Physicochemical interaction can be improved by changing surface energies, intermolecular interactions, surface-induced crystallizations, and growing oxygenated functionalities on the reinforcement surfaces. The number of traditional methods like plasma modification, high--energy irradiation, and chemical etching, etc., are used to treat the reinforcement surface; nonetheless, all the above-mentioned methods reduce single fiber strength properties [2].

Reinforcements have excellent mechanical (strength and modulus) properties due to the fewer number of flaws and stress concentration points at the surface. For advanced structural composites application the largely used reinforcement are: glass fiber (GF); carbon fibers (CF); aramid fibers (AF); and natural fibers (NF) such as jute and a few polymeric fibers such as nylon, poly-tetrafluoroethylene, polypropylene, etc. The van der Waals and hydrogen bond forces between the fiber and matrix are required during composite development [3,4]. For an ideal interface the fiber–matrix adhesion energy should be higher than the cohesion energy of the polymer matrix. The fiber surface treatment proved beneficial to improve the strength properties of composites, but the controlled designing of fiber–matrix interfacial/nanostructured interphase requires further research and investigations [2,5].

Fiber–matrix interface strengthening and its engineering have enhanced composites performance for several of the above-mentioned industrial applications. This chapter will highlight the importance of fiber–matrix strengthening and covers newly developed methods to create a strong and robust interface without sacrificing the vital fiber properties such as single-fiber strength. The later section of the chapter will briefly focus on the advanced nano-indentation and nano-mechanical characterization techniques used to investigate fiber–matrix interface/interphase at the nanoscale level.

5.2 Sizings to protect reinforcements and strengthen interface

Reinforcements need coating with polymers and/or other solvent-based materials for protected during manufacturing and handling. The process of fiber coating with such polymers to impart a surface finish and to protect from critical flaws/fluffiness is called sizing. During operations and mechanical loading these surface flaws or pits developed during the fiber manufacturing process act as a stress concentrator for crack propagation and cause early failure. Polymer sizings shield the reinforcing fibers from breaking and provide strand integrity. Compatible sizings also improve adhesion of reinforcement with matrix materials and improve composite processing.

Fig. 5.1 shows the role of polymer sizings for applying surface finish and protection from breakage [2]. The sizing does enhance the reinforcement wettability with the matrix materials. For advanced composites applications commercially supplied sizing are refinished or modified to improve fiber thermo oxidative stability and mechanical performance [6].

Appropriate sizings for reinforcement is vital to quantify the good interfacial adhesion. The diffusion of the sizing's polymer into the matrix materials affects the interfacial strength and other properties like abrasion resistance and bending strength. The sizing's materials adsorbed on the reinforcement surface obstructs its dissolution with matrix during composites processing, which reduces the strength properties.

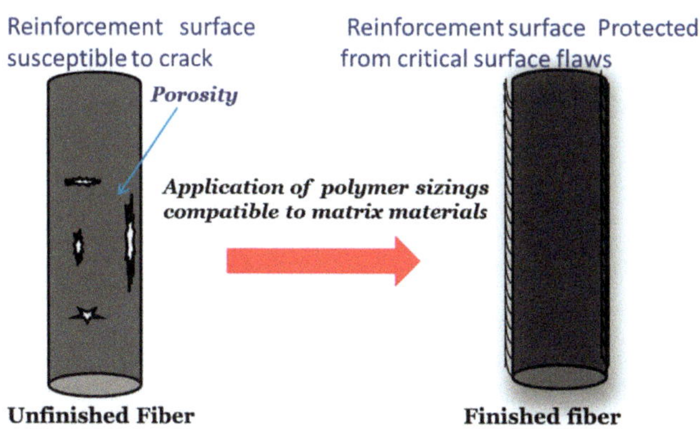

Figure 5.1 Polymer sizings to protect fiber surface during manufacturing process.
M. Sharma, S. Gao, E. Mader, H. Sharma, L.Y. Wei, J. Bijwe, Carbon fiber surfaces and composite interphases. Compos. Sci. Technol. 102 (2014) 35–50. With permission Elsevier 3971260625316.

The usage of suitable coupling agent along with the sizing materials creates the functionalities at the fiber surface that ease the further chemical interactions with the matrix materials [7]. A variety of alkaline and silanes coupling agent used to improve the fiber–matrix interfacial adhesion [8]. The sizing molecular weight is also an important criterion for the reinforcement surface energies and interfacial shear strength (IFSS) properties of epoxy composites [9]. Daia et al., explain how the reinforcement desizing improves the composites interfacial strength significantly but sacrifices single-fiber strength. The desized carbon fiber has fewer activated carbon atoms with high dispersive surface energy thus exhibiting an improved interfacial properties due to the reduced acidic parameter at the reinforcement surface [10]. Zinck and Gerard suggested that the hygrothermal-resistant interphase in a GF anhydride–cured epoxy doesn't need to have an organosilane coupling agent reactive towards the anhydride [11]. Fibers were subjected to surface modifications to remove the outer layer and introduce oxygen containing functionality [12]. The functionalities alter the surface energies and improves fiber matrix the wettability [13]. In composites fabrication and applications the interdiffusion of sizing and matrix creates an interphase region that protects the reinforcements and retards the crack initiation and propagation process. Hence the selection of appropriate sizing and its compatibility with polymer matrix is vital to achieve the strengthened fiber–matrix interface.

5.3 Strengthening of fiber–matrix interface by reinforcement modifications

Reinforcement treatment is an efficient method for improving fiber–matrix interface properties. Reinforcement treatment methods can be categorized as wet, dry, and multiscale modification methods. Plasma treatment, chemical treatment, high-energy irradiation, and nanoparticles attachments are the main methods currently adapted to increase the fiber–matrix wettability and its interfacial adhesion with polymeric matrices. Treatment generally increased roughness on the fiber surface. Fig. 5.2 shows the conventional and newly developed methods to carbon fiber surface modification [2].

5.3.1 Conventional methods for reinforcement modification

Acid etching corrodes the reinforcement surface and establishes pits, crevasses, expanded micro-voids, flaws, and perforations on the fiber surface to improve fiber–matrix mechanical interlocking but at the cost of reducing single-fiber strength [14,15]. Physicochemical changes on carbon fiber surface due to the acidic treatment were analyzed by surface-enhanced Raman scattering [16], Fourier transform infrared spectroscopy (FTIR), and X-ray photoelectron spectroscopy (XPS) [17]. The interfacial strengthens properties (shear strength, delamination, and fracture toughness) of CF-cyanate ester and CF-polyethersulfone composites enhanced

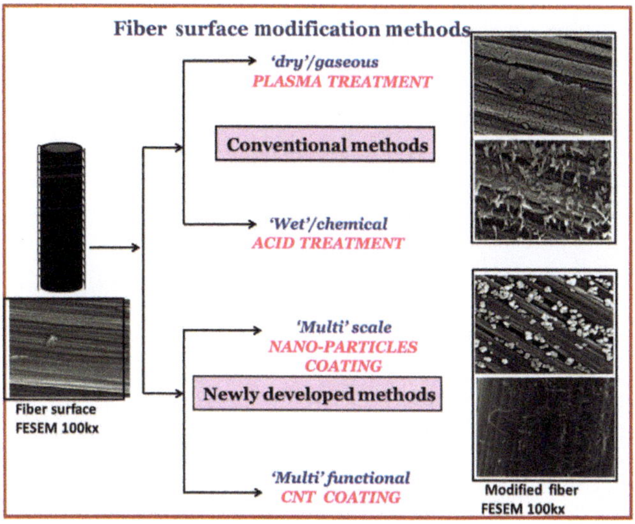

Figure 5.2 Various techniques for reinforcement surface modifications
M. Sharma, S. Gao, E. Mader, H. Sharma, L.Y. Wei, J. Bijwe, Carbon fiber surfaces and composite interphases. Compos. Sci. Technol. 102 (2014) 35–50; J. Bijwe, M. Sharma, Carbon fabric-reinforced polymer composites and parameters controlling tribological performance, in: J.P. Davim (Ed.), Wear of advanced materials. ISTE Wiley; 2011. pp. 1–59; M. Sharma, J. Bijwe, K. Singh, P. Mitschang, Exploring potential of Micro-Raman spectroscopy for correlating graphitic distortion in carbon fibers with stresses in erosive wear studies of PEEK composites. Wear 270 (11) (2011) 791–799. With permission Elsevier 3971260625316.

by acid treatment and plasma-oxidized fiber surface [18]. The interfacial strengthening is achieved by improved interlaminar shear strength properties.

Plasma/cold plasma treatment to the reinforcement influences the fiber–polymer interfacial properties [5]. Plasma modifications of fibers to improve composites properties have been studied since the 1980s. Due to the plasma treatment, the surface layers of the reinforcement alter physicochemical and influence the fiber–matrix adhesion and the interface. Fig. 5.3 explicates the strategies to promote fiber–matrix adhesion by growing compatible functionalities on the reinforcement surface [2]

The plasma modification influences the interfacial adhesion by including oxygenated functionalities at the reinforcement surface. Henceforth, these functionalities increase the surface energies and reactivity to increase intermolecular bonding and fiber wettability with hydrophilic polymer matrix. Scanning tunneling microscopy studies on plasma treated carbon fibers revel that structure disordering and concentration oxygen functionalities at the surface [19]. The plasma treatment to the fiber surface introduces pitting and deep ridges that are beneficial for improving the composites' strength by mechanical interlocking but reduces the strength of individual fiber strands because of the greater chance of stress propagation from the

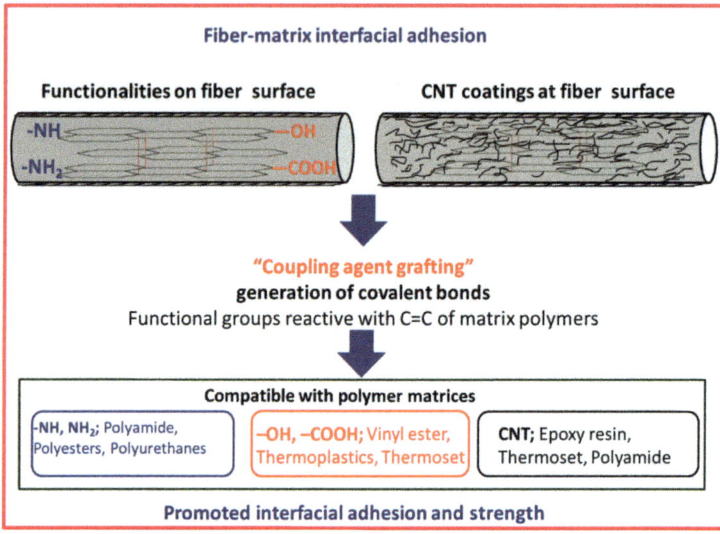

Figure 5.3 Promoted fiber–matrix interfacial adhesion via employing surface functionalities on the reinforcement surface.
M. Sharma, S. Gao, E. Mader, H. Sharma, L.Y. Wei, J. Bijwe, Carbon fiber surfaces and composite interphases. Compos. Sci. Technol. 102 (2014) 35–50. With permission Elsevier 3971260625316.

deeper channels at the surface. Chemical grafting, poly-condensation, and plasma polymerization are the other improved methods used to introduce polymeric coatings on the fiber surfaces.

Plasma polymerization is another way to modify the fiber surface to improve composites shear strength and toughness properties [20]. Plasma polymerization is the method to perform selectively coating on the fiber surface. The optimization of the polymerization process with difference sizes of coating may influence interface properties. The coating layer thickness introduced during plasma polymerization needs to be adjusted to the depth of penetration of the matrix materials. The optimum layer thickness (nano-meter range) has been observed on acrylic acid/1, 7-octadiene coated glass fibers by plasma polymerizations [1]. In few cases the plasma polymerization can substitute the different sizing's process and may save the cost for the additional coatings. Silane-based plasma polymerization on the GF improved the interfacial adhesion and interlaminar shear strength with unsaturated polyester resin [21]. Electrochemical polymerization is preferred over other methods because of the deposition of homogenous coating over a long range of surfaces. The plasma methods are favorable because of the introduction of chemical functionalities that modify the outermost layers of fibers. However, other electro polymerization methods modify the fiber sublayers thereby leading to additional reduction in the single-fiber strength properties in the latter process.

Electrochemical oxidation can be a potential method for functionalizing the fiber surface. The electrochemical oxidation alters the surface energy and roughness of the reinforcement materials and promotes its adhesion with the matrix polymer. The electrochemical process-to-fiber-surface treatment initiates an electrontransfer and alters the oxidation state, whereby the electrolyte type, its concentration, and exposure conditions become the essential parameters to achieve and strengthen the interface. The electrolyte adsorption increases surface reactivity by generating the extended surface areas via the formation of ultra-micro pores on the fiber surface. The number of electrolytes is used to induce functionalities to the reinforcement materials. Ammonium carbonate, sodium hydroxide, nitric acid, and sulfuric acid are the most commonly used electrolytes on the fiber surface [22]. Electrochemical treatment eliminates the weak boundary layers from the fiber surface and alters the reactivity because of the formation of acidic and basic moieties. Electrolytic treatment to the fiber surface improved with interfacial strength, however; higher electrolyte doses reduce the single-fiber strength. Therefore, an optimized dose of the electrochemical process parameters is essential.

Polymer grafting, with the aid of a silane-modified polypropylene copolymers coupling agent, is also a conventional method to modify the glass fiber surface [23]. The silane coupling creates chemical bonds with the glass fiber. As a result, however, the dangling chains participate in a cocrystallization process with the matrix materials.

High-energy irradiation is an environmental friendly method that treats the fiber surface and promotes fiber–matrix adhesion without reducing the strength properties [24,25]. High-energy irradiation methods alter the surface roughness by displacing the atoms and creating active sites for bonding with matrix materials. Preradiation and coradiation gamma-ray grafting methods adopted to alter the fiber's surface [24,26]. The Co_{60} gamma-ray irradiation increases the surface roughness of fibers after the optimized dose of 30 kGy; higher irradiation was unable to affect the interfacial strength properties [27]. Increased structural disorder and decreased surface crystallites size with the incremental treatment dose for carbon fiber has been suggested by Raman characterizations. Further, the gamma-modified fibers showed the enhanced friction and wear properties of CF/PEI composites with improved strength properties [15,28].

Cryogenic modifications, the pyrolyzed impurities introduced during carbonizations steps of carbon fiber manufacturing, act as stress concentrator and creates a weakening of the structure during the tensile loading. Pyrolyzed deposits on the fiber surface were removed with thermal treatments. The single-fiber strength incremented due to the removal of amorphous carbon at the surface but affects the fiber–matrix adhesion. The increase in interfacial strength of carbon fiber–epoxy composites containing thermal treated fibers is due to the high reactivity of the active sites on the oxidized fiber surface [29]. Kim et al. treated carbon fibers with plasma, nitric acid, and liquid nitrogen to improve the impact strength and tensile strength properties of nylon and rubber composites. The interfacial strength of the hybrid rubber composites improved the tensile and impact strength properties to noticeable amount [30].

Self-assembly is an indispensable tool to create monolayer polymer structures in the area of microelectronics, biomaterials, and coating technology. It is the procedure to convert randomly oriented molecules into an organized supramolecular structure by local interactions between them. The interactions can be electrostatic, weak van der Waals forces or strong hydrogen bonding [1]. Holmes et al., investigated such self-assembly structuring of silanized glass fiber beneficial to improve the interfacial properties of glass fiber–epoxy composites due to the steric hindrance due to the size of the EP molecules [31]. The self-assembly on a CF plated with silver reacted with thiols of a different chain length. The thiols were grafted onto the Ag plated CF via Ag–S bonds. The interfacial properties of Ag plated carbon fiber–epoxy composites increased up to 6 wt.% when compared to unassembled carbon fibers.

5.3.2 Recently developed treatment techniques: strategies to retain fiber strength properties

The nanoparticle attachment to fiber surface for the development of stable and homogenous coatings and further improvement in the fiber–matrix interfacial properties have been recently evaluated [32]. Materials in nanoscale range like carbon nano tubes (CNT) are explored in depth using different deposition techniques due to their unique physical, mechanical, thermal, electrical, and field emission [33–37].

In the advanced composites materials applications, the usages of nanoscale fillers for reinforcement modification are beneficial to alleviate the predominating matrix and associated problems. The inclusion of CNT form percolating networks on the fiber surface thus increases the surface-area-to-volume ratio; increases mechanical linking; as well as the local stiffening of fiber/matrix interface thereby strengthening the fiber–matrix interface by efficient stress transfer from polymer to the reinforcement materials.

Nano-fillers are a potential method for interface strengthening. There are a number of conventional as well as advanced methods to facilitate the attachment of functionalized nano-additives with the reinforcement such as chemical grafting, direct-grafting attachment process like chemical vapor deposition (CVD), and newly developed dip coating methods.

The high surface area to volume ratio of nano-additives is the key factor for creating large and homogenous surface on to the fibers to facilitate covalent and/or mechanical bonding with the matrix materials. CNT structurally are much entangled as compared to graphene sheets. There are methods by the aid of surfactant to create disentangled CNT structures [38].

Surface treatment of fibers is favorable, as established coating/sizing application methods can be scaled up to the industrial scale and mass customization. The nanoparticles attachment on the fiber surface is beneficial to enhance interface properties. Homogenous distribution of the CNT on fiber can be achieved by direct-grafting of these nano-additives using CVD methods, but the reduction in single-fiber strength is still the big problem. Judiciously, fiber surface treatment can be done with newly developed dip coating method by dipping the fiber in sonicated suspension of nano-additives in suitable solvent.

CNT grafted for carbon fibers is advantageous for providing conductive layers to the composites with enhanced fracture toughness and interfacial strength properties [39–41]. CVD grafting CNT/nano-filler attachments to the reinforcement enhanced the capillary wetting by the polymer matrix as well as mechanical interlocking (more pronounced in the case of thermosetting polymers), which contribute to a better stress transfer from the tough matrix to the high strength fibrous materials.

CVD method is used to modify carbon fiber; the nanotubes layer that has been grafted on CF improved the fiber surface area and the interfacial shear, compressive strength of its composites with epoxy composites. The single-fiber strength is reduced because of the adsorption of iron particles on the fiber surface during CVD of ferrocene [42]. The fragmentation test indicates the increase in interfacial strength of carbon fiber–epoxy composites due to the exact control on orientation and length of aligned CNT [43]. The mechanical properties of carbon fiber–polypropylene composites increased with grafted CNT [44]. Thermally evaporated CNT functionalized on woven carbon fibers fracture toughness and fatigue durability of its composites with epoxy matrix without scarifying the structural stiffness [45].

CNT were deposited on glass fiber by CVD technique using Ni/Fe particle catalyst. Before deposition the glass fiber was plasma treated to facilitate the CVD process. Nano-indentation tests had suggested a 35 percent higher stiffness of the interfacial region. The grafting of carbon nanoparticles on calcined glass fiber was facilitated by the CVD technique. The carbon nano-materials on the surface of glass fiber increased the electrical conductivity and dynamic mechanical storage modulus of glass fiber–epoxy composites [46,47].

Nano filler dip coating was employed on fibers surface using dip coating methods. The rare earth nanoparticles adsorbed on reinforcement surface to enhance its chemical reactivity by formulating the sulfonic, carbonyl, hydroxyl, carboxyl functionalities [2]. Nano particle coatings at fibers surface are preferred due to their capability to produce higher specific surfaces on the reinforcement. Nano YBF_3 attachment to the carbon fiber enhanced surface reactivity and improved strength and wear resistance properties of carbon fiber–polymer composites due to highly electronegative F-atom [48,49]. Fig. 5.4 shows the homogenous attached of the nano rare earth particle to the carbon fiber.

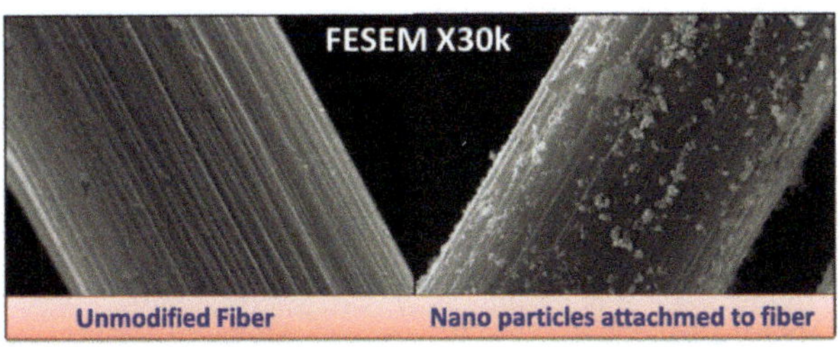

Figure 5.4 Fiber surface with nanoparticle attached by using dip coating methods.
M. Sharma, S. Gao, E. Mader, H. Sharma, L.Y. Wei, J. Bijwe, Carbon fiber surfaces and composite interphases. Compos. Sci. Technol.102 (2014) 35–50. With permission Elsevier 3971260625316.

The multiscale and multifunctional coating methods have an edge over the conventional treatment methods, as they are capable of improving the interfacial properties without reducing the individual fiber strength properties. The understanding of fiber–matrix interface and/or nanostructured interphase region, underlying properties, and its formation is a topic of deep research. It is needed to study the studies on properties of the nanoscale interphase/interfacial region and underlying mechanism by using advanced characterization techniques.

5.4 Interfacial design and characterization

Hierarchal reinforcement-matrix nanostructured interfaces/interphase behaves as a vast number of ways in construction and properties. At the reinforcement top layer the surface properties, such as the employment of chemical oxygenated functionalities, various polymer sizings are accounted. At the fiber–matrix subsurface the typical interlaminar shear strength (ILSS), IFSS properties are vital. The fiber–matrix pullout, fragmentation, and a three point bend test are the characterization techniques adopted to characterize the interface subsurface. The most crucial region that needs to be investigated are the details in the nanostructured interphase region, wherein the much advanced nano-mechanical, nano-tribological, and nano-indentation characterizations will be accounted.

5.4.1 Fiber/matrix interface characterization and failure mechanism

Fiber–matrix adhesion test: Fiber tow was separately dipped in the solutions of polymer or polymer melt to analyze the effect of polymer viscosity on the fiber–matrix adhesion. After 10 min the fibers were carefully taken out from all three solutions at the same time and allowed to dry in identical conditions. The difference in the layer of matrix adhering to the fiber strand was examined with SEM after sputtering. The SEM in Fig. 5.5 suggested the enhanced matrix intake in the latter case when the tows of fiber (untreated and treated) were dipped in the polymer melt at the same interval of time. Fig. 5.5 shows the well = impregnated treated carbon fiber with thermoplastic matrix in comparison to the untreated carbon fibers.

Interlaminar shear strength 0(ILSS) properties are evaluated according to ASTM D2344 using the formula:

$$ILSS = 0.75 \times \left(\frac{P}{bd}\right)$$

where P is maximum load, b is width and d is depth of sample.

A single-fiber fragmentation test is developed from the Kelly and Tyson model to calculate the IFSS of the fiber–matrix interface [28,51]. The single-fiber sample needed to be studied for the fragmentation test will be encapsulated and fixed in in a dogbone shape silicon mold. The polymer matrix (for thermosetting epoxy resins the pouring in needed however for thermoplastic polymer like polypropylene (PP)

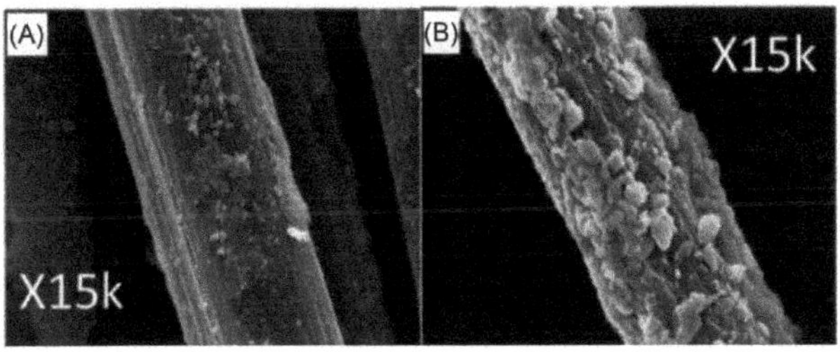

Figure 5.5 Single carbon fiber impregnated with polyether matrix: (A) before and (B) after plasma modification shows enhanced matrix attached for plasma modified fiber.
M. Sharma, S. Gao, E. Mader, H. Sharma, L.Y. Wei, J. Bijwe, Carbon fiber surfaces and composite interphases. Compos. Sci. Technol.102 (2014) 35–50; M. Sharma, J. Bijwe, Influence of fiber–matrix adhesion and operating parameters on sliding wear performance of carbon fabric polyethersulphone composites. Wear 271(11–12) (2011)2919–2927. With permission Elsevier 3971260625316.

the polymer powder is heat at melting temperature). The tensile test was performed on the dogbone-shaped coupons. During testing the tensile load was transferred to fiber by shear stresses in the polymer matrix through the interfacial region. Elongating in a tensile tester results in fiber breakage. The fibers fractured into fragments after the peak loading. The tensile experiment needs to be done in situ under optical microscope. Reinforcements embedded in the matrix materials breaks into fragments at areas where the fiber axial stress reaches the ultimate tensile strength. The resin chosen for the purpose should have a higher strain-to-failure than the reinforcement to facilitate the breakage. The tensile stress at the fracture location is null when fiber fragmented. The toughened matrix provides continues shearing and stress transfer. At some juncture, the tensile stress in the fiber increases roughly linearly. The higher the axial strain, the more fractures will be occurring till the fragment will be too short to transfer enough stresses into the fiber to cause further breakage. After the fragmentation, shape of the broken fiber and the debonding mechanism will determine the interlaminar properties. If the interfacial adhesion is high the fragments can propagate through the matrix and cause its deformation around the fragmentation. On the other hand, if the interface is feeble, the fiber–matrix debonding will occur to facilitate the fiber pullouts. As per Kelly and Tyson model the IFSS is calculated as per the following equation [50, 51]:

$$\mathscr{T}_{\mathrm{IFSS}} = \sigma_{\mathrm{F}} d_{\mathrm{f}} / 2lc$$

$$lc = \frac{4}{3} lav$$

where $\mathscr{T}_{\mathrm{IFSS}}$ is the interfacial shear strength and d_{f} is the fiber diameter, lc is the critical length, σ_{f} is the fiber strength at the critical fragment length. l_{c} is determined by average fragment length l_{av}

The fiber pullout micro bond is another method for measuring the interface properties of composites materials. The number of resin drops was cured on the single-fiber and the coated fibers are pulled out from the blades to evaluate the pullout force [52,53]. Using fiber pullout method the interfacial shear strength $\mathcal{T}$ determined by the equation:

$$T = \frac{f}{\pi D l}$$

where F is the maximum pullout force, D is the fiber diameter. $\boldsymbol{l}$ is the embedded length in the range of 40–60 μm.

A single fiber pullout test was performed to characterize the interfacial properties of acid-treated carbon fibers reinforced with maleic anhydride grafted polystyrene, CF-epoxy to improve IFSS [68,69,70cst]. Improved tribological properties with less frictional force, tribo-couple heating and improved wear resistance were suggested for treated carbon fiber composites with thermoplastic matrix [54–58]. ILSS of polyhedral oligomeric silsesquioxane coating fibers improved interfacial properties of vinyl ester matrix and carbon fiber-coated composites increased up to 38 percent [59]. Surface fluorination of reinforcement is a beneficial method to improve interfacial adhesion without affecting the single-fiber strength properties. The incremented surface polarity of oxi-fluorinated fiber enhanced its interfacial adhesion with epoxy [60].

5.4.2 Advanced techniques to characterize nanostructured interface/interphase

Recently developed atomic force microscopic (AFM) techniques are very promising for nanostructure interface characterizations. AFM-nano indentations, AFM infrared (AFM-IR), AFM-Lorentz Contact Resonance (AFM-LCR), and AFM-nano thermal analyses (AFM-nanoTA) are the advanced techniques needed to be adopted for analyzing the interfacial regions focusing on nanoscale interphases. The investigations on molecular rearrangement in between fiber-sizings-nano-additives-matrix network is an area of prime interest and to get complete understanding for interface/interphase and its strengthening for potential composites applications [61]. Fig. 5.6 explicates the typical interface region for glass fiber–epoxy composites.

5.5 Potential applications of strengthened fiber–matrix interfaces

The lightweight fiber-reinforced structural composites have advantages in aerospace, marine offshore, and the high performance automobile sector as a replacement for steel counterparts. Fiber-reinforced composite materials can achieve high specific strength and load bearing capabilities when used in combination with a reinforcement/matrix modification and further interface strengthening.

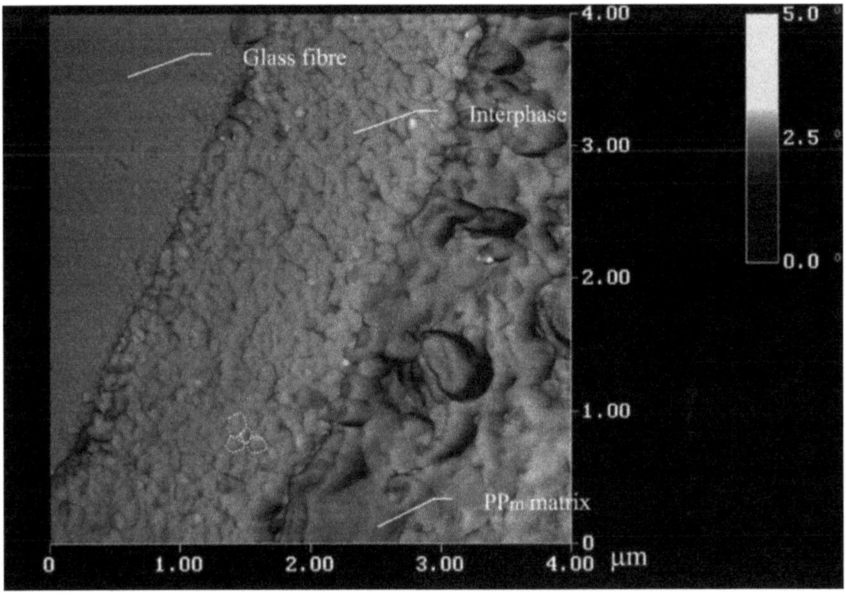

Figure 5.6 AFM image of glass fiber–polypropylene matrix interface.
S.L. Gao, E. Mäder, Characterization of interphase nanoscale property variation in glass fiber reinforced polypropylene and epoxy resin composites. Compos. Part A Appl. Sci. Manuf. 33 (4) (2002) 559–576. With permission Elsevier 3971260845248.

Fig. 5.7 highlights the schematic to develop hybrid composites with tailored composites surface with strengthened fiber–matrix interface. The figure explicates various steps which include the reinforcement surface modification followed by the impregnation with the thermoplastic matrix to develop the strengthened interface followed by tailoring at the top composites layers with the nano-additives/solid lubricants from submicron to nanoscale range. The final hybrid/tailored composites have improved tribological properties at the surface with strengthened interface at the bulk of composites. Interface strengthening is directly beneficial to explore the potential of reinforced composites for advanced applications like cryogenic propellant tanks for heavy-lift vehicles in space applications. NASA has tested a composites propellant tank with improved utilities, showing a 30% lighter and 25% cost reduction compared to the metallic counterparts.

5.6 Prospective

This chapter is a brief attempt to highlight the importance of fiber–matrix interface strengthening towards the composites lightweight applications. The acceptability of composites materials for numerous applications has already suggested their edge over state-of-the-art steel and other conventional materials. Reinforced composites can be tailored selectively as per the intended application. For example, in marine

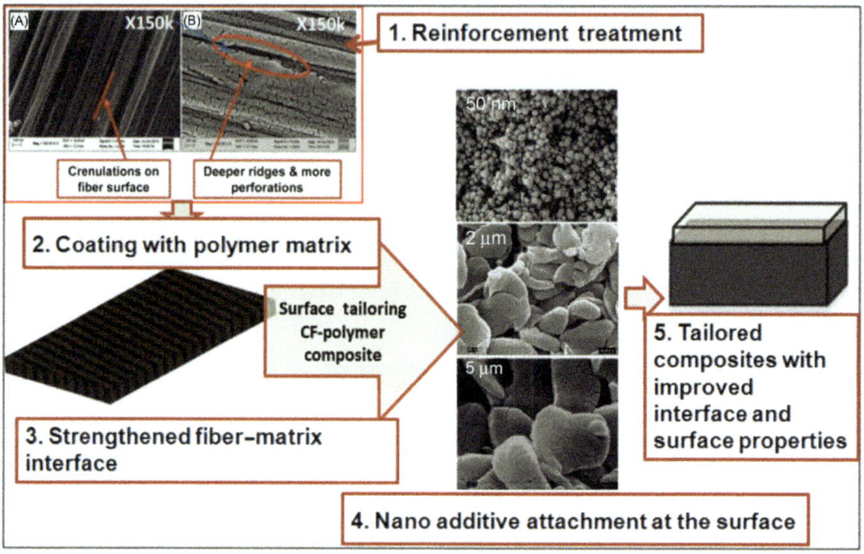

Figure 5.7 Schematic to fabricate hybrid/tailored polymer composites with strengthened interface.

and offshore applications, composites to develop a flexible riser for deep sea exploration required the excellent fluid tightness and protection from corrosive fluids and gasses, whereas the composites used for marine bearing need wear resistance and fatigue-resistance properties. Henceforth, the right choice of reinforcement for its surface modification and/or attachment of nano-additives with the selection of suitable processing techniques can achieve the strengthened interface and qualified composites for predetermined applications. The investigation on the fiber and/or matrix surface modification to strengthen the interface needs more depth analyses on nano-structure interfaces and underlying mechanisms using advance nano-mechanical and nano-indentation techniques.

References

[1] J. Karger-Kocsis, H. Mahmood, A. Pegoretti, Recent advances in fiber/matrix interphase engineering for polymer composites, Prog. Mater. Sci. 73 (2015) 1–43.
[2] M. Sharma, S. Gao, E. Mader, H. Sharma, L.Y. Wei, J. Bijwe, Carbon fiber surfaces and composite interphases, Compos Sci Technol 102 (2014) 35–50.
[3] E. Fitzer, K.H. Geigl, W. Huttner, R. Weiss, Chemical interactions between the carbon fiber surface and epoxy resins, Carbon. N. Y. 18 (1979) 389–393.
[4] H. Dvir, J. Jopp, M. Gottlieb, Estimation of polymer–surface interfacial interaction strength by a contact AFM technique, J. Colloid Interface Sci. 304 (2006) 58–64.
[5] U. Zielke, K.J. Huttinger, W.P. Hoffman, Surface-oxidized carbon fibers: I. Surface structure and chemistry, Carbon. N. Y. 34 (1996) 983–998.

[6] R.E. Allred, S.P. Wesson, E.E. Shin, L. Inghram, L. McCorkle, D. Papadopoulos, et al., The influence of sizings on the durability of high-temperature polymer composites, High Perform. Polym. 15 (2003) 395–419.
[7] Z. Dai, B. Zhang, F. Shi, M. Li, Z. Zhang, Y. Gu, Chemical interaction between carbon fibers and surface sizing, J Appl. Polym. Sci. 124 (2012) 2127–2132.
[8] M.A. Fuqua, S. Huo, C.A. Ulven, Natural fiber reinforced composites, Polym. Rev. 52 (2012) 259–320.
[9] R.L. Zhang, Y.D. Huang, D. Su, L. Liu, Y.R. Tang, Influence of sizing molecular weight on the properties of carbon fibers and its composites, Mater. Des. 34 (2012) 649–654.
[10] Z. Daia, F. Shi, B. Zhang, M. Li, Z. Zhang, Effect of sizing on carbon fiber surface properties and fibers/epoxy interfacial adhesion, Appl. Surf. Sci. 257 (2011) 6980–6985.
[11] P. Zinck, J. Gerard, Thermo-hydrolytic resistance of polyepoxide–glass fibres interfaces by the microbond test, Compos. Sci. Technol. 68 (2008) 2028–2033.
[12] K.K. Chee Ho, H. Qian, A. Bismarck, Carbon fiber: surface properties, in: L. Nicolais, A. Borzacchiello (Eds.), Wiley Encyclopedia of Composites, John Wiley & Sons, Hoboken, NJ, USA, 2012.
[13] F. Hoecker, J. Karger-Kocsis, Surface energetics of carbon fibers and its effects on the mechanical performance of CF/EPcomposites, J. Appl. Polym. Sci. 59 (1996) 139–153.
[14] J. Bijwe, M. Sharma, Carbon fabric-reinforced polymer composites and parameters controlling tribological performance, in: J.P. Davim (Ed.), Wear of Advanced Materials, ISTE Wiley, 2011, pp. 1–59.
[15] M. Sharma, J. Bijwe, K. Singh, P. Mitschang, Exploring potential of Micro-Raman spectroscopy for correlating graphitic distortion in carbon fibers with stresses in erosive wear studies of PEEK composites, Wear 270 (11) (2011) 791–799.
[16] B. Xu, X. Wang, Y. Lu, Surface modification of polyacrylonitrile-based carbon fiber and its interaction with imide, Appl. Surf. Sci. 253 (2006) 2695–2701.
[17] Z. Wv, C.U. Pittman, Nitric acid oxidation of carbon fibers and the effects of subsequent treatment in refluxing aqueous NaOH, Carbon. N. Y. 33 (5) (1995) 597–605.
[18] C. Marieta, E. Schulz, I. Mondragon, Characterization of interfacial behaviour in carbon-fibre/cyanate composites, Compos. Sci. Technol. 62 (2002) 299–309.
[19] A.K. Stark, L.A. Berglund, M. Tagawa, N. Ohmae, Effect of atomic oxygen on the mechanical properties of highly graphitized carbon fibers, Carbon. N. Y. 32 (1994) 641–644.
[20] C. Jones, Effects of electrochemical and plasma treatments on carbon fibre surfaces, Surf. Interface Anal. 20 (1993) 357–367.
[21] V. Cech, Plasma-polymerized organosilicones as engineered interlayers in glass fiber/polyester composites, Compos. Interface 14 (2007) 321–334.
[22] J. Gulyas, E. Foldes, A. Lazar, B. Pukanszky, Electrochemical oxidation of carbon fibres: surface chemistry and adhesion, Compos. Part A 32 (2001) 353–360.
[23] J.F. Feller, Y. Grohens, Coupling ability of silane grafted poly(propene) at glass fibers/poly(propene) interface, Compos. A 35 (2004) 1–10.
[24] R.L. Clough, High-energy radiation and polymers: a review of commercial processes and emerging applications, Nucl. Instrum. Methods. Phys. Res. B. 185 (2001) 8–23.
[25] Z. Xu, Y. Huang, C. Zhang, L. Liu, Y. Zhang, L. Wang, Effect of gamma-ray irradiation grafting on the carbon fibers and interfacial adhesion of epoxy composites, Compos. Sci. Technol. 67 (2007) 3261–3270.

[26] J.Q. Li, Y.D. Huang, S.Y. Fu, L.H. Yang, H. Qu, G. Wu, Study on the surface performance of carbon fibres irradiated by c-ray under different irradiation dose, Appl. Surf. Sci. 256 (2010) 2000–2004.
[27] J. Li, Y. Huang, Z. Xu, Z. Wang, High-energy radiation technique treat on the surface of carbon fiber, Mater. Chem. Phys. 94 (2005) 315–321.
[28] M. Sharma, J. Bijwe, E. Mäder, K. Kunze, Strengthening of CF/PEEK interface to improve the tribological performance in low amplitude oscillating wear mode, Wear 301 (1) (2013) 735–739.
[29] T. Ramanathan, A. Bismarck, E. Schulz, K. Subramanian, Investigation on the influence of acidic and basic surface groups on carbon fibres on the interfacial shear strength in an epoxy matrix by mean of single fibre pull-out test, Compos. Sci. Technol. 61 (2001) 599–605.
[30] S.Y. Kim, S.J. Baek, J.R. Youn, New hybrid method for simultaneous improvement of tensile and impact properties of carbon fiber reinforced composites, Carbon. N. Y. 49 (2011) 5329–5338.
[31] G. Holmes, E. Feresenbet, D. Raghavan, Using self-assembled monolayer technology to probe the mechanical response of the fiber interphase-matrix interphase interface, Compos. Interface 10 (2003) 515–546.
[32] H. Qian, E.S. Greenhalgh, M.S.P. Shaffer, A. Bismarck, Carbon nanotube-based hierarchical composites: a review, J. Mater. Chem. 20 (2010) 4729–4756.
[33] H. Dai, Carbon nanotubes: opportunities and challenges, Surf Sci 500 (2002) 218–241.
[34] M. Terrones, Science and technology of the twenty first century: synthesis, properties, and applications of carbon nanotubes, Ann Rev Mater Res 33 (2003) 419–501.
[35] J. Vavro, M.C. Llaguno, B.C. Satishkumar, D.E. Luzzi, J.E. Fischer, Electrical and thermal properties of C60 filled single-wall carbon nanotubes, Appl. Phys. Lett. 80 (2002) 1450–1452.
[36] H. Sharma, V. Kaushik, P. Girdhar, V.N. Singh, A.K. Shukla, V.D. Vankar, Enhanced electron emission from titanium coated multiwalled carbon nanotubes, Thin. Solid. Films. 518 (2010) 6915–6920.
[37] H. Sharma, D.C. Agarwal, M. Sharma, A.K. Shukla, D.K. Avasthi, V.D. Vankar, Tailoring of structural and electron emission properties of CNT walls and graphene layers using high-energy irradiation, J. Phys. D: Appl. Phys. 46 (2013) 315301–315309.
[38] J. Rausch, R.C. Zhuang, E. Mader, Surfactant assisted dispersion of functionalized multi-walled carbon nanotubes in aqueous media, Compos. Part A 41 (2010) 1038–1046.
[39] J.D.H. Hughes, H. Morley, E.E. Jackson, Aligned carbon fibre composite which approaches theoretical strength, J Phys. D: Appl. Phys. 13 (1980) 921–936.
[40] Q. Zhang, J. Liu, R. Sager, L. Dai, J. Baur, Hierarchical composites of carbon nanotubes on carbon fiber: influence of growth condition on fiber tensile properties, Compos. Sci. Technol. 69 (2009) 594–601.
[41] R.J. Sager, P.J. Klein, D.C. Lagoudas, Q. Zhang, J. Liu, L. Dai, et al., Effect of carbon nanotubes on the interfacial shear strength of T650 carbon fiber in an epoxy matrix, Compos. Sci. Technol. 69 (2009) 898–904.
[42] Z.G. Zhao, L.J. Ci, H.M. Cheng, J.B. Bai, The growth of multi-walled carbon nanotubes with different morphologies on carbon fibers, Carbon. N. Y. 43 (2005) 663–665.
[43] P. Lv, Y. Feng, P. Zhang, H. Chen, N. Zhao, W. Feng, Increasing the interfacial strength in carbon fiber/epoxy composites by controlling the orientation and length of carbon, nanotubes grown on the fibers, Carbon. N. Y. 49 (2011) 4665–4673.

[44] A.R. Suraya, S.M.Z. Sharifah, S.R. Yunus, I.R. Azowa, Growth of carbon nanotubes on carbon fibres and the tensile properties of resulting carbon fibre reinforced polypropylene composites, J. Eng. Sci. Technol. 4 (2009) 400–408.
[45] K.L. Kepple, G.P. Sanborn, P.A. Lacasse, K.M. Gruenberg, W.J. Ready, Improved fracture toughness of carbon fiber composite functionalized with multiwalled carbon nanotubes, Carbon. N. Y. 46 (2008) 2026–2033.
[46] C.D. Wood, M.J. Palmeri, K.W. Putz, G. Ho, R. Barto, L. Catherine Brinson, Nanoscale structure and local mechanical properties of fiber-reinforced composites containing MWCNT-grafted hybrid glass fibers, Compos. Sci. Technol. 72 (2012) 1705–1710.
[47] A. Rahaman, K.K. Kar, Carbon nanomaterials grown on E-glass fibers and their application in composite, Compos. Sci. Technol. 101 (2014) 1–10.
[48] S. Tiwari, J. Bijwe, S. Panier, Role of Nano-YbF3 treated carbon fabric on improving abrasive wear performance of polyetherimide composites, Tribol. Lett. 42 (2011) 293–300.
[49] S. Tiwari, J. Bijwe, S. Panier, Enhancing the adhesive wear performance of polyetherimide composites through nano-particle treatment of the carbon fabric, J. Mater. Sci. 47 (2012) 2891–2898.
[50] M. Sharma, J. Bijwe, Influence of fiber–matrix adhesion and operating parameters on sliding wear performance of carbon fabric polyethersulphone composites, Wear 271 (11–12) (2011) 2919–2927.
[51] A. Kelly, W.R. Tyson, Tensile properties of fibre-reinforced metals: copper/ tungsten and copper/molybdenum, J. Mech. Phys. Solids. 13 (1965) 329–350.
[52] B. Miller, P. Muri, L. Rebenfeld, A microbondmethod for determination of the shear-strength of a fiber-resin interface, Compos. SciTechnol. 28 (1) (1987) 17–32.
[53] B. Yu, Z. Jiang, Xi Tang, Y.C. Yoon, J. Yang, Enhanced interphase between epoxy matrix and carbon fiber with carbon nanotube-modified silane coating, Compos. Sci. Technol. 99 (2014) 131–140.
[54] M. Sharma, J. Bijwe, K. Singh, Studies for wear property correlation for carbon fabric-reinforced PES composites, Tribol Lett 43 (3) (2011) 267–273.
[55] M. Sharma, J. Bijwe, Influence of molecular weight on performance properties of polyethersulphone and its composites with carbon fabric, Wear 274–275 (2012) 388–394.
[56] M. Sharma, J. Bijwe, P. Mitschang, Abrasive wear studies on composites of PEEK and PES with modified surface of carbon fabric, Tribol. Int. 44 (2) (2011) 81–91.
[57] M. Sharma, J. Bijwe, P. Mitschang, Wear performance of PEEK-carbon fabric composites with strengthened fiber–matrix interface, Wear 271 (9–10) (2011) 2261–2268.
[58] M. Sharma, S. Tiwari, J. Bijwe, Optimization of material parameters for development of polyetherimide composites, Mater. Sci. Eng. B 168 (1–3) (2010) 55–59.
[59] H. Mahfuz, F. Powell, R. Granata, M. Hosur, M. Khan, Coating of carbon fiber with polyhedral oligomeric silsesquioxane (POSS) to enhance mechanical properties and durability of carbon/vinyl ester composites, Materials 4 (2011) 1619–1631.
[60] M.K. Seo, S.J. Ramanathan, Surface characteristics of carbon fibers modified by direct oxyfluorination, J. Colloid Interface Sci. 330 (1) (2009) 237–242.
[61] S.L. Gao, E. Mäder, Characterization of interphase nanoscale property variation in glass fiber reinforced polypropylene and epoxy resin composites, Compos. Part A Appl. Sci. Manuf. 33 (4) (2002) 559–576.

Hybrid ceramic/polymer composites for bone tissue regeneration

6

Daniela Iannazzo, Alessandro Pistone, Marina Salamò and Signorino Galvagno
University of Messina, Messina, Italy

Chapter Outline

6.1 Introduction

Bone tissue engineering and regeneration therapies have seen a tremendous growing interest with major progress over the last thirty years, thanks to the collaborative efforts of scientists, engineers, and surgeons to achieve the ultimate goal of creating bone grafts able to enhance bone repair and regeneration [1–3]. Musculoskeletal problems including bone and joint pathologies fractures related to osteoporosis, back pain, serious injuries, and different kind of bone diseases and disabilities affect million of people across the world; this number, with the rise in trauma victims and musculoskeletal disorders often associated with the increase in life expectancy, is expected to increase in the coming years [4].

The engineering of bone tissues can provide bone grafts for clinical use or offer solutions to be implemented in bone loss that occurs because of degenerative, surgical, or traumatic processes. Promises of tissue engineering are also brought to the bone regeneration of the cranium-maxillofacial skeleton area, including oral maxillofacial surgery, plastic surgery, otolaryngology, neurosurgery, general surgery, and head and neck oncology. The area is also of vital interest to most specialties of dentistry including periodontics, orthodontics, endodontics, and even general dental

Hybrid Polymer Composite Materials: Processing. DOI: http://dx.doi.org/10.1016/B978-0-08-100789-1.00006-X

practices [5]. Bone regeneration is a complex, well-organized physiological process in which different cell types and their activated signaling pathways are involved.

In this context, mesenchymal stem cells (MSC) play a pivotal role, and their differentiation is regulated by specific signaling molecules such as growth factors, cytokines and hormones, and their activated intracellular networks [6]. The development of controlled release systems for the regeneration of bone, cartilage, and the osteochondral interface is an emerging field of regenerative medicine, and a wide range of therapeutic agents in combination with a great variety of tissue engineering approaches have been evaluated for bone regeneration [7]. Among the different classes of therapeutic drugs for bone tissue regeneration, bisphosphonates (BP) are well-established drugs, used in the development of metabolic bone disorders-related therapies such as osteoporosis, tumor-induced hyperkalaemia, Paget's disease, and inflammation-related bone losses. Increasing evidence on the advantages of using BP in combination with biocompatible scaffolds for the delivery and sustained release of these therapeutic drugs strategies are reported in the literature [8].

In a parallel way, the functionalization of biomaterials with growth factors, nucleic acids, or cells can enhance and accelerate the bone formation, aiding in recapitulating signals present in bone development and healing, regenerating interfaces of bone with other connective tissues, and enhancing vascularization of bone tissues [9]. Both synthetic strategies must be designed with programmed release kinetics in order to deliver the bioactive molecules in a spatio-temporal manner, mimicking the natural wound healing process and promoting osteoprogenitor cellular infiltration and proliferation for integrative tissue repair [10].

In this scenario, the development of new biomaterials for bone tissue engineering, suitable for scaffolds preparation, represents a significant challenge to current regenerative medical research. In general, an ideal scaffold to be used for bone tissue regeneration should possess biocompatibility and osteoconductivity, namely, the ability to stimulate the regeneration of the bone tissue. Moreover, an ideal material must be nonantigenic, resistant to infection, easily adaptable, and of course, readily and sufficiently available to trigger osteogenesis. Moreover, it should closely mimic the natural bone extracellular matrix and allow the osteogenic cells to lay down the bone tissue matrix and the osteogenic differentiation promoted by morphogenic signals. Finally, it should have a sufficient vascularization to meet the growing tissue nutrient supply and clearance needs.

The development of new medical technologies and the recent achievements in material science, biochemistry, molecular biology, and genetic engineering allowed for the creation of newly combined synthetic materials for bone grafting and bone drug delivery systems (DDS). Among the different synthetic scaffolds, CaP-based ceramics such as hydroxyapatite (HA) and beta tricalcium phosphate (β-TCP), as well as polymers, microsponges and other nanomaterials, have been shown to increase the therapeutic efficacy and reduce adverse effects.

The modification of the materials bulk structure with drugs, cytokines, growth factors, and cells provide synthetic materials with not only osteoconductive but also osteoinductive properties as well as the control of the speed of biodegradation, thereby bringing it closer to the kinetics of osteogenesis. Moreover, the possibility to treat bone pathologies locally, at the targeted bone injury and for longer time, can improve a patient's compliance [11,12].

The search of bulk biomaterial to be used in bone tissue engineering has been mainly focused on developing materials able to mimic native bone tissue in order to minimize immunological and disease transfer risks from allogeneic bone. Thus alloplastic bone substitutes are predominantly based on CaP-based ceramics (Fig. 6.1) such as synthetic HA with chemical formula $Ca_5(PO_4)_3(OH)$, β-TCP, biphasic calcium phosphates, consisting of a mixture of TCP and HA (BCP), amorphous calcium phosphate (ACP), calcium sulfates, and multiphasic bio-glasses with osteoconductive and bioactive properties [13,14].

These biomaterials are similar in composition and structure with the mineral phase of bone, are FDA-approved, and were among the most investigated materials for scaffold composition for over three decades [15,16]. The osteoconductive properties of CaP-based ceramics support tissue ingrowth, osteoprogenitor cell growth, and the development of bone formation. Moreover, their surfaces also allow for a direct and strong bond with the bone tissue, thus favoring the exchange of $Ca^{2+}$and P ions between cell matrix and the substrate [17].

Since HA is present in the form of nano-sized crystals in natural bone tissues, great scientific interest was also devoted in developing HA nanocrystals that are endowed with a high level of mechanical properties and expected to be in a dynamic equilibrium with the biological environment in the resorption/mineralization cycle [18]. However, despite the good biological properties of CaP-based ceramics, the major disadvantage of CaP-based materials is their inability to be used as load-bearing bioceramics, because of brittleness and poor fatigue resistance [19].

With the advancement of nanotechnology and sintering technology, it was possible to obtain high strength bioceramics with the required enforcements or combinations, such as ceramic/polymer, ceramic/ceramic, or ceramic/metal composites, with the optimal density and porosity for the specific application or implant site [20]. Next-generation biomaterials for bone tissue regeneration should exhibit the

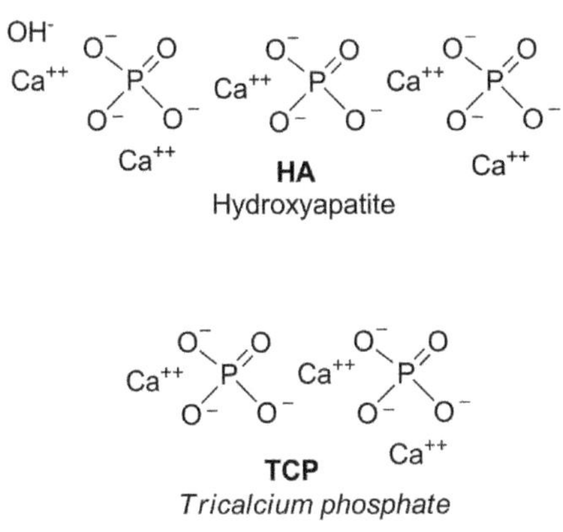

Figure 6.1 Examples of Ca-P ceramics for bone tissue regeneration.

appropriate combination of mechanical support and morphological guidance for cell proliferation and attachment while at the same time serving as matrices for the sustained delivery of drugs and/or biomolecular signals.

The so-called ceramic/polymer scaffolds can achieve this goal by combining the tunable macro/microporosity and osteoinductive properties of ceramic materials with the mechanical/physical properties of biodegradable polymers [21].

The type of polymer and its composition, hydrophobicity, crystallinity, and degradability can positively affect the rate of drug release as well as the rate of tissue ingrowth [22]. Natural or synthetic biodegradable polymers offer great potential in controlled delivery of drugs and bioactive molecules that can be covalently bound to polymers or physically entrapped inside the polymer matrix and released as the polymer degrades in the physiological environment. Their degradation products are normal metabolites of the body or products that can be completely eliminated from the body with or without further metabolic transformations, thus avoiding tissue rejection phenomena [23]. The most used polymers for bone tissue engineering belong to polylactic/polyglycolic acid derived polymers and natural protein and carbohydrates polymers such as chitosan, alginate, hyaluronic acid, and collagen [24].

A lot of research is nowadays focused on hybrid ceramic/polymer (CPC) scaffolds as matrices for the sustained delivery of therapeutic drugs and/or biomolecular signals. Composites made of synthetic HA in the forms of powders, granules, and gels in combination with synthetic or natural polymers and conjugated with drugs, peptides, growth factors, and embryonic stem cells have been proposed, expanding the possibilities of reconstructing pathologically modified mineralized tissues [25,26].

The development of composite scaffolds Ca-P ceramics/polymer-based for bone tissue regeneration involves the optimization of several key parameters such as the volume fraction, size, and shape of the inorganic phase, porosity, and suitable bonding formation at the polymer/ceramicinterface [27]. The therapeutic agent can be loaded in the scaffold by chemical interaction to the composite surface or by physical entrapment within the scaffold. There are three main methods for drug loading: (1) pre-encapsulation of the agent (by using micro- or nanospheres, etc.), followed by the loading of the encapsulating system into the CPC scaffold; (2) the surface immobilization of the agent by noncovalent interaction mechanisms such as hydrophobic, electrostatic, or van der Waals interactions or (3) by specific covalent interaction, which may be afforded by introducing functional groups on the molecular agent or in the biomaterial in order to achieve a better control of the binding as well as a stable incorporation of the molecular agent within the scaffold [28].

In this chapter, hybrid composites constituted by CaP-based ceramic materials with several synthetic or natural polymers will be discussed, summarizing and analyzing their chemical and mechanical properties as well as the results of in vitro or in vivo release studies and their biocompatibility and ability to induce bone tissue regeneration. Ceramic/polymer nanocomposites will be also discussed for their advanced biomedical applications, including tissue engineering and the design of drug delivery system for bone tissue regeneration.

6.2 Ceramic/polymer composites

The combination of ceramics with polymers for the synthesis of hybrid scaffolds able to mimic bone tissue native structures has been widely investigated [29,30]. The most commonly used ceramics for the preparation of polymer/ceramic hybrid composites are β-TCP and HA since their mineral compositions are similar to those found in human bones. Among the different organic counterparts, synthetic polymers have shown many beneficial properties, such as high flexibility and good shape-forming ability. Polymers derived from natural sources have the potential advantage of biological recognition but also the drawback of potential immunogenicity, the presence of pathogenic impurities, and less control over their mechanical properties and batch-to-batch consistency. Natural polymers proposed in tissue-engineered constructs can be further classified in polysaccharide-based or protein-based polymers.

6.2.1 Ceramic/synthetic polymer composites

Properly designed synthetic polymer/ceramic composites to improve scaffold bioactivity and tailored physicochemical properties for specific biomedical applications have been widely investigated [31,32] (Table 6.1). Polyamide (PA), a polymer containing monomers of amides joined by peptide bonds, has been reported to promote adhesion and functional osteoblasts expression because of its structural similarity to collagen; moreover, it shows exceptional hardness, good impact strength, and abrasion resistance [48].

Shen et al., investigated a porous biocomposite choosing PA6 as a polymer matrix and biphasic calcium phosphate (BCP) ceramic as an inorganic phase to enhance the mechanical properties and osteoconduction [49]. The synthesized bioceramic/PA6 scaffolds were cultured with bone marrow stromal stem cells (BMSC) to investigate the in vitro cytocompatibility as well as the in vivo histocompatibility, by implanting the biomaterial in subcutaneous sites of mice for four and eight weeks. The results of these studies demonstrated that the composite scaffolds provided a favorable environment for initial cell adhesion, maintained cell viability and cell proliferation, and had good tissue compatibility. Polylactic/polyglycolic acid-derived polymers have been deeply investigated, and many ceramic hybrid composites have been proposed. This group of polymers include polylactic acid (PLA), polyglycolic acid (PGA), poly-(D/L-Lactic Co-Glycolic) acid (PLGA), and polycaprolactone (PCL). These synthetic polymers are highly biocompatible, and their degradation properties, due to hydrolysis reactions, can be tuned by varying changing different parameters such as the molecular weight [50] (Fig. 6.2).

The most used ceramic particles added to the polymers to improve the osteoconductivity and/or to increase the mechanical properties are porous or dense of structures of HA, ACP, and BCP [51–53]. Of particular interest are PLA/HA-based composites which are fully resorbable in vivo.

Russias et al. investigated from a micromechanistic perspective, the in vitro degradation behavior of such composites manufactured using a simple hot-pressing

Table 6.1 Examples of synthetic polymer/ceramic composites for bone tissue regeneration

Synthetic polymer/ ceramic composites		Bioactive agent	Biological study	Reference
Ceramic	Polymer			
				[33]
BCP	PA 6	–	In vitro and in vivo	[34]
HA	PLGA	Dexamethasone	In vitro and in vivo	[35]
HA	PCL	Clodronate	In vitro	[11]
HA	PLGA	Alendronate	In vitro	[36]
TCP/DCP/ $CaCO_3$/HA	PLGA	rhBMP-2	In vitro and in vivo	[37]
TCP	PCL	BSA	In vitro	[38]
TCP	PCL	rhBMP-2	In vitro	[39]
TCP	PCL	Doxorubicin	In vitro	[40]
HA	PCL	Tetracycline	In vitro	[41]
HA/TCP	PLA	Gentamycin	In vitro	[42]
DCP/$CaCO_3$	PLGA	Gentamycin	In vitro	[43]
HA	PU	Ceftazidime	In vitro	[44]
HA/TCP	PLA	Ciprofloxacin	In vivo	[45]
TCP	PCL	Gatifloxacine	In vivo	[46]
HA	PCL	Vancomycin	In vitro	[47]

route for two different HA particles with an average particle size of 5 μm or ~25–30 μm long and ~5 μm in diameter [54]. The authors observed that composites with ceramic contents ranging between 70 and 85 wt% have mechanical properties that match reasonably those of human cortical bone. However, these properties deteriorate with immersion in Hanks' balanced salt solution due to the degradation of the polymer phase. This degradation resulted was found to be more pronounced in samples with larger ceramic content because of the dissolution of the smaller amount of polymer between the ceramic particles.

Son et al., investigated the inductive effect in osteogenic culture of dexamethasone (DEX) loaded PLGA microspheres on HA scaffold surfaces [55]. In this study,

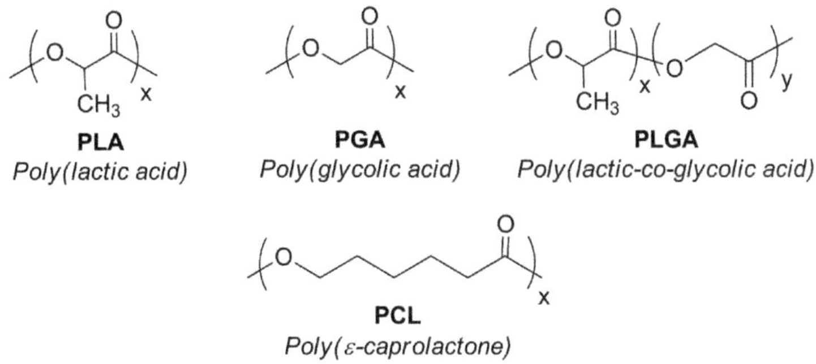

Figure 6.2 Examples of synthetic polymers used for ceramic/polymer scaffolds.

PLGA microsphere surfaces were coated with polyethylenimine (PEI) in order to give a net positive-charged surface and thus to allow the DEX-loaded PLGA microspheres to be immobilized on the negatively charged HA scaffold surfaces. In vitro drug release studies from the encapsulated microspheres were performed prior to the implantation in the femur defects of beagle dogs. A release profile of DEX over a four-week immersion study indicated an initial burst release followed by a sustained release. The in vivo evaluation of the defects filled with the hybrid composite scaffolds indicated enhanced volume and quality of new bone formation when compared to defects that were either unfilled or filled with HA scaffolds alone. Various bioactive molecules for bone regeneration were efficiently incorporated with calcium phosphate-based bioceramics using biodegradable synthetic polymers.

Among the different drugs used for bone tissue diseases, BP can greatly improve the performance of tissue engineering scaffolds in inhibiting bone resorption associated with different bone diseases, such as tumor-associated-altered bone metabolism or osteoporosis [56].

Puppi et al. investigated the release of BP from ceramic/polymer composites using a composite bioactive scaffolds made of biodegradable three-arm branched star PCL, HA, and the clodronate (CD), a bisphosphonate drug that has demonstrated efficacy in the treatment of various bone diseases and as an antiinflammatory drug [57]. The authors developed these hybrid composites with different structural scales and textures by applying either electrospinning or wet-spinning techniques and consequenty investigated the different drug release kinetics through the tuning of fiber dimensions and mesh porosity.

Wang et al., investigated the in situ release of the bisphosphonate drug, alendronate (AL), and DEX from PLGA/HA sintered microspherical scaffold. In vitro osteogenesis was successfully achieved with synovium-derived mesenchymal stem cells (SMSC), which possess strong chondrogenic tendency, as indicated by high yields of alkaline phosphatase (ALP) and bone calcification [58]. The SMSC, after treatment with AL- and DEX-releasing PLGA-HA, hybridized microsphere scaffolds have been successfully differentiated into osteogenic cells thus indicating that the system is a promising scaffold for osteogenesis in situ and bone repair therapy.

PLGA microsphere/calcium phosphate cement composites have been also investigated as possible carrier of growth factors.

The in vivo release of rhBMP-2 loaded PLGA/Ca-P cement consisting of a mixture of TCP, dicalcium phosphate (DCP), calcium carbonate, and HA was investigated [59]. In vitro release studies with rhBMP-2 showed a sustained, small release (5–10%) after four weeks in medium while the in vivo release characteristics with low and high molecular weight PLGA microspheres showed higher release efficiencies (25–50%). The discrepancy between in vitro and in vivo results could be referred to a more dynamic flow in vivo and to the presence of high amounts of salts and proteins that positively influenced the release properties.

Scaffolds for bone tissue regeneration should have high capacity for protein loading and should be able to maintain a controlled and sustained release. To accomplish these goals, an experimental matrix to test if different amounts of protein can be loaded on highly porous sintered TCP was investigated, where strength was improved due to PCL coating and model protein bovine serum albumin (BSA) release kinetics was studied [60]. BSA was encapsulated efficiently within the PCL coating without significant denaturation and the release kinetics of protein from PCL coated porous TCP scaffold was studied in vitro. The amount of loaded protein is controlled easily by varying protein composition in PCL coating. The authors demonstrated that the protein release from PCL coated TCP is dependent on the degradation behavior of PCL. The mechanical strength, investigated before and after coating with PCL showed an increase in compressive strength with increasing PCL concentration.

Rai et al., investigated the drug loading and release behavior of PCL/TCP hybrid composite scaffold, synthesized by fused deposition modeling for bone regeneration [61]. Fibrin sealant was used to promote the loading efficiency of rhBMP-2 onto the PCL and PCL-TCP composites. The authors reported that PCL/TCP-fibrin retained rhBMP-2 longer than PCL-fibrin composites, and interpreted that this was likely due to the formation of intermolecular linkages between rhBMP-2 and TCP. Bone tissue engineering implants with sustained local drug delivery provide an opportunity for better postoperative care for bone tumor patients because these implants offer sustained drug release at the tumor site and reduce systemic side effects.

A macroporous PCL scaffold embedded with a porous matrix composed of chitosan, nanoclay, and TCP, by freeze-drying was developed [62]. This composite scaffold was evaluated on its ability to deliver the anticancer agent doxorubicin and to promote formation of a mineralized matrix in vitro. The authors reported that the scaffold scaffolds comprising nanoclay released up to 45% of the drug for up to 2 months, while the scaffold without nanoclay released 95% of the drug within 4 days. The results of this study suggest that the scaffold can be used clinically in reconstructive surgery after bone tumor resection.

Ceramic/polymer scaffolds have been widely studied for the release of antibiotics after surgical interventions aimed at the implantation of prosthesis or for the prevention from bacterial infections. In fact, the use of these drugs either orally or intravenously, because of the very little accessibility of the site of infection, often prolongs the treatment of bone infections.

HA scaffolds synthesized by polyurethane foam reticulate method have been coated with a mixture of HA powder and PCL [63]. Then, the antibiotic drug tetracycline hydrochloride (TCH) was mixed with the HA/PCL coating. The authors reported an increase in compressive strength and elastic modulus by this coating scaffold, where the dissolution rate was highly dependent on the HA/PCL ratio; the higher concentration and HA amount caused the increased biodegradation. The drug release rate was sustained for prolonged periods and was highly dependent on the degree of coating dissolution, suggesting the possibility of a controlled drug release in the porous scaffold with HA/PCL coating.

A calcium phosphate/PLA blend was used for the release of gentamycin. The drug, incorporated in the polymer, showed a burst release followed by a slower sustained release, typical for PLA polymers [64]. Also, the controlled release of gentamycin from the CaP-ceramics (DCP/$CaCO_3$) and PLGA composite was reported [65]. In this study, scaffolds containing up to 20 wt% of microspheres were loaded with gentamycin (10–30 wt%) and showed a sustained in vitro release pattern, while the release pattern from microspheres alone exhibited a burst release.

Liu et al., developed a biodegradable scaffold for bone regeneration with the capacity of controlled drug delivery, using ceftazidime as a model drug [42]. The drug encapsulated in ethyl cellulose (EC) microspheres was incorporated in a HA/polyurethane (HA/PU) composite scaffold to generate an antibiotic drug delivery system. The incorporation of microspheres into scaffolds also in this case significantly reduced the initial burst release, and the system exhibited a sustained release of the model drug for up to 60 days.

A ciprofloxacin implant formulation composed of HA, TCP, PLA, and 40% ciprofloxacin was characterized in vivo for use in treatment of multibacterial bone infection [66]. The implant, inserted in the femur of rabbits, showed approximately 90% of the total ciprofloxacin released within eight weeks, maintaining therapeutic levels in the femur and tibia.

A hybrid composite of TCP/PCL loaded with gatifloxacine by a solvent-free process was developed [67]. The composite released the drug for four weeks in Hanks' balanced salt solution, and demonstrated a sustained bactericidal activity against *Streptococcus milleri* and *Bacteroides fragilis* for at least one week. The composite, implanted in an osteomyelitis lesion of rabbit mandible, showed to be effective in controlling the infection and supported bone tissue reconstruction at the bone defect.

Highly porous HA scaffolds, fabricated by a polyurethane foam reticulate method, were coated with hybrid coating solution, consistituted by PCL, HA powders, and the antibiotic vancomycin [68]. The encapsulated drug within the coated scaffold was released in a highly sustained manner as compared to the rapid release of drugs directly adsorbed on the pure HA scaffold.

6.2.2 Ceramic/natural polymer composites

Polymers derived from natural sources, used for the development of hybrid ceramic/polymer composites for bone tissue regeneration, with respect to the synthetic polymers above discussed, have the potential advantage of biological recognition that may positively support cell adhesion and function. However, as with many natural products, they may exhibit immunogenicity and contain pathogenic impurities. Moreover, with respect to the synthetic counterparts it may also exhibit less control over their mechanical properties, biodegradability, and batch-to-batch consistency [43]. Natural polymers proposed in tissue-engineered constructs can further be polysaccharide-based, such as chitin, chitosan, alginate, hyaluronan, and cellulose or protein-based such as collagen and gelatin.

6.2.2.1 Ceramic/carbohydrate-based polymer composites

Biodegradable carbohydrate-based polymers used for the development of hybrid ceramic/polymer composites for bone tissue regeneration are mainly chitin and chitosan, a derivative of chitin, alginate, hyaluronan, and cellulose. Different biopolysaccharides/ceramics combinations tested as templates for bone regeneration (Fig. 6.3; Table 6.2).

Torres et al., evaluated the incorporation of alginate, a polymer comprising 1,4′-linked β-D-mannuronic acid and α-L-guluronic acid blocks in the production of tunable TCP/HA porous scaffold [70]. The coating with alginate conferred to the scaffold an improved compressive strength, fracture toughness and Young's

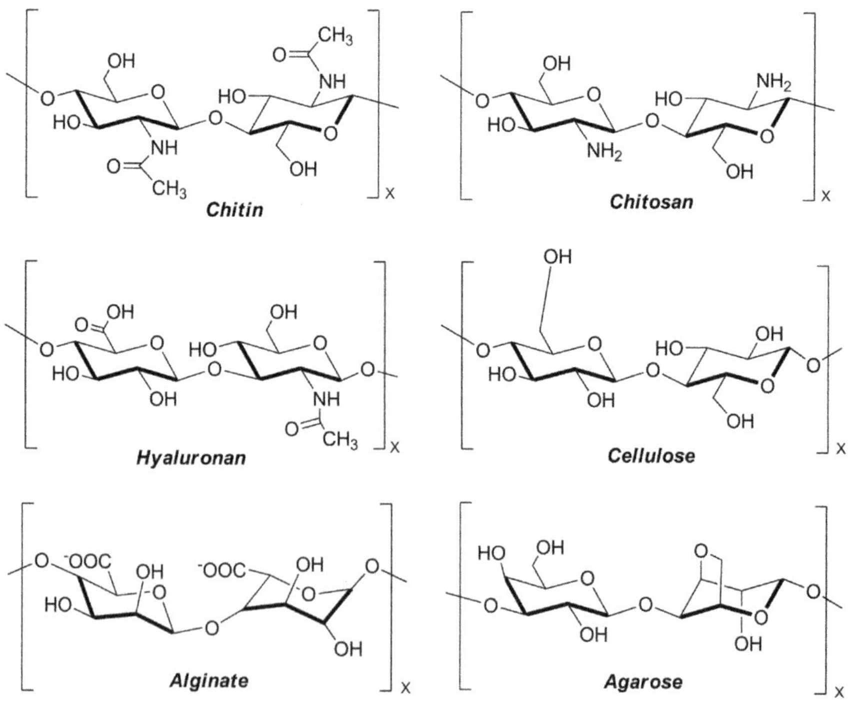

Figure 6.3 Examples of biopolysaccharides used for ceramic/polymer scaffolds.

Table 6.2 Examples of biopolysaccharides/ceramic composites for bone tissue regeneration

Biopolysaccharides/ceramic composites		Bioactive agent	Biological study	Reference
Ceramic	Polymer			
TCP/HA	Alginate	–	In vitro	[32]
TCP/HA	Alginate	TGF-β	In vivo	[67]
TCP/DCP	Chitosan	Flomoxef	In vitro	[6]
TCP/ACP	Chitosan	Gentamicin	In vitro	[69]
HA	Chitosan	Tetracycline	In vitro	[14]
TCP	Chitosan	PDGF	In vitro and in vivo	[16]
DCP	Hydroxypropyl methylcellulose	–	–	[4]
TCP	Amylopectin	rhTGF-β	In vitro and in vivo	[2]
TCP	Agarose	Vancomycin	In vitro	[19]
HA	Hyaluronan	–	In vivo	[57]
HA	Hyaluronan	Collagenase I	In vitro and in vivo	[61]

modulus, to values similar to those of native bone. The hybrid composites demonstrated valuable characteristics such as biocompatibility, biodegradability, hydrophilicity, low toxicity, and gelling capacity with divalent cations; moreover, they supported osteoblast adhesion, maturation, and proliferation.

Seol et al., also reported alginate hydrogel 3-D scaffolds, fabricated using injection molding, for cartilage regeneration. In this study, the authors assembled the hydrogel with the ceramic scaffolds consisting of HA and TCP. The transforming growth factor-beta (TGF-β) was added to the alginate scaffolds to promote cartilage tissue regeneration [71]. In vivo experiments were then performed and articular cartilage tissue regeneration was examined in the knee joints of rabbits, using histological and immunohistochemical analyses. The results of these biological studies demonstrated that the regeneration of osteochondral tissue, especially articular cartilage tissue regeneration, was better in hybrid scaffolds than in the hydrogel scaffolds alone, thus highlighting the key role of ceramic counterparts for articular cartilage tissue regeneration.

Chitosan is often applied together with CaP-ceramics because of its biodegradability by lytic enzymes and the improved mechanical/physical properties of ceramic/chitosancomposites [72]. Moreover, chitosan can be added to CaP-ceramics as an adjuvant to make the cement more injectable without substantially modifying the setting reaction [73]. The in vitro antibiotic release from TCP/DCP/chitosan composites using flomoxef sodium as a model drug was investigated [74]. Results of in vitro drug release studies showed the typical profile observed in a skeleton type drug delivery which was characterized by an initial burst, followed by a more sustained release. Moreover, the authors reported that changing the concentration of chitosan, can control the rate of drug release from the system.

The opposite construction of chitosan/ceramic composites, namely, the insertion of ceramic particles in a porous chitosan composite, was also formulated [75]. The added ceramic particles gave extra osteogenic potential giving to the composite materials enhanced osteoblast proliferation and differentiation.

Macroporous chitosan scaffolds reinforced by TCP/ACP ceramics were reported by Zhang et al. using a thermally induced phase separation technique [76]. Then, the antibiotic gentamicin—sulfate was loaded and the controlled drug release on these scaffolds was investigated. The authors reported an initial burst release from macroporous chitosan that was diminished by addition of the ceramic particles and a total release of 90% after 3 weeks. The more sustained drug release observed from the hybrid composite was suggested to be due to the higher extend of chitosan cross-linking. The authors also investigated the morphology of osteosarcoma MG63 cells cultured on the scaffolds. Scanning electron microscopy (SEM) micrographs showed no apparent morphological differences for osteoblastic cells grown on the pure chitosan scaffolds and those grown on composite scaffolds; the cells, attached and migrated on these scaffolds, exhibited a biological appearance, thus suggesting a good cellular compatibility.

Teng et al., explored chitosan/HA composites as controlled drug delivery systems for the antibiotic tetracycline hydrochloride [77]. The authors developed functional scaffolds with a different gradient of structure and drug concentration demonstrating the high effectiveness of the hybrid scaffolds in regulating the release of drugs, and hence their capability to serve as a temporary drug carrier in tissue regeneration.

These functional scaffolds also have potential application to the delivery of some bioactive molecules such as growth factors. Thus, the release of platelet-derived growth factor (PDGF) to chitosan/TCP scaffolds fabricated by freeze-drying a mixture of chitosan solution and TCP powder and soaking in a PDGF-BB solution was investigated [37]. The authors measured the in vitro release by ^{125}I labeled PDGF and in vivo bone regeneration in calvarial defects in rats. The release test showed an initial burst, followed by a slower maintained release. The histologic examination demonstrated the ability of chitosan/TCP sponge to promoted osseous healing of the rat calvarial defects, as compared to controls. The addition of the growth factor further enhanced the bone regeneration.

Hydroxypropyl methylcellulose (HPMC) a carbohydrate derived from cellulose was also applied in combination with ceramic particles or as an addition to calcium phosphate cement. These hybrid composites showed poor mechanical properties but a fast in vivo dissolution, followed by replacement with newly formed bone [78]. Injectable hybrid composites with a high solubility for rapid setting by incorporating DCP with the gelling agent HMCP have been reported [79]. The authors reported that the addition of HPMC dramatically increased the paste injectability. This effect could be particularly relevant for minimally invasive surgical techniques or for procedures involving defects with limited accessibility.

Injectable pastes comprised of amylopectin a soluble highly branched polysaccharide polymer of glucose found in plants, with β-TCP granules and loaded with rhTGF-β, by absorbing the growth factor on the granules, have been prepared [80]. The in vitro results showed a release of 80% after 24 h incubation in serum. The in vivo efficacy also was assessed, in comparison to a sham control group and a placebo treated group, using a rabbit unilateral segmental defect model. Radiographs of defect sites demonstrated a positive bone response of the added TGF-β was observed after 56 days. Moreover, the in vivo pharmacokinetics of the growth factor evaluated in the same rabbit model suggested that the growth factor remained intact at the defect site for more than 21 days.

The use of agarose for the fabrication of TCP-based composites was reported. The scaffolds were prepared by a shaping technique that allows obtained pieces at a temperature low enough to simultaneously include active substances susceptible to heat degradation, such as the antibiotic vancomycin [81]. The authors reported the controlled drug release was also controlled by the drying procedures employed to process and preserve the obtained scaffolds, which resulted in the generation of different pore architectures and certain chemical interactions that yielded different drug release patterns.

Hyaluronan and hyaluronic acid are other carbohydrate-based polymers for tissue engineering, inherently recognizable by cells and exhibiting favorable cell-material interactions. The polysaccharide hyaluronan was investigated by Liljensten et al. for bone grafts consisting of HA granules in a hyaluronan carrier [39]. The bone response to porous ceramic HA granule/hyaluronan composites was evaluated in an experimental bone defect model in rabbits and the results of this in vivo study suggested the hybrid composite properties as an interesting bone substitute.

In a more recent work, it was reported a HA, calcium sulfate hemihydrates (CS) and hyaluronic acid laden collagenase (HA/CS/HyA-Col) as a bone substitute for the alveolar bone regeneration [82]. The composite material was mechanically tested and the biocompatibility was evaluated by a water-soluble tetrazolium salt-1 (WST-1) assay. The in vivo bone formation was assessed in rats with alveolar bone defects, and the bone augmentation by the hybrid composite was confirmed by micro-computed tomography (micro-CT) images and by histological examination. The composite showed excellent biocompatibility and the results of this study showed an improved new bone formation with matured bone morphology in bone defects implanted with HA/CS/HyA composite containing type I collagenase with respect to the HA/CS/HyA composite that lacks the collagenase and the porous HA granules.

Other biopolysaccharides have shown the ability to improve cement cohesion. Khairounet et al. investigated the cohesion properties of a commercial HA after the addition of a wide range of these polymers such as hydroxyethyl starch, starch, sodium dextran sulphate, $\alpha/\beta/\lambda$-cyclodextrins, alginic acid, hyaluronic acid, and chondroitinesulphate [83]. Results for all these composites show a reduced cohesion time and small changes in setting time and mechanical properties when compared to the original cement.

6.2.2.2 Ceramic/protein-based polymer composites

Composites of CaP-ceramics with protein-based polymers have been extensively investigated because of the possibility to produce organic–inorganic composites with mechanical/physical and biological characteristics, similar to human bone. Among these, collagen and gelatin have been proposed in tissue-engineered constructs (Table 6.3).

Collagen is the main structural protein present in the extracellular space of various animal connective tissues. Depending upon the degree of mineralization, collagen tissues may be rigid, like in bones, compliant like in tendon, or may have a

Table 6.3 Examples of protein-based polymers/ceramic composites for bone tissue regeneration

Protein-based polymers/ceramic composites		Bioactive agent	Biological study	Reference
Ceramic	Polymer			
HA	Collagen	rhBMP-2	In vivo	89
TCP/DCP	Collagen	VEGF	In vitro	90
HA	Collagen	–	In vivo	91
HA	Gelatin	–	In vitro	95
TCP	Gelatin	–	In vitro	96
TCP	Gelatin	BMP-2	In vitro and in vivo	97
HA	Gelatin	Gentamycin	In vitro	98
TCP	Gelatin	Vancomycin	In vitro	99
TCP/HA	Fibrin glue	–	–	100
BCP	Fibrin glue	–	In vitro and in vivo	101
TCP	Soybean	–	–	103
TCP/DCP	Polypeptide graft copolymers	–	–	104

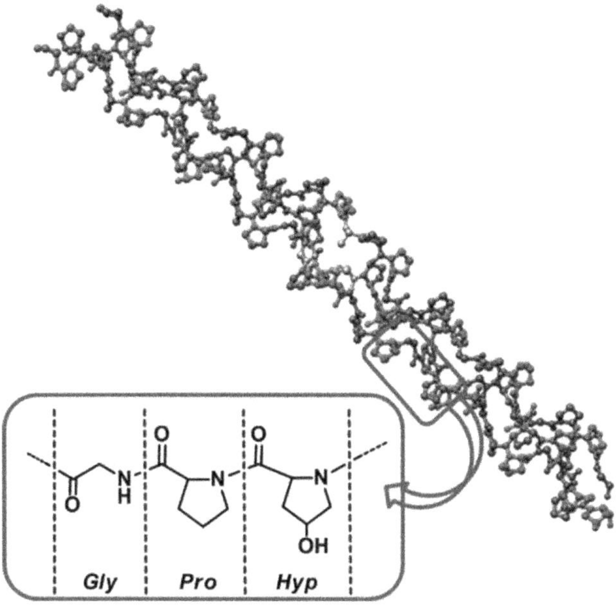

Figure 6.4 Common amino acid sequence motif of collagen.

gradient from rigid to compliant like in cartilage. From a chemical point of view collagen is composed of a triple helix, which generally consists of two identical polypeptide chains and an additional chain that slightly differs in its chemical composition. This protein is unique in terms of amino acid composition, repeating sequence pattern, high degree of posttranslational modification, and characteristic intermolecular cross-links [84]. The most common motifs in the amino acid sequence of are glycine-proline-AA and glycine-AA-hydroxyproline, where AA is any amino acid other than glycine, proline, or hydroxyproline (Fig. 6.4).

Gelatin, which is used in food and industry, derives from the irreversibly hydrolyzed collagen and as collagen, contains many glycine, proline, and 4-hydroxyproline residues. Higher levels of these pyrrolidines result in stronger gels. Chemical cross-links can be introduced, using transglutaminase, in order to link lysine to glutamine residues or using glutaraldehyde to link the amino groups of two lysine residues [27]. Both collagen and gelatin materials are degradable biopolymers by the proteolysis reaction of collagenase or gelatinase, respectively.

An HA/collagen composite with structure, chemical composition, and crystallinity similar to natural bone by a coprecipitation method followed by cold isostatic pressure shaping process was reported [85]. The authors investigated an anterior fusion model of the cervical spine in beagle dogs, the influence of the cross-linking degree, as well as the effect of recombinant human bone morphogenetic protein-2 (rhBMP-2) preadsorption to the composite on the bone substitution rate. The results of this study highlighted the importance of cross-linking of the HA/collagen composite in controlling both the mechanical strength and bioresorbability.

Radiographic and histological results suggested that a combined treatment of cross-linking of the composite with preadsorption of rhBMP-2 growth factor may be a very valuable replacement of existing ceramic systems in the anterior fusion of the cervical spine, as well as inlay grafting of bone defects in weight-bearing sites.

Lode et al. reported a strategy to accelerate vascularization of the implant region, developing a collagen/ceramic composite functionalized with the vascular endothelial growth factor (VEGF), which is known to mediate angiogenesis in vivo [44]. The authors studied the VEGF release following physical adsorption to the composite and demonstrated a high VEGF binding capacity of composite and a sustained drug release with a moderate initial burst. Moreover, a proliferation assay using endothelial cells revealed maintenance of biological activity of VEGF after release from the composite; the biological efficacy of released VEGF was even higher than that of nonreleased control VEGF.

Three-dimension scaffolds for bone tissue regeneration were produced, combining nanometric HA, synthesized by precipitation within collagen fibrils following a biologically inspired mineralization process and PCL, which was employed to give to the material a 3-D structure [86]. The chemical-physical analysis carried on the hybrid composite revealed a high similarity in composition and morphology with biologically mineralized collagen fibrils; moreover, the scaffold degradation pattern was fond to be suitable for physiological processes. The scaffolds were implanted in sheep mandibles using prototyped surgical guides and customized bone plates. Then, after three months healing, SEM analysis of the explanted scaffold revealed a massive cell seeding of the scaffold, with cell infiltration within the scaffold's interconnected pores. Histology confirmed cell penetration and neoangiogenesis within the scaffold.

Several porous ceramic-based gelatin/collagen scaffolds have been formulated; among these, the so-called gelatin sponges are prepared by freeze-drying a gelatin/calcium phosphate mixture [87]. A ceramic/gelatin porous scaffold with drug delivery function was prepared by macroporous HA ceramic soaked in gelatin solution that penetrated in the bulk of the ceramic have been prepared. Gelatine, reproducing the proteinaceous part of bone was cross-linked in order to modulate its solubility in the physiologic fluids and the kinetic of gelatine release from ceramic matrix was evaluated [88].

The introduction of 18% of gelatin inside a TCP matrix was also reported and the chemical and physical properties of the composite as well as the biological response was investigated [89]. The authors reported an increase in compression strength due to the high density of the gelatin with respect to the control cement. The in vitro studies performed with primary culture of osteoblasts derived from healthy and pathological bones demonstrated that the gelatin-enriched cement improves osteoblasts' activity and differentiation thus suggesting the application of the composite as bone substitute also in the presence of osteopenic bone.

Gelatin sponges containing TCP particles, loaded with the growth factors BMP-2 have been investigated [90]. The BMP-2 loading was performed by simply dropping a solution of the growth factor to freeze-dried sponges prepared by chemical cross-linking of the gelatin in the presence of different amounts of TCP. Drug release studies were performed in vivo by labeling theBMP-2 with ^{125}I. The in vivo release tests revealed that the growth factor was released at a similar time profile, irrespective of the ceramic content. The osteoinduction activity of gelatin or gelatin/TCP sponges incorporating BMP-2 was investigated following the implantation into the back subcutis of rats in terms of histological and biochemical examinations. Results of this study demonstrated for the gelatin sponge a significantly higher osteoinduction activity with respect to the gelatin/TCP sponges. However, in vitro collagenase digestion experiments revealed that the hybrid sponge collapsed easier than the gelatin sponge, without ceramic incorporation.

Antibiotic drugs have been also incorporated in ceramic/gelatin sponges and release studies have been performed [68,91]. Kim et al., loaded gentamycin to HA/gelatin sponges, fabricated in a foam type via a novel freeze-drying and cross-linking technique and investigated the in vitro drug entrapment and release as a function of cross-linking density of the gelatin, the amount of HA, and of drug loading. The authors showed that the increasing of cross-linking density, lead to an higher drug entrapment and a more sustained release pattern. Biological studies performed in vitro with osteoblast-like human osteosarcoma showed that cells spread and grew actively on all the foams. The cell proliferation rate, quantified indirectly on the cells cultured on titanium disks coated with gelatin and with gelatin/HA composites using MTT assay, exhibited an up-regulation with gelatin coating compared with bare Ti substrate and a slight decrease on the composite coatings; however, the alkaline phosphatase activities expressed by the cells cultured on composites were significantly enhanced compared with those on pure gelatin foam.

Zhou et al., reported the controlled release of vancomycin in gelatin/TCP composite scaffolds for the treatment of osteomyelitis. In this study, biodegradable gelatin sponges containing different contents of TCP were loaded with an aqueous solution containing vancomycin and genipin, leading to a final concentration of gelatin of 10%. Results of drug release studies showed that the composite scaffolds could be used as a vancomycin sustained release system, achieving local therapeutic drug levels over an extended duration.

Other than collagen and gelatin, also other protein-based structures have been used for the preparation of ceramic/protein-based polymer composites. Fibrin, and especially fibrin glue, was applied for the introduction of ceramic particles or mixed together with calcium phosphate cement [35,36,92].

Le Nihouannen et al. combined fibrin glue, a mixture of fibrinogen and thrombin, often used in surgery due to its hemostatic, chemotactic, and mitogenic properties, with calcium phosphate ceramic granules composed by HA and TCP. The authors reported that thrombin influenced the properties of the composite by inducing the nucleation of crystalline precipitate at the ceramic/fibrin glue interface. However, they suggest that combining fibrin sealant and calcium phosphate ceramics could lead to new scaffolds for bone tissue engineering with the synergy of the properties of the two biomaterials.

In a similar work, Jegoux et al. combined fibrin glue with biphasic calcium phosphate granules consisting of a mixture of TCP and HA (BCP) in order to facilitate the handling of biomaterials.

The authors compared the bioactivity of two composites combining calcium phosphate granules with two different types of fibrin glue, one with natural aprotinin and the other with the synthetic tranexamic acid. The in vivo tests, performed using New Zealand rabbits, showed no adverse reactions either samples. Moreover, the bone ingrowth and resorption was found to be qualitatively and quantitatively equivalent with the two types of fibrin sealant.

Several studies have been performed using TCP ceramics in a casein and soybean protein scaffold and polypeptides into CaP-cements, because of their biodegradability, biomimetic properties, and mechanical reinforcement [40,93]. Vaz et al. the developed meltable hybrid composites based on casein and soybean proteins and TCP ceramic reinforcement. The authors reported that the incorporation of TCP into the soybean thermoplastic decreased its mechanical properties but lead to the nucleation of a bioactive calcium phosphate film on their surface when immersed in a simulated body fluid solution.

Lin et al. developed water-based calcium phosphate cements composed of equimolar TCP and DCP ceramics, reinforced by polypeptide graft copolymers, and micelles that were formed by polypeptide copolymers. The authors reported that the molecular structure of the polypeptide graft polyethylene glycol (PEG) copolymers with peptides such as poly-γ-benzyl L-glutamate (PBLG-gPEG), poly-γ-ethyl L-glutamate (PELG-g-PEG) and poly-γ-methyl L-glutamate (PMLG-g-PEG) as well as the association forms of the polypeptide copolymers, exhibited a marked effect on the mechanical properties of their hybrid composite. Moreover, it was reported that polypeptide copolymers with more hydrophilic side chains and with core-shell micelle forms gave more effective reinforcement effect.

6.3 Ceramic/polymer nanocomposites

Ever growing interest is given to the use of nanoscale structures for tissue engineering applications. Materials converted into nano size, provide unique surface properties different from those of the bulk materials or single molecules, which critically may influence their interaction with the biological systems [94]. Thus, biocompatible and biodegradable nanomaterials have been used as support matrices or as a scaffold for the delivery of bioactive agents and cultured cells to targeted tissues and to promote three-dimensional tissue reconstructions [30,41].

The use of nanostructured biomaterials in bone regeneration is inspired by the native bone architecture. The inorganic phase of bones mainly consists of nonstoichiometric HA crystals with lengths of about 100 nm, widths of 20–30 nm. and thicknesses of 3–6 nm, which are embedded between the collagen fibers. Several reports showed that nano-CaP-based biomaterials exhibited physicochemical and biological characteristics better than conventional sized CaP-ceramics, being more similar to natural bone nanocrystals [95]. The hybridization of nano-sized HA with synthetic or natural polymers represents a promising approach to facilitate the preparation of HA nano-scaffold nanomaterials [96].

Polymer/ceramic nanocomposites used in bone tissue regeneration for the delivery of drugs or growth factors have shown to offer new opportunities to provide more focused and fine-tuned treatment of diseases at a molecular level, enhancing their therapeutic efficacy and reducing the side effects. The opportune design of the nanosystems can make them independent in the normal tissue environments and selective at the diseased pharmacological site [97].

Nano-HA/polyamide (n-HA/PA) composite scaffolds utilizing thermally induced phase inversion processing technique were prepared and seeded on the scaffold MSC derived from bone marrow of neonatal rabbits [98]. The results of the biological studies confirmed that n-HA/PA scaffolds were biocompatible and no negative effects on the MSC was observed in vitro. Moreover, the in vivo biocompatibility and osteogenesis of the composite scaffolds was investigated implanting the scaffolds in rabbit mandibles. The results of histological and microradiographical studies demonstrated a good biocompatibility and extensive osteoconductivity with the host bone. In addition, the authors observed that the introduction of MSC to the scaffolds dramatically enhanced the efficiency of new bone formation, especially at the initial stage after implantation.

More recently, it was investigated the in vitro and in vivo biological responses to strontium-containing nano-structured carbonated HA/sodium alginate (SrCHA) spheres that were used for sinus lifts in rabbits [99]. The authors evaluated the cytocompatibility after exposing murine preosteoblasts to the extracts of these materials. The microscopic and histomorphometric analyses, revealed an overall viability of the cells exposed to carbonated HA and SrCHA extracts while the in vitro tests demonstrated the cytocompatibility of the extracts. Bone formation was observed in both groups; moreover, a greater bioresorption of SrCHA with respect to CHA was observed thus indicating a great biocompatibility, osteoconductivity, and bioresorption ability of SrCHA nanostructured material.

Significative changes in the biological or mechanical properties of bio-nanocomposites can be achieved with the use of only a small amount of nanofillers. A significant part of the literature reports the use of nanofillers in the form of nanoparticles (NP) such as silica or iron oxide NP, nanofibers, or carbon-based nanomaterials to control the bioactivity of ceramic-based nanocomposites for bone tissue regeneration [100].

In order to overcome a long-standing problem in the use of these conventional scaffolds, namely, the impossibility of reloading the scaffold with the bio-agents after implantation, some research groups developed nanosystems ceramic and/or biopolymers that are able to attract and deliver bio-agents in a controlled manner by the use of magnetic nanoparticles (MNP) [33]. These nanoparticles also function with bio-agents such as drugs, growth factors, and stem cells, and can act as carriers for the delivery of bio-agents toward and inside the magnetic scaffold under the effect of a magnetic field, thus improving the bone regeneration processes. The interest in magnetic nanoparticles is also justified by the increasing evidence that that magnetic fields alone and magnetic responsive scaffolds can play unique roles in promoting bone repair and regeneration [101]. On the basis of these evidences, the synergistic effects of magnetic scaffolds was highlighted in response to external magnetic fields on the bone regeneration in situ by transforming standard commercial scaffolds made of HA and collagen in magnetic scaffolds [102]. The magnetic nanocomposites were developed by dip-coating the scaffolds in aqueous ferrofluids containing iron oxide nanoparticles coated with various biopolymers. The obtained magnetization values have shown to be suitable for generating magnetic gradients and to enable magnetic guiding in the vicinity and inside the scaffold. Biological studies indicated the ability of the magnetic scaffolds to support adhesion and proliferation of human bone marrow stem cells in vitro.

Magnetic HA bone substitutes to enhance tissue regeneration have also been developed, incorporating magnetite at three different ratios to porous HA ceramic composites [69]. The authors evaluated in vitro the ability of these magnetic HA-based scaffolds to enhance tissue regeneration using human osteoblast-like cells cultures. Results of these biological studies indicated high biocompatibility with no negative effects arising from the presence of magnetite or by the use of a static magnetic field. Moreover, the implantation of the nanocomposites HA/Fe_3O_4 (90/10) in vivo in a critical size lesion of rabbit condyle have shown a good level of histocompatibility, thus opening new perspectives for the application of a magnetic field in a clinical setting of bone replacement.

In order to minimize the formation of potentially cytotoxic magnetic phases such as magnetite or other iron oxide phases, the same research group proposed a bioinspired mineralization process to develop biomimetic hybrid scaffolds made of (Fe^{2+}/Fe^{3+})-doped HA nanocrystals nucleated on self-assembling collagen fibers and endowed with super-paramagnetic properties [103]. The in vitro biological studies, performed using human osteoblast-like cells cultures on magnetic and nonmagnetic materials demonstrated a great biocompatibility of the synthesized nanomaterials. Moreover, the magnetization of the super-paramagnetic scaffolds, induced applying an external static magnetic field, have shown to improve cell proliferation in comparison to the nonmagnetic scaffold.

Magnetic nanofibrous scaffolds of PCL have been developed by incorporating magnetic nanoparticles; their physicochemical, mechanical, and biological properties for bone regeneration purposes have been investigated [104]. The authors reported that the incorporation of magnetic NP greatly improved the hydrophilicity of the nanofibers; in addition, the tensile mechanical properties were significantly enhanced. The synthesized nanocomposites exhibited magnetic behaviors, which increased gradually with MNP content. The in vitro evaluation of the bone-forming ability in simulated body fluids confirmed the substantial improvement gained by the addition of magnetite. The in vivo biological studies by implanting the nanocomposite subcutaneously in rats showed minimal adverse tissue reactions, while inducing substantial neo-blood vessel formation, which was greatly limited in pure PCL.

De Santis et al., also reported 3-D fully biodegradable and magnetic nanocomposite scaffolds for bone tissue engineering, consisting of a PCL matrix reinforced with iron-doped HA nanoparticles (FeHA) for bone regeneration [105]. The in vitro biological studies showed that the cell growth in the magnetized scaffolds was 2.2-fold greater than that in nonmagnetized ones. The in vivo testing, performed in a rabbit animal model, suggested that the PCL/FeHA scaffolds were completely filled with newly formed bone after four weeks, thus proving a good level of histocompatibility.

A nanofibrous composite scaffold composed of super-paramagnetic γ-Fe_2O_3 NP, HA nanoparticles and PLA, prepared using electrospinning technique, was reported [106]. The scaffolds, implanted in a white rabbit model of lumbar transverse defects and provided with static magnetic field post-surgery, showed accelerated new bone tissue formation and remodeling in the rabbit defect thus providing a new strategy for scaffold-guided bone repair.

In the search of new biocompatible nanomaterials for tissue engineering applications, carbon-based nanomaterials, such as carbon nanotubes and graphene, emerged as promising nanofillers for casting bone scaffolds [107]. Carbon nanotubes (CNT) offer potential advantages over the more widely studied nanoparticle systems including their ability to cross cellular membranes and shuttle drugs, biomolecules, including DNA, siRNA and proteins, into various types of cells such as cancer cells and T cells so as to prevent biofouling and their application in biosensing [30,46,108–111]. The opportunity offered by the multimodal conjugation of carbon nanotubes, which allows the insertion of more than one type of functional group to the nanotube surface, is a key property that establishes the superiority over other agents [111,112]. Thus, functionalized CNT, because of their high mechanical strength and biocompatibility, emerged as a promising nanofiller for nanocomposites to be used as bone scaffolds [113–116].

Haddon's group investigated the possibility of injecting CNT into a bone fracture to support the growth of new tissues and to heal fractures. The authors functionalized single-walled carbon nanotubes (SWCNT) with phosphonates and poly (aminobenzene sulfonic acid) in the solution phase and as films on substrates. Microscopy studies showed that HA nucleated and crystallized on the surface of the functionalized SWCNT, and the authors reported that the negatively charged functional groups present on SWCNT attracted the calcium cations leading to self-assembly of HA. The thickness of the HA layers was found to be a function of the mineralization time.

In a follow-up investigation, the same research group evaluated the ability of SWCNT and multiwalled carbon nanotubes (MWCNT) functionalized with different organic groups to control cell growth in osteosarcoma ROS 17/2.8 cells. The authors reported that CNT carrying neutral electric charge sustained the highest cell growth and production of plate-shaped crystals. Thus, the results of these studies demonstrated that the presence of CNT in a bone scaffold can improve the cell adhesion, which is crucial for cell growth, proliferation, differentiation, and migration within the scaffold.

A different method of nucleation and growth of HA by using MWCNT was reported [38]. Inspired by self-assembly of nano-hydroxyapatite (nHA) on collagen associated with the 67 nm periodic microstructure of collagen, the authors used MWCNT with approximately 40 nm bamboo periodic microstructure as a template for nHA deposition to form a nHA–MWCNT composite, thus creating a self-assembled HA on MWCNT.

CNT-reinforced porous polyurethane nanocomposite scaffold for osteoblast growth and mineralization was also reported [117]. The nanocomposite foams, synthesized by thermally induced phase separation and loaded with CNT, did not cause any osteoblast cytotoxicity or detrimental effects on osteoblast differentiation or mineralization. Moreover, it was observed that the production of VEGF increased in proportion to CNT loading, thus revealing the potential of the nanocomposite scaffolds to influence cellular behavior.

Marrs et al., have investigated the mechanical properties of MWCNT incorporated in polymethyl methacrylate, the common polymer material for bone cement and dental prostheses [47]. The authors reported that the incorporation of MWCNT favorably alters the static and fatigue mechanical properties of the polymer offering thermal benefits and improving the longevity of the implants; this effect was attributed to the high thermal conductivity of the carbon nanotubes.

MWCNT particles were implanted into mouse skull subperiosteum and tibial bones and found that the MWCNT did not cause any major inflammatory reaction [118]. Moreover, from these studies emerged that MWCNT showed high bone tissue compatibility, permit bone repair, become integrated into new bone, and accelerate bone formation stimulated by recombinant human bone morphogenetic protein-2 (rhBMP-2).

It was also demonstrated the biocompatibility of CNT for bone regeneration [98]. MWCNT-polycarbosilane composites fabricated by the spark plasma sintering (SPS) method were implanted in the subcutaneous tissue and femur of rats at 1 and 4 weeks after implantation. The results of histological investigations showed that there was little inflammatory response in the subcutaneous tissue, and newly formed bone tissue was observed in the femur.

In a similar study, MWCNT-chitosan composite scaffolds, adsorbed with rhBMP-2 and implanted in a muscle tissue, were investigated for their ability to induce ectopic formation of bone tissue [119]. The biocompatible character of the synthesized scaffolds was remarkable according to in vitro and in vivo experiments.

In order to combine the ability of both magnetic nanoparticles and carbon nanotubes to promote bone formation, Cunha et al. reported the synthesis of specifically tailored Fe_3O_4/carboxylated MWCNT and evaluated in vitro their biocompatibility and their potential for tissue engineering applications. The synthesized magnetic hybrid nanocomposites were analyzed in vitro by incubation with MSC for 1, 3, and 7 days, either in the presence or in absence of a static magnetic field. The results of this study demonstrated that the introduction of magnetite into the MWCNT structure increases biocompatibility of oxidized MWCNT. Moreover, the authors reported that the presence of a static magnetic field further increases the Fe_3O_4/MWCNT influence on the cell behavior.

The same research group also investigated magnetic HA-based nanomaterials as bone-specific systems for controlled drug delivery. The authors investigated the biological behavior of the synthesized HA, decorated with magnetite NP by a deposition method (HA/Fe_3O_4) and the nanocomposite system obtained using magnetic MWCNT as a nanofiller for HA (HA/MWCNT/Fe_3O_4). The synthesized nanosystems were also doped with clodronate in order to combine the effect of bone biomineralization induced by HA-based composites with the decrease of osteoclast formation induced by the drug. The in vitro analysis of the preosteoclastic RAW

264.7 cell proliferation by MTT assay confirmed the high biocompatibility of the nanosystems. The clodronate-doped systems were able to release the drug in vitro, and the analysis of TRAP staining of RAW 264.7 conditioned with sRAKL to induce osteoclastogenesis, showed a higher reduction of the osteoclast formation, compared to the parent ones without clodronate.

More recently, also graphene-based materials have been explored for wound healing, stem cell engineering, regenerative medicine, and tissue engineering [120]. This nanomaterial is considered today the most amazing and versatile substance available to mankind, and since its discovery in 2004, it is without any doubt the most intensively studied material [38] [30]. Although it is just one atom thick, graphene possesses outstanding mechanical, electronic, optical, thermal, and chemical properties and the research on biomedical applications of graphene has seen dramatic progress and is expanding rapidly. Potential applications in biomedicine and in particular in drug delivery, cancer therapy, and biological imaging have been reported [121]. Moreover, the research on graphene and graphene oxide (GO)-based materials for cell culture is of particular interest. Studies in this field demonstrated that graphene and GO are able to accelerate the growth, differentiation, and proliferation of stem cells and therefore hold great promise in tissue engineering, regenerative medicine, and other biomedical fields [34,122].

Graphene-reinforced chitosan films have shown enhanced mechanical properties and no toxicity when tested on murine fibrosarcoma L929 cell culture [123]. Also, graphene oxide (GO) chitosan hydrogel scaffolds prepared by covalent conjugation of chitosan amino groups with the carboxylic groups of graphene showed a significant improvement in cell adhesion, differentiation, proliferation, and calcium phosphate deposition by mouse preosteoblast MC3T3-E1 cells [124].

Graphene-based composite materials were investigated for wound healing by preparing chitosan–PVA nanofibrous scaffolds (CS-PVA) containing graphene [45]. In this study three materials have been investigated: chitosan–PVA–graphene, electrospun fibers, and chitosan–PVA fibers without graphene, in order to assess the wound healing potential of graphene-based composite in mice and rabbit. The results of this study showed that the samples containing graphene, healed completely and at a faster rate as compared to others in both mice and rabbit.

3-D composite scaffolds were fabricated using gelatin methacrylate (GelMA) and GO [125]. The authors reported that the incorporation of GO inside the GelMA hydrogels enhanced their mechanical properties and reduced the UV-induced cell damage while preserving their favorable characteristics for 3-D cell encapsulation. Moreover, no adverse effect on encapsulated NIH-3T3 fibroblast cells was observed, demonstrating the excellent cellular viability, proliferation, spreading, and alignment of the nanosystem.

Besides wound healing, a number of studies have been also performed by exploring the use of graphene for stem cell engineering and musculoskeletal tissue engineering [62]. Moreover, the graphene functionalization with protein/peptides is expected to be very useful for tissue engineering applications and several micro- and nano-fabrication approaches can be employed to achieve spatial patterning of cells and/or proteins [126–129].

6.4 Conclusions

Hybrid composites constituted by CaP-based ceramic materials with several synthetic or natural polymers have been discussed in this chapter, summarizing and analyzing their chemical and mechanical properties as well as the results of in vitro or in vivo release studies and their biocompatibility and ability to induce bone tissue regeneration.

The next-generation ceramic/polymer nanocomposites used in bone tissue regeneration, have shown to offer new opportunities to provide more focused and fine--tuned treatment of diseases at a molecular level, enhancing their therapeutic efficacy and reducing the side effects.

By optimizing the interplay between the material used as bone scaffold and the bioactive components, the field of tissue engineering has the potential to revolutionize the treatment of bone degenerative disorders. With improved experimental animal models that are able to mimic not only a human defect but a human response, these hybrid scaffolds are able to mimic bone tissue native structures and will certainly provide efficient methods to clinically regenerate bone in the future. The close collaboration among chemists, engineers, and biologists will allow the development of highly efficient release systems and to study how the kinetics of bioactive agents released from these systems can affect bone repair.

References

[1] A.R. Amini, C.T. Laurencin, S.P. Nukavarapu, Bone tissue engineering: recent advances and challenges, Crit. Rev. Biomed. Eng. 40 (2012) 363–408.

[2] J.R. Porter, T.T. Ruckh, K.C. Popat, Bone tissue engineering: a review in bone biomimetics and drug delivery strategies, Biotechnol. Prog. 25 (2009) 1539–1560.

[3] A.J. Salgado, O.P. Coutinho, R.L. Reis, Bone tissue engineering: state of the art and future trends, Macromol. Biosci. 4 (8) (2004) 743–765.

[4] J.A. Plannell, M. Navarro, Challenges of bone repair, in: J.A. Planell, S.M. Best, D. Lacroix, A. Merolli (Eds.), Bone Repair Biomaterials, Elsevier Academic Press, 2009, pp. 1–24.

[5] M.E. Elsalanty, D.G. Genecov, Bone grafts in craniofacial surgery, Craniomaxillofacial Trauma Reconstr. 2 (3) (2009) 125–134.

[6] A.L. Oliveira, A.J. Pedro, C.S. Arroyo, J.F. Mano, G. Rodriguez, J. San Roman, et al., Biomimetic Ca-P coatings incorporating bisphosphonates produced on starch-based degradable biomaterials, J. Biomed. Mater. Res. B Appl. Biomater. 92 (2010) 55–67.

[7] V.E. Santo, M.E. Gomes, J.F. Mano, R.L. Reis, Controlled release strategies for bone, cartilage, and osteochondral engineering—part ii: challenges on the evolution from single to multiple bioactive factor delivery, Tissue Engineering: Part B 19 (4) (2013) 327–352.

[8] J.P. Cattalini, A.R. Boccaccini, S. Lucangioli, V. Mouriño, Bisphosphonate-based strategies for bone tissue engineering and orthopedic implants, Tissue. Eng. Part. B. Rev. 18 (2012) 323–340.

[9] T.N. Vo, F.K. Kasper, A.G. Mikos, Strategies for controlled delivery of growth factors and cells for boneregeneration, Adv. Drug. Deliv. Rev. 64 (2012) 1292–1309.

[10] J.E. Samorezov, E. Alsberg, Spatial regulation of controlled bioactive factor delivery for bone tissue engineering, Adv. Drug. Deliv. Rev. 84 (2015) 45–67.

[11] R.Z. Le Geros, Calcium phosphate-based osteoinductive materials, Chem. Rev. 108 (2008) 4742–4753.

[12] D. Bellucci, A. Sola, M. Gazzarri, F. Chiellini, V. Cannillo, A new hydroxyapatite-basedbiocompositeforbonereplacement, Mater. Sci. Eng. C, Mater. Biol. Appl. 33 (2013) 1091–1101.

[13] S. Bansal, V. Chauhan, S. Sharma, R. Maheshwari, A. Juyal, S. Raghuvanshi, Evaluation of hydroxyapatite and beta-tricalcium phosphate mixed with bone marrow aspirate as a bone graft substitute for posterolateral spinal fusion, Indian J. Orthopaedics 43 (2009) 234–239.

[14] W. Paul, C.P. Sharma, Ceramic drug delivery: a perspective, J. Biomater. Appl. 17 (2003) 253–264.

[15] W.E. Teo, S. Liao, C. Chan, S. Ramakrishna, Fabrication and characterization of hierarchically organized nanoparticle-reinforced nanofibrous composite scaffolds, Acta. Biomater. 7 (1) (2011) 193–202.

[16] M. Pilia, T. Guda, M. Appleford, Development of composite scaffolds for load-bearing segmental bone defects, BioMed Res. Int. 458253 (2013), pp. 15.

[17] J.F. Osborn, H. Newesely, The material science of calcium phosphate ceramics, Biomaterials. 1 (2) (1980) 108–111.

[18] S.V. Dorozhkin, Nanodimensional and nanocrystallineapatites and other calcium orthophosphates in biomedical engineering, biology and medicine, Materials 2 (2009) 1975–2045.

[19] M. Prakasam, J. Locs, K. Salma-Ancane, D. Loca, A. Largeteau, L. Berzina-Cimdina, Fabrication, properties and applications of dense hydroxyapatite: a review, J. Funct. Biomater. 6 (2015) 1099–1140.

[20] S.V. Dorozhkin, Calcium orthophosphate-containing biocomposites and hybrid biomaterials for biomedical applications, J. Funct. Biomater. 6 (2015) 708–832.

[21] W.J. Habraken, J.G. Wolke, J.A. Jansen, Ceramic composites as matrices and scaffolds for drug deliveryin tissue engineering, Adv. Drug. Deliv. Rev. 59 (2007) 234–248.

[22] D.W. Hutmacher, Scaffolds in tissue engineering bone and cartilage, Biomaterials. 21 (2000) 2529–2543.

[23] J.A. Burdick, M.N. Mason, A.D. Hinman, K. Thorne, K.S. Anseth, Delivery of osteoinductive growth factors from degradable PEG hydrogels influences osteoblast differentiation and mineralization, J. Controlled Release 83 (2002) 53–63.

[24] T.A. Holland, A.G. Mikos, Biodegradable polymeric scaffolds. Improvements in bone tissue engineering through controlled drug delivery, Adv. Biochem. Eng. Biotechnol. 102 (2006) 161–185.

[25] S.Y. Ivanov, R.F. Mukhametshin, A.A. Muraev, D.V. Solodkaya, Synthetic materials used for the substitutionof bone defects: Critical review, Ann. Oral Maxillofacial Surg. 1 (1) (2013) 1–4.

[26] I.Y. Kim, S.J. Seo, H.S. Moon, M.K. Yoo, I.Y. Park, B.C. Kim, et al., Chitosan and its derivatives for tissue engineering applications, Biotechnol. Adv. 26 (2008) 1–21.

[27] V. Mouriño, J.P. Cattalini, J.A. Roether, P. Dubey, I. Roy, A.R. Boccaccini, Composite polymer-bioceramic scaffolds with drug delivery capability for bone tissue engineering, Expert. Opin. Drug. Deliv. 10 (2013) 1353–1365.

[28] W.M. Saltzman, W.L. Olbricht, Building drug delivery into tissue engineering, Nat. Rev. Drug. Discov. 1 (2002) 177–186.

[29] H.S. Yu, G.Z. Jin, J.E. Won, I. Wall, H.M. Kim, Macrochanneled bioactive ceramic scaffolds in combination with collagen hydrogel: a new tool for bone tissue engineering, J. Biomed. Mater. Res. A. 100 (2012) 2431–2440.

[30] D. Iannazzo, A. Pistone, S. Galvagno, S. Ferro, L. De Luca, A.M. Monforte, et al., Synthesis and anti-HIV activity of carboxylated and drug conjugated multi-walled carbon nanotubes, Carbon. N. Y. 82 (2015) 548–561.

[31] Z. He, L. Xiong, Assessment of Cytocompatibility of bulk modified poly-l-latic acid biomaterials, Polymer-Plastics Technol. Eng. 48 (10) (2009) 1020–1024.

[32] D. Menon, Anoop, K. Anand, V.C. Anitha, S. Nair, Hydroxyapatitereinforced polyamide 6,6 nanocomposites through melt compounding, Int. J. Polym. Mater. Polym. Biomater. 59 (7) (2010) 498–509.

[33] J.J. Kim, R.K. Singh, S.J. Seo, T.H. Kim, J.H. Kim, E.J. Lee, et al., Magnetic scaffolds of polycaprolactone with functionalized magnetite nanoparticles: physicochemical, mechanical, and biological properties effective for bone regeneration, RSC Adv. 4 (2014) 17325–17336.

[34] M. Kalbacova, A. Broz, J. Kong, M. Kalbac, Graphene substrates promote adherence of human osteoblasts and mesenchymal stromal cells, Carbon. N. Y. 48 (15) (2010) 4323–4329.

[35] U. Kneser, A. Voogd, J. Ohnolz, O. Buettner, L. Stangenberg, Y.H. Zhang, et al., Fibrin gel-immobilized primary osteoblasts in calcium phosphate bone cement: in vivo evaluation with regard to application as injectable biological bone substitute, Cells. Tissues. Organs. 179 (2005) 158–169.

[36] D. Le Nihouannen, L.L. Guehennec, T. Rouillon, P. Pilet, M. Bilban, P. Layrolle, et al., Micro-architecture of calcium phosphate granules and fibrin glue composites for bone tissue engineering, Biomaterials. 27 (2006) 2716–2722.

[37] Y.M. Lee, Y.J. Park, S.J. Lee, Y. Ku, S.B. Han, P.R. Klokkevold, et al., The bone regenerative effect of platelet-derived growthfactor-BB delivered with a chitosan/tricalcium phosphate sponge carrier, J. Periodontol. 71 (2000) 418–424.

[38] S. Liao, G. Xu, W. Wang, F. Watari, F. Cui, S. Ramakrishna, et al., Self-assembly of nano-hydroxyapatite on multi-walled carbon nanotubes, Acta. Biomater. 3 (2007) 669–675.

[39] E.L. Liljensten, A.G. Attaelmanan, C. Larsson, H. Ljusberg-Wahren, N. Danielsen, J. M. Hirsch, et al., Hydroxyapatite granule/carrier composites promote new bone formation in cortical defects, Clin. Implant. Dent. Relat. Res. 2 (2000) 50–59.

[40] J. Lin, S. Zhang, T. Chen, C. Liu, S. Lin, X. Tian, Calcium phosphate cement reinforced by polypeptide copolymers, J. Biomed. Mater. Res. Part B Appl. Biomater. 76 (2006) 432–439.

[41] C. Liu, Z. Xia, J.T. Czernuszka, Design and development of three-dimensional scaffolds for tissue engineering, Chem. Eng. Res. Design 85 (2007) 1051–1064.

[42] H. Liu, L. Zhang, P. Shi, Q. Zou, Y. Zuo, Y. Li, Hydroxyapatite/polyurethane scaffold incorporated with drug-loaded ethyl cellulose microspheres for bone regeneration, J. Biomed. Mater. Res. Part B Appl. Biomater. 95 (2010) 36–46.

[43] X. Liu, P.X. Ma, Polymeric scaffolds for bone tissue engineering, Ann. Biomed. Eng. 32 (3) (2004) 477–486.

[44] A. Lode, C. Wolf-Brandstetter, A. Reinstorf, A. Bernhardt, U. König, W. Pompe, et al., Calcium phosphate bone cements, functionalized with VEGF: release kinetics and biological activity, J. Biomed. Mater. Res. A. 81 (2007) 474–483.

[45] B. Lu, T. Li, H. Zhao, X. Li, C. Gao, S. Zhang, et al., Graphene-based composite materials beneficial to wound healing, Nanoscale 4 (2012) 2978–2982.

[46] S. Marchesan, M. Prato, Under the lens: carbon nanotube and protein interaction at the nanoscale, Chem. Commun. 51 (2015) 4347–4359.
[47] B. Marrs, R. Andrews, T. Rantell, D. Pienkowski, Augmentation of acrylic bone cement with multiwall carbon nanotubes, J. Biomed. Mater. Res. A 77 (2006) 269–276.
[48] Y. Xiong, C. Ren, B. Zhang, H. Yang, Y. Lang, L. Min, et al., Analyzing the behavior of a porous nano-hydroxyapatite/polyamide 66 (n-HA/PA66) composite for healing of bone defects, Int. J. Nanomedicine. 9 (2014) 485–494.
[49] J. Shen, Y. Li, Y. Zuo, Q. Zou, L. Cheng, P. Shi, et al., Bioactive ceramic/polyamide 6 scaffold for bone regeneration: in vitro and in vivo evaluation, Polymer-Plastics Technol. Eng. 50 (2011) 1367–1374.
[50] J. Wei, S.J. Heo, D.H. Kim, S.E. Kim, Y.T. Hyun, J.W. Shin, Comparison of physical, chemical and cellular responses to nano- and micro-sized calcium silicate/poly(e-caprolactone) bioactive composites, J. R. Soc. Interface 5 (23) (2008) 617–630.
[51] M.J. Mondrinos, R. Dembzynski, L. Lu, V.K.C. Byrapogu, D.M. Wootton, P.I. Lelkes, et al., Porogen-based solid freeform fabrication of polycaprolactone-calcium phosphate scaffolds for tissue engineering, Biomaterials. 27 (2006) 4399–4408.
[52] M. Navarro, M.P. Ginebra, J.A. Planell, S. Zeppetelli, L. Ambrosio, Development and cell response of a new biodegradable composite scaffold for guided bone regeneration, J. Mater. Sci. Mater. Med. 15 (2004) 419–422.
[53] Y. Shikinami, Y. Matsusue, T. Nakamura, The complete process of bioresorption and bone replacement using devices made of forged composites of raw hydroxyapatite particles/poly l-lactide (F-u-HA/PLLA), Biomaterials. 26 (2005) 5542–5551.
[54] J. Russias, E. Saiz, R.K. Nalla, K. Gryn, R.O. Ritchie, A.P. Tomsia, Fabrication and mechanical properties of PLA/HA composites: a study of in vitro degradation, Mater. Sci. Eng. C 26 (2006) 1289–1295.
[55] J.S. Son, M. Appleford, J.L. Ong, J.C. Wenke, J.M. Kim, S.H. Choi, et al., Porous hydroxyapatite scaffold with three-dimensional localized drug delivery system using biodegradable microspheres, J. Controlled Release 153 (2011) 133–140.
[56] A. Pistone, D. Iannazzo, S. Panseri, M. Montesi, A. Tampieri, S. Galvagno, Hydroxyapatite-magnetite-MWCNT nanocomposite as a biocompatible multifunctional drug delivery system for bone tissue engineering, Nanotechnology. 25 (425701) (2014), pp. 9.
[57] D. Puppi, A.M. Piras, F. Chiellini, E. Chiellini, A. Martins, I.B. Leonor, et al., Optimized electro- and wet-spinning techniques for the production of polymeric fibrous scaffolds loaded with bisphosphonate and hydroxyapatite, J. Tissue. Eng. Regen. Med. 5 (2011) 253–263.
[58] Y. Wang, X. Shi, L. Ren, Y. Yao, D.A. Wang, In vitro osteogenesis of synoviummesenchymal cells induced by controlled release of alendronate and dexamethasone from a sintered microspherical scaffold, J. Biomater. Sci. Polym. Ed 21 (2010) 1227–1238.
[59] P.Q. Ruhé, O.C. Boerman, F.G. Russel, P.H. Spauwen, A.G. Mikos, et al., Controlled release of rhBMP-2 loaded poly(dl-lactic-co-glycolic acid)/calcium phosphate cement composites in vivo, J. Control Release 106 (2005) 162–171.
[60] W. Xue, A. Bandyopadhyay, S. Bose, Polycaprolactone coated porous tricalcium phosphate scaffolds for controlled release of protein for tissue engineering, J. Biomed. Mater. Res. Part B Appl. Biomater. 91 (2) (2009) 831–838.
[61] B. Rai, S. Teoh, D. Hutmacher, T. Cao, K. Ho, Novel PCL-based honeycomb scaffolds as drug delivery systems for rhBMP-2, Biomaterials. 26 (2005) 3739–3748.
[62] M. Chen, D.Q. Le, S. Hein, P. Li, J.V. Nygaard, M. Kassem, et al., Fabrication and characterization of a rapid prototyped tissue engineering scaffold with embedded

multicomponent matrix for controlled drug release, Int. J. Nanomedicine. 7 (2012) 4285–4297.

[63] H.W. Kim, J.C. Knowles, H.E. Kim, Hydroxyapatite/poly(-caprolactone) composite coatings on hydroxyapatite porous bone scaffold for drug delivery, Biomaterials. 25 (2004) 1279–1287.

[64] I. Soriano, C. Evora, Formulation of calcium phosphates/poly (d,l-lactide) blends containing gentamicinfor bone implantation, J. Controlled Release 68 (2000) 121–134.

[65] J. Schnieders, U. Gbureck, R. Thull, T. Kissel, Controlled release of gentamicin from calcium phosphatepoly (lactic acid-co-glycolic acid) composite bone cement, Biomaterials. 27 (2006) 4239–4249.

[66] C. Castro, C. Evora, M. Baro, I. Soriano, E. Sánchez, Two-month ciprofloxacin implants for multibacterialbone infections, Eur. J. Pharmaceut. Biopharmaceut. 60 (2005) 401–406.

[67] T. Miyai, A. Ito, G. Tamazawa, T. Matsuno, Y. Sogo, C. Nakamura, et al., Antibiotic-loaded poly-epsilon-caprolactone andporous beta-tricalcium phosphate composite for treating osteomyelitis, Biomaterials. 29 (2008) 350–358.

[68] H.W. Kim, J.C. Knowles, H.E. Kim, Hydroxyapatite porous scaffold engineered with biological polymerhybrid coating for antibiotic Vancomycin release, J. Mater. Sci. Mater. Med. 16 (2005) 189–195.

[69] S. Panseri, C. Cunha, T. D'Alessandro, M. Sandri, A. Russo, G. Giavaresi, et al., Magnetic hydroxyapatite bone substitutes to enhance tissue regeneration: evaluation in vitro using osteoblast-like cells and in vivo in a bone defect, PLoS. ONE. 7 (6) (2012) e38710.

[70] A.L. Torres, V.M. Gaspar, I.R. Serra, G.S. Diogo, R. Fradique, A.P. Silva, et al., Bioactive polymeric–ceramic hybrid 3D scaffold for application in bone tissue regeneration, Mater. Sci. Eng. C 33 (2013) 4460–4469.

[71] Y.J. Seol, J.Y. Park, W. Jeong, T.H. Kim, S.Y. Kim, D.W. Cho, Development of hybrid scaffolds using ceramic and hydrogel for articular cartilage tissue regeneration, J. Biomed. Mater. Res. A 103A (2015) 1404–1413.

[72] H.H. Xu, C.G. Simon, Fast setting calcium phosphate-chitosan scaffold: mechanical properties and biocompatibility, Biomaterials. 26 (2005) 1337–1348.

[73] S. Takagi, L.C. Chow, S. Hirayama, F.C. Eichmiller, Properties of elastomeric calcium phosphate cementchitosan composites, Dental Mater. 19 (2003) 797–804.

[74] M. Takechi, Y. Miyamoto, Y. Momota, T. Yuasa, S. Tatehara, M. Nagayama, et al., The in vitro antibiotic release from anti-washout apatite cement using chitosan, J. Mater. Sci. Mater. Med. 13 (2002) 973–978.

[75] Y. Zhang, M. Zhang, Cell growth and function on calcium phosphate reinforced chitosan scaffolds, J. Mater. Sci. Mater. Med. 15 (2004) 255–260.

[76] Y. Zhang, M. Zhang, Calcium phosphate/chitosan composite scaffolds for controlled in vitro antibioticdrug release, J. Biomed. Mater. Res. 62 (2002) 378–386.

[77] S.H. Teng, E.J. Lee, P. Wang, S.H. Jun, C.M. Han, H.E. Kim, Functionally gradient chitosan/hydroxyapatitecomposite scaffolds for controlled drug release, J. Biomed. Mater. Res. Part B Appl. Biomater. 90 (2009) 275–282.

[78] B.H. Fellah, P. Weiss, O. Gauthier, T. Rouillon, P. Pilet, G. Daculsi, et al., Bone repair using a new injectable self cross linkable bone substitute, J. Orthopedic Res. 24 (2006) 628–635.

[79] E.F. Burguera, H.H. Xu, M.D. Weir, Injectable and rapid-setting calcium phosphate bone cement withdicalcium phosphate dihydrate, J. Biomed. Mater. Res. B Appl. Biomater. 77 (2006) 126–134.

[80] B. Ongpipattanakul, T. Nyugen, T.F. Zioncheck, R. Wong, G. Osaka, L. De Guzman, et al., Development of tricalcium phosphate/amylopectin paste combined with recombinant human transforming growth factor beta 1 as abone defect filler, J. Biomed. Mater. Res. 36 (1997) 295–305.
[81] M.V. Cabañas, J. Peña, J. Román, M. Vallet-Regí, Tailoring vancomycin release from beta-TCP/agarose scaffolds, Eur. J. Pharmaceut. Sci. 37 (2009) 249–256.
[82] S. Subramaniam, Y.H. Fang, S. Sivasubramanian, F.H. Lin, C.P. Lin, Hydroxyapatite-calcium sulfate-hyaluronic acid composite encapsulated with collagenase as bone substitute for alveolar bone regeneration, Biomaterials. 74 (2015) 99–108.
[83] G. Tozzi, A. De Mori, A. Oliveira, M. Roldo, Composite hydrogels for bone regeneration, Materials 9 (2016) 267, (pp. 24).
[84] M.D. Shoulders, R.T. Raines, Collagen structure and stability, Annu. Rev. Biochem. 78 (2009) 929–958.
[85] S. Itoh, M. Kikuchi, Y. Koyama, K. Takakuda, K. Shinomiya, J. Tanaka, Development of a hydroxyapatite/collagen nanocomposite as a medical device, Cell. Transplant. 13 (2004) 451–461.
[86] L. Ciocca, I.G. Lesci, O. Mezini, A. Parrilli, S. Ragazzini, R. Rinnovati, et al., Customized hybrid biomimetic hydroxyapatite scaffold for bone tissue regeneration, J. Biomed. Mater. Res. B Appl. Biomater. 00B (2015), pp. 12.
[87] A. Ayvazyan, N. Morimoto, N. Kanda, S. Takemoto, K. Kawai, Y. Sakamoto, et al., Collagen-gelatin scaffold impregnated with bFGF accelerates palatal wound healing of palatal mucosa in dogs, J. Surg. Res. 171 (2011) e247–e257.
[88] A. Tampieri, G. Celotti, E. Landi, M. Montevecchi, N. Roveri, A. Bigi, et al., Porous phosphate-gelatine composite as bone graft with drug delivery function, J. Mater. Sci. Mater. Med. 14 (2003) 623–627.
[89] A. Bigi, S. Panzavolta, L. Sturba, P. Torricelli, M. Fini, R. Giardino, Normal and osteopenic bone-derived osteoblast response to a biomimetic gelatin-calcium phosphate bone cement, J. Biomed. Mater. Res. A. 78 (2006) 739–745.
[90] Y. Takahashi, M. Yamamoto, Y. Tabata, Enhanced osteoinduction by controlled release of bone morphogenic protein-2 from biodegradable sponge composed of gelatin and ß-tricalcium phosphate, Biomaterials. 26 (2005) 4856–4865.
[91] J. Zhou, T. Fang, Y. Wang, J. Dong, The controlled release of vancomycin in gelatin/Î²-TCP composite scaffolds, J. Biomed. Mater. Res. A 100 (2012) 2295–2301.
[92] F. Jegoux, E. Goyenvalle, M. Bagot D'arc, E. Aguado, G. Daculsi, In vivo biological performance of composites combining micro-macroporous biphasic calcium phosphate granules and fibrin sealant, Arch. Orthop. Trauma. Surg. 125 (2005) 153–159.
[93] C.M. Vaz, M. Fossen, R.F. van Tuil, L.A. de Graaf, R.L. Reis, A.M. Cunha, Casein and soybean protein-based thermoplastics and composites as alternative biodegradable polymers for biomedical applications, J. Biomed. Mater. Res. A. 65 (2003) 60–70.
[94] R. Langer, N.A. Peppas, Advances in biomaterials, drug delivery, and bionanotechnology, AIChE J. 49 (2003) 2990–3006.
[95] P. Wang, L. Zhao, J. Liu, M.D. Weir, X. Zhou, H.H.K. Xu, Bone tissue engineering via nanostructured calcium phosphate biomaterials and stem cells, Bone Res. 2 (2014) 14017, (pp. 22).
[96] Y. Sasaki, S. Yamane, K. Kurosu, S.I. Sawada, K. Akiyoshi, Templated formation of hydroxyapatite nanoparticles from self-assembled nanogels containing tricarboxylate groups, Polymers 4 (2012) 1056–1064.
[97] T. Dvir, B.P. Timko, D.S. Kohane, R. Langer, Nanotechnological strategies for engineering complex tissues, Nature Nanotechnol. 6 (2011) 13–22.

[98] W. Wang, F. Watari, M. Omori, S. Liao, Y. Zhu, A. Yokoyama, et al., Mechanical properties and biological behavior of carbon nanotube/polycarbosilane composites for implant materials, J. Biomed. Mater. Res. Part B Appl. Biomater. 82 (2007) 223–230.
[99] H. Valiense, M. Barreto, R.F. Resende, A.T. Alves, A.M. Rossi, E. Mavropoulos, et al., In vitro and in vivo evaluation of strontium-containing nanostructured carbonated hydroxyapatite/sodium alginate for sinus lift in rabbits, J. Biomed. Mater. Res. Part B Appl. Biomater. 104 (2) (2015) 274–282.
[100] F. Zhao, D. Yao, R. Guo, L. Deng, A. Dong, J. Zhang, Composites of polymer hydrogels and nanoparticulate systems for biomedical and pharmaceutical applications, Nanomaterials 5 (4) (2015) 2054–2130.
[101] X.U. Hai-Yan, G.U. Ning, Magnetic responsive scaffolds and magnetic fields in bone repair and regeneration, Front. Mater. Sci. 8 (2014) 20–31.
[102] N. Bock, A. Riminucci, C. Dionigi, A. Russo, A. Tampieri, E. Landi, et al., A novel route in bone tissue engineering: magnetic biomimetic scaffolds, Acta. Biomater. 6 (2010) 786–796.
[103] A. Tampieri, M. Iafisco, M. Sandri, S. Panseri, C. Cunha, S. Sprio, et al., Magnetic bioinspired hybrid nanostructured collagen-hydroxyapatite scaffolds supporting cell proliferation and tuning regenerative process, ACS Appl. Mater. Interfaces 6 (2014) 15697–15707.
[104] R.K. Singh, K.D. Patel, J.H. Lee, E.J. Lee, J.H. Kim, T.H. Kim, et al., Potential of magnetic nanofiber scaffolds with mechanical and biological properties applicable for bone regeneration, PLoS. ONE. 9 (2014) e91584.
[105] R. De Santis, A. Russo, A. Gloria, U. D'Amora, T. Russo, S. Panseri, et al., Towards the design of 3D fiber-deposited poly(-caprolactone)/iron-doped hydroxyapatite nanocomposite magnetic scaffolds for bone regeneration, J. Biomed. Nanotechnol. 11 (2015) 1236–1246.
[106] J. Meng, B. Xiao, Y. Zhang, J. Liu, H. Xue, J. Lei, et al., Super-paramagnetic responsive nanofibrous scaffolds under static magnetic field enhance osteogenesis for bone repair in vivo, Sci. Rep. 3 (2013) 2655.
[107] J.V. Veetil, K. Ye, Tailored carbon nanotubes for tissue engineering applications, Biotechnol. Prog. 25 (2009) 709–721.
[108] D. Iannazzo, A. Piperno, A. Ferlazzo, A. Pistone, C. Milone, M. Lanza, et al., Functionalization of multi-walled carbon nanotubes with coumarin derivatives and their biological evaluation, Org. Biomol. Chem. 10 (2012) 1025–1031.
[109] D. Iannazzo, A. Piperno, A. Pistone, G. Grassi, S. Galvagno, Recent advances in carbon nanotubes as delivery systems for anticancer drugs, Curr. Med. Chem. 20 (2013) 1333–1354.
[110] S. Baldo, S. Buccheri, A. Ballo, M. Camarda, A. La Magna, M.E. Castagna, et al., Carbon nanotube-based sensing devices for human Arginase-1 detection, Sensing Bio-Sensing Res. 7 (2016) 168–173.
[111] A. Pistone, A. Piperno, D. Iannazzo, N. Donato, M. Latino, D. Spadaro, et al., Fe_3O_4–MWCNT PhCOOH composites for ammonia resistive sensors, Sensors Actuat. B 186 (2013) 333–342.
[112] D. Tasis, N. Tagmatarchis, A. Bianco, M. Prato, Chemistry of carbon nanotubes, Chem. Rev. 106 (3) (2006) 1105–1136.
[113] L.P. Zanello, B. Zhao, H. Hu, R.C. Haddon, Bone cell proliferation on carbon nanotubes, Nano. Lett. 6 (2006) 562–567.
[114] B. Zhao, H. Hu, S.K. Mandal, R.C. Haddon, A bone mimic based on the self-assembly of hydroxyapatite on chemically functionalized single-walled carbon nanotubes, Chem. Mater. 17 (2005) 3235–3241.

[115] C. Cunha, S. Panseri, D. Iannazzo, A. Piperno, A. Pistone, M. Fazio, et al., Hybrid composites made of multiwalled carbon nanotubes functionalized with Fe_3O_4 nanoparticles for tissue engineering applications, Nanotechnology. 23 (2012) 465102.

[116] G. Grassi, A. Scala, A. Piperno, D. Iannazzo, M. Lanza, C. Milone, et al., A facile and ecofriendly functionalization of multiwalled carbon nanotubes by an old mesoionic compound, Chem. Commun. 48 (2012) 6836–6838.

[117] G. Jell, R. Verdejo, L. Safinia, M.S.P. Shaffer, M.M. Stevens, A. Bismarck, Carbon nanotube-enhanced polyurethane scaffolds fabricated by thermally induced phase separation, J. Mater. Chem. 18 (2008) 1865–1872.

[118] Y. Usui, K. Aoki, N. Narita, N. Murakami, I. Nakamura, N. Ishigaki, et al., Carbon nanotubes with high bone-tissue compatibility and bone-formation acceleration effects, Small. 4 (2008) 240–246.

[119] A. Abarrategi, M.C. Gutiérrez, C. Moreno-Vicente, M.J. Hortigüela, V. Ramos, J.L. López-Lacomba, et al., Multiwall carbon nanotube scaffolds for tissue engineering purposes, Biomaterials. 29 (2008) 94–102.

[120] S. Goenka, V. Sant, S. Sant, Graphene-based nanomaterials for drug delivery and tissue engineering, J. Controlled Release 173 (2014) 75–88.

[121] L.M. Zhang, J.G. Xia, Q.H. Zhao, L. Liu, Z. Zhang, Functional graphene oxide as a nanocarrier for controlled loading and targeted delivery of mixed anticancer drugs, Small. 6 (4) (2010) 537–544.

[122] T.R. Nayak, H. Andersen, V.S. Makam, C. Khaw, S. Bae, X. Xu, Graphene for controlled and accelerated osteogenic differentiation of human mesenchymal stem cells, ACS Nano 5 (6) (2011) 4670–4678.

[123] H. Fan, L. Wang, K. Zhao, N. Li, Z. Shi, Z. Ge, et al., Fabrication, mechanical properties, and biocompatibility of graphene-reinforced chitosan composites, Biomacromolecules. 11 (2010) 2345–2351.

[124] D. Depan, B. Girase, J.S. Shah, R.D.K. Misra, Structure process property relationship of the polar graphene oxide-mediated cellular response and stimulated growth of osteoblasts on hybrid chitosan network structure nanocomposite scaffolds, Acta. Biomater. 7 (2011) 3432–3445.

[125] S.R. Shin, B. Aghaei-Ghareh-Bolagh, T.T. Dang, S.N. Topkaya, X. Gao, S.Y. Yang, et al., Cell-laden microengineered and mechanically tunable hybrid hydrogels of gelatin and graphene oxide, Adv. Mater. 25 (2013) 6385–6391.

[126] H. Bae, H. Chu, F. Edalat, J.M. Cha, S. Sant, A. Kashyap, et al., Development of functional biomaterials with micro- and nanoscale technologies for tissue engineering and drug delivery applications, J. Tissue. Eng. Regen. Med. 8 (2014) 1–14.

[127] S. Sant, S.L. Tao, O.Z. Fisher, Q. Xu, N.A. Peppas, A. Khademhosseini, Microfabrication technologies for oral drug delivery, Adv. Drug. Deliv. Rev. 64 (2012) 496–507.

[128] D. Iannazzo, A. Pistone, S. Galvagno, Functionalization Methods of Graphene, in: Vijay Kumar Thakur, ManjuKumari Thakur (Eds.), Chemical Functionalization of Carbon Nanomaterials: Chemistry and Applications, CRC Press, Boca Raton, 2015, pp. 510–537.

[129] D. Iannazzo, A. Pistone, C. Espro, S. Galvagno, Drug Delivery Strategies for Bone Tissue Regeneration, in: Silvia Panseri, Dr. Francesca Taraballi, Dr. Carla Cunha (Eds.), Biomimetic Approaches for Tissue Healing, Published by OMICS Group eBooks, 731 Gull Ave, Foster City, CA 94404, USA, 2015, pp. 1–39.

Natural and synthetic fillers for reaching high performance and sustainable hybrid polymer composites

7

Daniela de França da Silva Freitas, Sibele P. Cestari and Luis C. Mendes
Universidade Federal do Rio de Janeiro (UFRJ), Rio de Janeiro, RJ, Brazil

Chapter Outline

7.1 Introduction

During the past 50–70 years, a significant advance was made in the petrochemical industry that led to the development of different polymers and copolymers. They represent a hydrocarbon material class that is similar to other materials in their high molecular weight and low-density (0.90–2.2 g/cm^{-3}) but much smaller than conventional materials, such as steel, wood, iron, etc. Whether natural or synthetic, there is a significant diversity in the polymer origin, number of monomers, method of preparation, chemical structure, chain configuration, and tacticity of the polymeric chain. There are still differences on fusibility, solubility, and mechanical behavior. From the middle of the 1970s, the polymeric blends were introduced into the market. They tried to combine the qualities of two or more polymers in applications performed previously by homopolymers and copolymers.

Although these polymers have a low specific weight and fewer issues with regard to processability as compared to other metals, most of the polymers have mechanical properties inferior to other materials when it comes to high performance application. In an attempt to replace traditional engineering materials, and considering the qualities and deficiencies of polymers, researchers in academia and industry have developed a new material: a so-called polymer composite.

Hybrid Polymer Composite Materials: Processing. DOI: http://dx.doi.org/10.1016/B978-0-08-100789-1.00007-1

According to IUPAC [1], a composite is a multicomponent material comprising multiple different (nongaseous) phase domains in which at least one type of phase domain is a continuous phase. Also, it was defined that a polymer composite is a composite in which at least one component is a polymer.

Generally speaking, polymer composites represent a class of heterogeneous materials, multiphase, in which one of the components, discontinuous, provides the primary resistance to stress (structural component) and the other, usually solid, acts as a means of stress transfer (matrix component). The set of properties is characterized by the contribution of each individual component [2].

Independent of the chemical structure, a thermoplastic or thermoset polymer must be considered as a matrix component. The mechanical requirements of the polymer are related to the application and performance of the final artifact. It should not be forgotten that the polymers are susceptible to the action of chemical and physical agents that can cause premature breakdown of properties.

Concerning the structural component, there is a large spectrum of material that may be applied in polymer composites. With respect to its chemical nature, it can be organic (cellulose, carbon, and graphite fibers) and inorganic (glass, steel, and metals). As regards the shape, they may be fibers, flakes, or powders. In the case of fibers, the properties of polymer composites is related to the dimensions, short or long, and if they were manipulated for the production of woven and nonwoven fibers. The main requirements for the structural materials are strength, stiffness, and ductility. These properties, together or individually, are responsible for supporting the artifact to prevent deformities beyond an acceptable limit.

It must be kept in mind that the higher the interaction between polymer matrix and the structural component is the better the transmission of shear forces and avoiding breakdown of the artifact. A more intimate interfacial region is necessary to ensure high performance. The interfacial region may be reduced by chemical modification of the matrix or structural component, or both. Another alternative is the addition of a component of low or high molar mass called a compatibilizing agent.

In general terms, the reinforced polymer composites may be classified according to the form of the structural component: particulate, fibrous, and laminate. Fibrous composites are based on natural/biofibers or synthetic fibers and they are most commonly applied. Green composites are referred to the materials in which both structural and matrix component are biodegradable. Low cost and simple fabrication methods are attributes of polymer matrix composites. If nonreinforced polymers are the material matrix, the level of strength, modulus, and impact resistance is limited. If a strong fibrous network is used, polymer composites are characterized by high specific strength and stiffness, high fracture resistance, good abrasion, impact, corrosion, and fatigue resistances [3].

When a rigid particulate is added to a polymer matrix, according to Ahmed and Jones [4], material with increased stiffness, a reduction of the coefficient of thermal expansion, and an improvement in creep resistance and fracture toughness are achieved. With respect to the modulus, it comprises a complex relationship among the properties of individual constituent phases: resin, filler, and the interfacial region. Parameters, such as size, shape, aspect ratio and distribution of the reinforcing particles, play a very pronounced role on the mechanical properties. Particularly, the applied stress is strongly dependent on the orientation degree for nonspherical particles [4].

For decades, the study of polymer composites loaded with inorganic or organic filler was limited to macrocomposites where filler length was in micrometer scale. From the middle of the 1980s, nanocomposites appeared: materials that in at least one dimension are nanoscale. Toyota Central Research was the first group to produce a nanocomposite based on polyamide-6 containing a small amount of the lamellar clay and a quaternary ammonium salt with a long aliphatic chain. The work served as the basis for triggering the search for nanocomposites. A generation of various commercial products with mechanical properties and chemical-resistance has stimulated interest in academia and industry to develop other polymer/nanoclay systems [5,6,7]. The modified clay was recognized as a nanoparticle. Its dimensional scale and large surface area per unit volume allowed higher molecular interaction between polyamide-6 and modified clay nanoparticle. This induced an interaction that produced a hybrid polymeric nanocomposite with properties much higher than the polymer alone and any conventional macrocomposite containing the same polymer [8,9].

The research has focused on the use of one-dimensional lamellar fillers (generally inorganic), mainly recognized by anisotropic properties. These lamellae compounds generally are hydrophilic and needs prior chemical modification to make them compatible with the polymers [10].

The intercalation of organic molecules in the galleries of lamellar fillers is a procedure of modification in order to control chemical and physico-chemical products derived from the intercalation reactions. The properties of the lamellar charges are modified by changing the electron density between the species involved [11,12]. The applicability of the process consists in obtaining nanocomposite—inorganic filler/polymer—where the dispersibility of the polymeric matrix within the filler lamellae promotes a large effect on the mechanical properties as well as the thermal and permeability properties of nanomaterial [11].

In contrast to microcomposites, where the filler is in micrometers scale, the polymer nanocomposites may comprise fillers in the form of leaves—nanolayers; only one dimension in the nanoscale, tube–nanotube—two dimensions are in the nanometer range and nanoparticle in which three dimensions are on the nanometer scale. To be considered a polymer nanocomposite, the material should contain one component in dimension at least 100 nm. Due to a relatively low quantity of filler, there is no loss in processability and there is also greater dispersibility of the nanoclay in the polymeric matrix. This leads to changes in the polymer matrix that gives rise to new structures and properties [13].

Some research papers have also reported on the development of hybrid polymer composites and nanocomposites loaded with organic and/or inorganic fillers. Many of the works are related to environmental conservation: the reuse of materials discarded in some industrial process as municipal solid waste.

7.2 Hybrid polymer composites with natural fillers

According to the Cambridge Dictionary, filler might be a substance that is used to fill small holes and cracks, especially in wood and walls. It also could be understood as a substance that's added to any formulation—polymer, a polymer composite, ceramics, paint, varnish, etc.—that results in the increasing of bulk, weight, viscosity, opacity, strength, and so on. Natural fillers comprise a series of substance: inorganic or organic. Because of their intrinsic qualities they have been widely used in the formulation of hybrid polymer composites.

In the dentistry area, restorative hybrid polymeric composites find large applications. The influence of the polymerization modes on the degree of conversion (DC) and Vickers microhardness (VMH) is usually investigated in order to assess the performance of the composite. Those properties were evaluated using commercial dual-curing self-adhesive resin cement. The specimens were separated in four different groups, changing the light irradiation time and mode of initiation reaction between them and also applying a device as a light barrier. DC and VMH were higher for specimens polymerized with dual-polymerization (light and chemical polymerization) than other specimen in that only a chemical polymerization (dark cure) was applied. The presence of a light barrier device did not affect the DC but reduced the VMH. The variation of light exposure did not change DC but raised slightly VMH indicating its dependence on the incident energy [14].

Four commercial dental restorative hybrid polymeric materials (resin composite, polyacid-modified resin composite, ormocer, and resin-modified glass ionomer cement) were submitted to the action of acid medium in order to evaluate the degradation products by high performance liquid chromatography (HPLC) and contact angle (θ) measurements. All materials degraded on a caries-simulated medium. It was suggested that a great effort should be made to disseminate oral health information since a high-caries challenge environment (low pH) can lead to dental composite degradation with potential toxic risks to the patients [15].

The concept of sustainability must be in our mind in order to preserve the planet for the next generations. In Brazil, the sugarcane industry releases byproducts from sugar and alcohol production. One of these, sugarcane bagasse (SCB), was used as an organic filler to prepare a hybrid composite based on high density polyethylene (HDPE). The optical microscopy (OM) and scanning electron microscopy (SEM) showed that the cellulosic material was homogeneously embedded within the HDPE matrix. Contact angle measurement using different liquids—water-based inks, ethanol, and ink for ink-jet printers—and printing tests were performed viewing its application as a synthetic newsprinter. The best result was achieved for the composite with the highest content of SCB as well as the shortest drying time [16].

Still following the path of sustainability, it was proposed the using of coffee dregs (COFD) in four different types (integral, extracted, major size, and minor size) and proportions (90–10%) in composites with high density polyethylene (HDPE). The action of the particle size and soluble extraction over the polyolefin properties was then evaluated. Concerning the size, the performance was similar, but the extracted fraction had a worse behavior. Thermal stability of HDPE was improved. After extrusion in a two-stage cokneader system extruder, the compound was injected as cylindrical specimens. The SEM revealed that all composites had a fine dispersion of the COFD into the polymeric matrix. The compressive moduli of the composites resulted in similar to the neat polymer one. In this outlook COFD polymeric composites can be used as building material [17,18].

Viewing the application in composites, natural graphite was expanded with octadecylamine using chemical way (acid mixture H_2SO_4/HNO_3, 4:1) and sonication. The expansion of the graphite layers was showed using SEM images (Fig. 7.1). Infrared spectroscopy detected the presence of CH absorption. An additional diffraction peak at a lower 2θ angle was noticed. The relaxometry showed that the relaxation time was a function of the frequency and the curve of the graphite/octadecylamine presented two peaks—approximately before 10^6 Hz and after 10^7 Hz. The results permitted inferring that organically modified graphite was achieved. The relaxometry results are presented in the Fig. 7.2. The acid mixture (AM sample) showed a straight line parallel to the frequency axis, being T_1H quasi-constant for all frequencies. For octadecylamine, the variation of relaxation time versus frequency presented a sigmoidal curve (S shape). For the graphite modified with acid mixture (AM-G sample), the curve revealed a rise from 10^5 Hz to 10^7 Hz, after which it remained constant. The profile was due to the presence of carboxyl and hydroxyl groups inside the graphite galleries, which can be chemically linked or adsorbed on the surface. The relaxometry curve of graphite modified with octadecylamine (ODA-AM-G) also showed that the relaxation time was dependent on the frequency. The profile of the graphite/octadecylamine curve indicated two rises. The first started at 10^5 Hz and finished at approximately 10^6 Hz. There was a plateau between 10^6 and 10^7 Hz, but the point near 10^9 Hz suggested that a second rise occurred. It could be assumed that at least two kinds of structures were present in the graphite/octadecylamine sample. The first rise could represent the exfoliated structure, where the octadecylamine molecules are less constricted, while the second one could be attributed to the delaminated structure—octadecylamine molecules are more squeezed between the graphite layers [19].

Morphology, thermal stability, flammability, water absorption, and compression strength of sustainable hybrid HDPE composites filled with gypsum waste were carried out. As the amount of filler increased in the composites, an improvement of thermal stability and flammability was noticed, but the compression strength was similar to neat HDPE. For all the compressed specimens, the matrix was HDPE and dispersed phase was constituted by gypsum encapsulated for HDPE Fig. 7.3 (a, c, and e). There is good adhesion between matrix and dispersed phase. In Fig. 7.3 (b, d, and f) show the presence of cellular structure as it was achieved in the injected specimens. It was also observed that the gypsum crystallized in different forms—as prismatic and anisotropic needles—as a function of its percentage in the composite [20].

The inorganic composite containing cement, water, sand, and aggregate is known as concrete. The use of recycled material based on debris, waste concrete, and others can be considered for use as a sustainable hybrid composite. From this

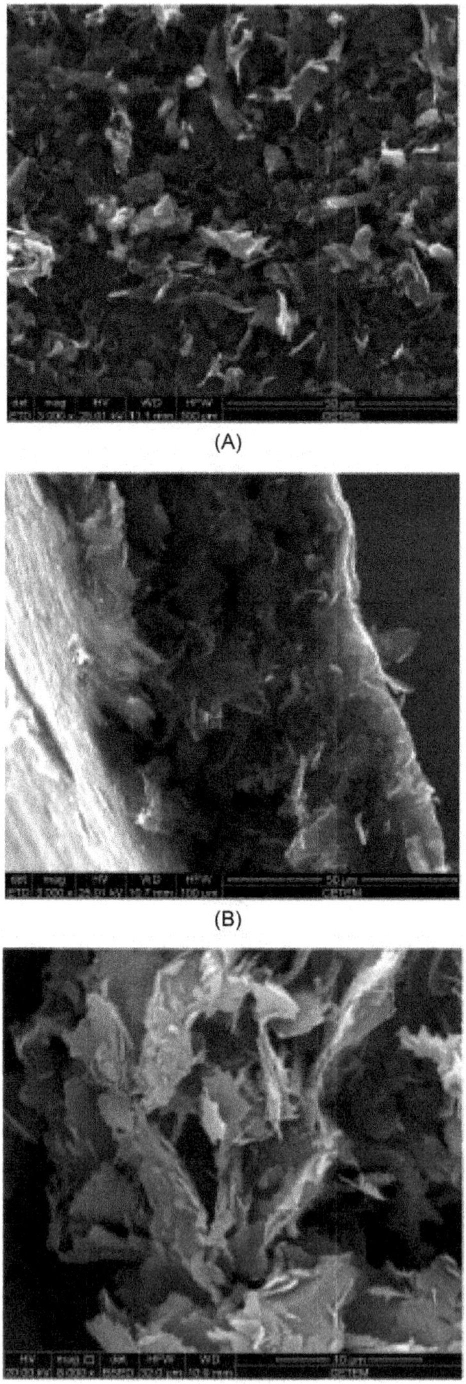

Figure 7.1 SEM images for NG (A), AM-G (B), and ODA-AMG (C) samples.

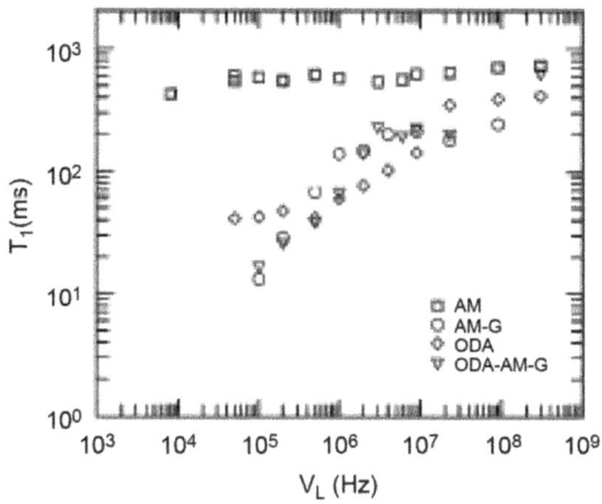

Figure 7.2 LFNMR for AM, AM-G, ODA-AM-G, and ODA samples.

premise, concrete, a as primary waste emerged from the concrete industry, was chosen as filler for preparing polymeric composites for application in the building industry. In order to recognize its chemical composition, concrete was submitted to heat treatment at high temperature (700°C). The residue was characterized by EDX analysis. The main components were calcium oxide, silicon oxide, aluminum oxide, and iron oxide. Because of its heterogeneity and in order to improve compatibility with polymeric matrix, the concrete waste was chemically modified with oleic acid. Infrared spectrum, SEM image, and thermal stability indicated that there was the chemical anchoring of oleic acid on the surface concrete particles [21].

Polyethylene terephthalate (PET) and polycarbonate (PC) presents a series of applications as commodity or engineering plastics due mainly to their mechanical properties. Both can be fully recycled. Its blend (PET/PC) is commercially available. In this view and thinking on the subject of environmental preservation, a blend of rPET/PC (80/20 wt/wt%) was filled with nanoparticles of zinc oxide (nZnO, 0–3 wt%) in order to produce nanocomposites as a shield for ultraviolet (UV) radiation. The PET and PC carbonyl indices and ^{1}H LFNMR results suggested that the UV damage was attenuated in the nanocomposite filled with the highest content of nZnO [22]. Recycled polycarbonate (rPC) filled with nano-zinc oxide particles and irradiated with gamma-radiation was prepared in order to produce nanocomposites (rPC/nZnO) to act as barrier against ultraviolet light. Before exposure to UV light the PC carbonyl index (CI) show trend to decrease—corroborating the scission of the rPC carbonate bonds. After a long exposure time, an increase of CI was noticed. This which was attributed to the free radicals recombination and esterification reaction. The combination of nZnO nanoparticles and gamma-radiation revealed some effect as barrier to UV light on the rPC matrix [23].

The agribusiness—sugar and alcohol—is an important branch of Brazil's economy. Until recently, sugarcane bagasse was considered as a waste and therefore harmful for environmental preservation. In order to provide added value this kind of waste can be designed for application in other sectors of economic activity.

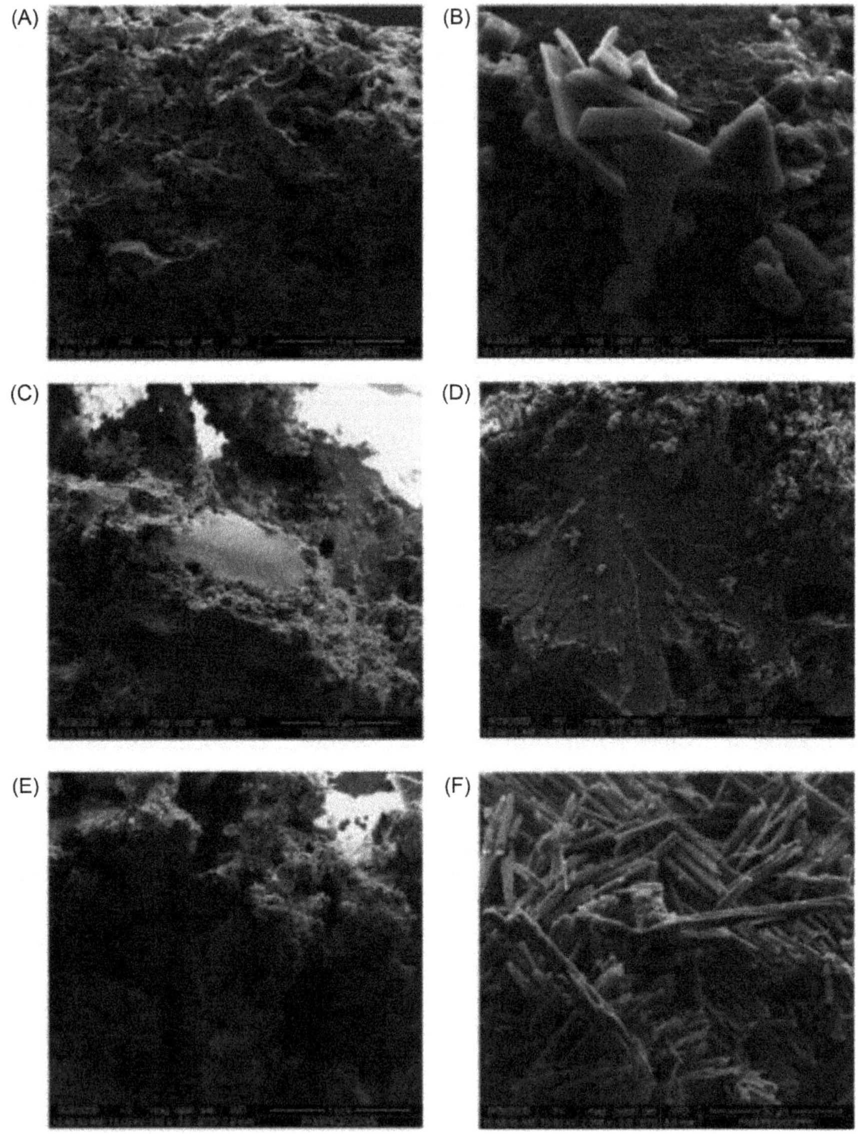

Figure 7.3 SEM photomicrograph adhesion crystals in the HDPE matrix (A, C, and E) of the composites and (B, D, and F) gypsum crystallized in different forms and cellular structure, 50/50, 40/60, and 30/70, respectively [20].

Then, viewing to create an upcycled material to produce masonry bricks, composites of recycled high density polyethylene (HDPE) and micro fibers of sugarcane bagasse were studied. The polymer/filler ratio ranged from 100–0 to 60–40%. Several conventional characterization techniques were applied. Good dispersion and adhesion of the filler into the polymeric matrix were noticed. No significant variation on the melting and crystallization temperatures (T_c and T_m) of HDPE in the composites led to conclusion that the average size of the polymeric crystals did not change substantially. As T_c was invariable and the degree of crystallinity of the composites increased to all filler contents, it was assumed that some transcrystallization had happened. Sugarcane bagasse (SCB) crystalline arrangement acted as a substrate for the additional crystallization. There was a rise in the compressive moduli of the composites, ranging from 37% to 63%. The modulus at the crossover point and the complex viscosity also rose due to the fiber ratio. The results show that the combination of these materials led to the reinforcement of the polymer compressive strength, making it a potential candidate for building material. The SEM images of the composites (Fig. 7.4) showed a fine dispersion between polymer and filler. The same good dispersion between HDPE and SCB as mentioned in previous work was seen; it was attributed it to the very fine fibers, with a dimension inferior to the #100 mesh sieve. This fraction represented around 55% of the total SCB loaded in the composites. The images showed that most of the SCB particles were fractured instead of pulled (Fig. 7.5). This can be taken as an indication of good adhesion between the materials. It was believed that the presence of lignin in the SCB acted as compatibilizing agent between polymer and filler. Considering that we did not use any delignification process (NaOH, H_2SO_4) or compatibilizing agent as is usual in natural fiber reinforced composites (NFRC), the interface resulted adequately for the intended purpose. The SEM images on Figs. 7.4 and 7.5 show broken fibers in the cryogenically fractured section, denoting a strong polymer/filler interaction. The increase in the filler content tends to decrease the adhesion as shown by some voids in the polymeric matrix in Fig. 7.4C. Most of them were caused by pulled fibers that were positioned parallel to the fracture section. As the intended application is compressive stress, this pulling effect on the SCB fibers will not be a problem [24].

The overlay of the complex viscosity curves (η^*) of the composites is presented in (Fig. 7.6). The viscosity rose with the increase of SCB content, and decreased with the increase of frequency, showing the pseudoplastic behavior of the materials. The presence of fibers has changed the average flow of the rHDPE, and restrained the mobility of the molten polymer. The composites became more resistant to deformation than the neat polymer in the melt state. According to Lozano et al. [25], the decrease in viscosity at higher frequencies can be a plus because it means that it will probably not be a problem to process these composites at high shear rates (average processing conditions of polyolefin) [24].

7.3 Hybrid polymer composites with synthetic fillers

Currently, the biomaterials represent a very important market segment for human health. The production of synthetic material that can serve as repairer and applied in the human body without causing side effects should be stimulated. In this line, composites of poly(vinyl pyrrolidone)/hydroxyapatite (PVP/HA, 100/0; 80/20; 50/50; 20/80 wt/wt %) were prepared and characterized by

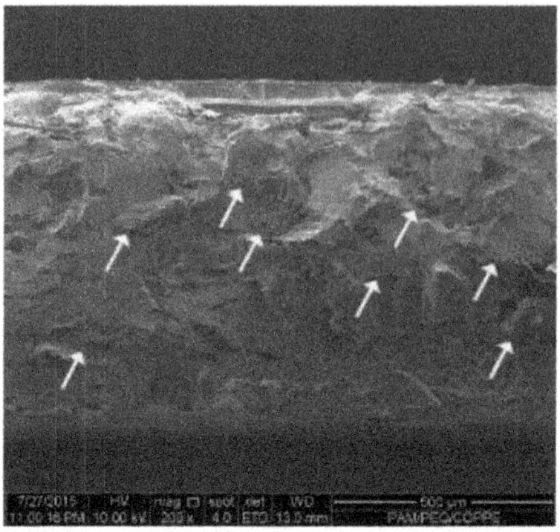

(A)

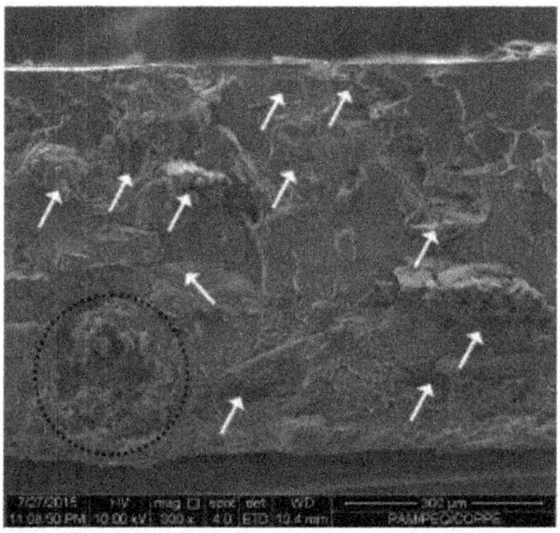

(B)

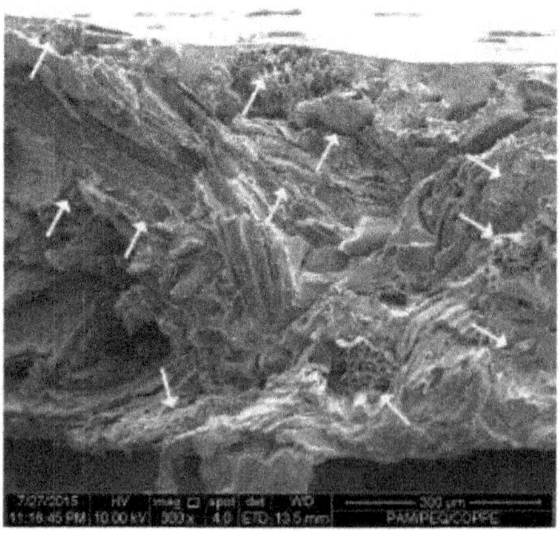

(C)

Figure 7.4 SEM photomicrographies of the composites 80–20 (A), 70–30 (B), and 60–40 (C), magnified at 300 × [24].

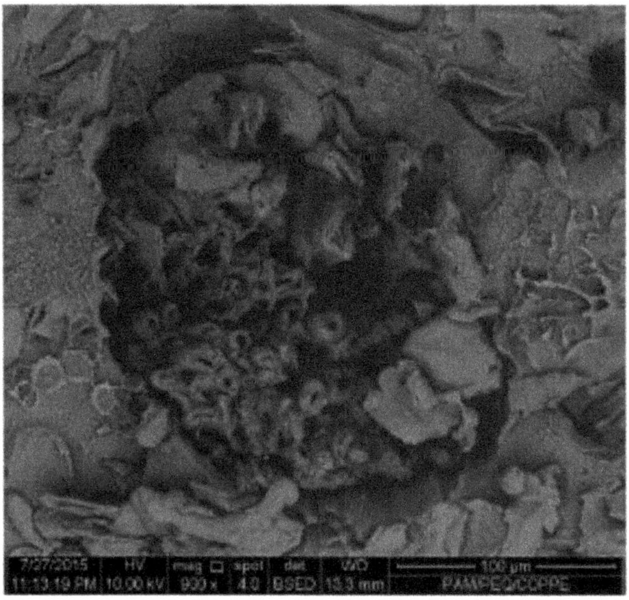

Figure 7.5 SEM photomicrographies with 900 × magnification of the dashed circle in Fig. 7.4B [24].

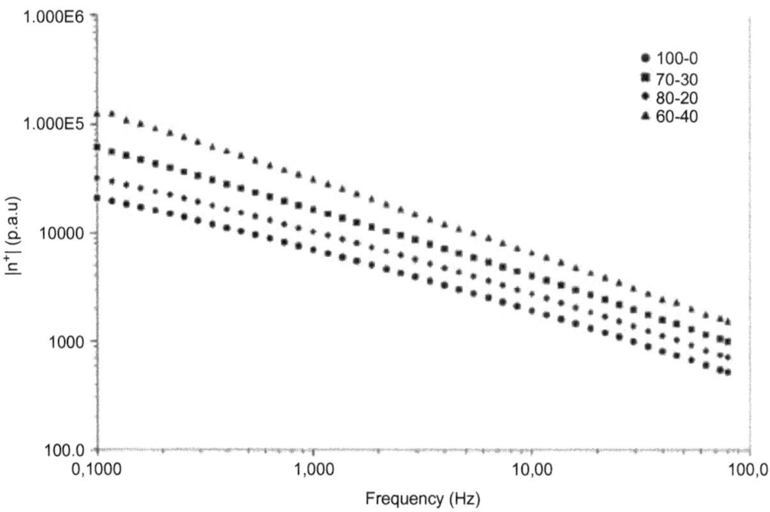

Figure 7.6 Overlay of the complex viscosity (η^*) of the composites [24].

means of standard techniques. Infrared observation revealed the formation of hydrogen bonding with HA hydroxyl groups—a slight displacement of PVP carbonyl stretching to lower frequency was detected. PVP enthalpy relaxation peak was displayed and it was dependent on its amount in the composite. TG/DTG curves showed that pyrrolidone pendant groups in PVP degraded sharply in low temperature. It was proposed that HA molecules exerted a catalytic action on the PVP degradation [26].

In order to produce a hybrid polymer composite—osteoinductive biomaterial—able to stimulate the migration of cells to tissue formation—bone regeneration, in situ hydroxyapatite (HA) was synthesized by the precipitation method, with and without the presence of collagen (COLL). The influence of HAs on structural and morphological characteristics were investigated. The Ca/P molar ratio was influenced by collagen addition: 1.89 and 2.38 for samples without and with collagen, respectively. In the presence of collagen, HA showed higher crystallinity degree and WAXD pattern with better resolution and intensity. Also, collagen influenced strongly the HA morphology as revealed by SEM images [27].

Several natural inorganic substances, particularly the phyllosilicates family, has been gaining importance as filler for preparing hybrid polymeric composites. The chemical reactions open a number of possibilities in order to obtain synthetic substances that, to some extent, have special characteristics. Having that in mind, nanolayered lamellar α-zirconium phosphate was synthesized. In order to increase the d-spacing, it was organically modified with a long-chain amine (octadecylamine) at different amine:phosphate ratios (0.5:1, 1:1, and 2:1) being dispersed in 2:1 ethanol/water solution. Observations by infrared spectroscopy (IR), thermogravimetry (TG/DTG), differential scanning calorimetry (DSC), and scanning electron microscopy (SEM) strongly indicated that amine:phosphate ratio regulated the amine insertion inside the zirconium phosphate lamellae. The occurrence of adsorbed and bonded amine molecules in the phosphate lamellae was evidenced by TG/DTG. Calorimetric data suggested that octadecylamine has different crystal arrangements in the phosphate galleries. A crystallographic diffraction pattern of zirconium phosphate was changed due to the octadecylamine [28].

Once modified, the organically modified lamellar zirconium phosphate was incorporated to linear low-density polyethylene (LLDPE) viewing to prepare nanocomposite, in the molten state, using a counter-rotating twin-screw extruder, controlled with temperature window between 170°C and 190°C at 100 rpm. Thermal results showed a decreasing of LLDPE melting temperature and crystallinity degree. The lamellar α-zirconium phosphate induced to increase of thermal stability suggested that some polyethylene chains entered into the filler's spacing. Both storage and loss moduli increased implying that there was an interaction between LLDPE and phosphate filler. The diffraction peak occurred in 2θ angle beneath 12° strongly suggested that partial intercalation of polyethylene chains among filler interlamellar spacing was achieved. The investigation on molecular relaxation revealed that the shift of two low-intensity relaxation time peaks to higher values. It could be considered as an additional indication of great interaction between polymer matrix and organically modified phosphate inside filler galleries. It was postulated that a partially intercalated and/or exfoliated nanostructured composite was achieved [29].

Considering the importance of high density polyethylene (HDPE) as commodity plastic, organically modified lamellar zirconium phosphate was added. Molten nanocomposites were prepared at selected conditions viewing to investigate the filler action on the HDPE properties. The presence of organically modified phosphate increased the thermal stability of the polymer matrix, but a reduction of its crystallinity degree was observed. The phosphate lamellar spacing increased (3.3 times higher). Low field nuclear magnetic resonance (LFNMR) revealed an increase of molecular mobility. The reduction of elastic modulus, elongation at break, and storage modulus were noticed. All results induced to infer that the presence of octadecylamine made easier the entrance of HDPE in the organically modified lamellar zirconium phosphate galleries. Therefore, it was postulated that a partially intercalated and/or exfoliated nanocomposite was achieved [30].

In the universe of the polymers, the polyamides family comprises a series of polymers with applications as commodity and engineering plastics. Seeking to insert polyamide-6 (PA-6) as engineering material, it was prepared melt extrusion nanocomposites filled with organointercalated zirconium phosphate. Thermal and thermo-mechanical, crystallographic and molecular mobility characteristics of the nanocomposites were evaluated concerning the influence of filler amine:phosphate ratio. For all nanocomposites, thermal degradation temperatures showed a slight decrease. Calorimetric data revealed changes on PA-6 crystalline phase, melting temperature, and crystallinity degree as function of the ratio amine:phosphate. Glass transition temperature (T_g), storage modulus (E'), and loss (E") decreased for 0.5:1 and 1:1 amine:phosphate ratios. Filler domains with size between 60 and 300 nm coexisted by SEM images. A crystallographic diffraction pattern of nanocomposites showed that diffraction peaks were shifted to lower angles according to amine:phosphate ratios. These represent sharp evidence that nanocomposites with different degree of intercalation and/or exfoliated could have been achieved [31].

7.4 Conclusions

From the 1950s until almost the end of the 1980s, polymer composites were mainly used in applications with high-performance requirements. In general, the matrix component used was constrained to a small number of thermosetting polymers: phenolic resin, epoxy resin, and unsaturated polyester. The reinforcing fillers were either organic or inorganic in nature. Clays, calcium and aluminum silicates, magnesium carbonate, asbestos, glass fiber, among others, belong to a group of inorganic filler most used. Carbon black from several origins and also different shapes and dimensions jointly cellulose—powder or fiber—were the best examples of organic reinforcing filler.

Near the end of the 1980s, the first scientific information related to production of nanocomposite occurred. It could be considered a turning point in modern science and technology. Hence, the concept of nanoscience and nanotechnology arose. Both involve and correlate various areas of human knowledge (engineering, physics, chemistry, biology, electronics, computing, and medicine). Nanotechnology has allowed the development of nanocomposites because of the inclusion of various other polymers and fillers in that the aspect ratio and special features might be precisely evidenced by their nanoscale dimension. It also represents something related to sustainability.

References

[1] W. Work, et al., Definition of terms related to polymer blends, composites, and multiphase polymeric materials (IUPAC Recommendations 2004), Pure Appl. Chem. 76 (11) (2004) 1985–2007.

[2] Mano, E. Polimeros como materiais de engenharia. São Paulo (SP): E. Blucher, 1991.

[3] S. Thomas, Polymer Composites., Wiley-VCH, Weinheim, 2012.

[4] S. Ahmed, F. Jones, A review of particulate reinforcement theories for polymer composites, J. Mater. Sci. 25 (12) (1990) 4933–4942.

[5] D. Merinska, et al., Polymer/clay nanocomposites based on MMT/ODA intercalates, Composite Interfaces 9 (6) (2002) 529–540.

[6] L. Zhang, et al., Preparation and properties of nylon 6/carboxylic silica nanocomposites via in situ polymerization, J. Appl. Polym. Sci. 122 (2) (2011) 1316–1324.

[7] A. Somwangthanaroj, M. Tantiviwattanawongsa, W. Tanthapanichakoon, Mechanical and gas barrier properties of nylon 6/clay nanocomposite blown films, Eng. J. 16 (2) (2012) 93–106.

[8] I. Isik-Gulsac, U. Yilmazer, G. Bayram, Mechanical and rheological properties, and morphology of polyamide-6/organoclay/elastomer nanocomposites, J. Appl. Polym. Sci. v. 125 (5) (2012) 4060–4073.

[9] J. Koo, Polymer Nanocomposites, McGraw-Hill, New York, 2006.

[10] A. Esteves, A. Barros-Timmons, T. Trindade, Nanocompósitos de matriz polimérica: estratégias de síntese de materiais híbridos, Química Nova 27 (2004) 798–806.

[11] M. Herrera-Alonso, et al., Intercalation and stitching of graphite oxide with diaminoalkanes, Langmuir. v. 23 (21) (2007) 10644–10649.

[12] W. Cardoso, Y. Gushikem, Preparation of lamellar compounds: synthesis of the crystalline zirconium hydrogenphosphate and its intercalation with amines. An experiment for undergraduate students, Quim. Nova. v. 28 (4) (2005) 723–726.

[13] A. Kumar, et al., Nanoscale particles for polymer degradation and stabilization—Trends and future perspectives, Progr. Polym. Sci. 34 (6) (2009) 479–515.

[14] L. Mendes, et al., Dual-curing, self-adhesive resin cement: influence of the polymerization modes on the degree of conversion and microhardness, Mat. Res. 13 (2) (2010) 171–176.

[15] M. Borges, et al., Degradation of polymeric restorative materials subjected to a high caries challenge, Dental Mater. v. 27 (3) (2011) 244–252.

[16] L.C. Mendes, S.P. Cestari, Printability of HDPE/natural fiber composites with high content of cellulosic industrial waste, MSA 02 (09) (2011) 1331–1339.

[17] S. Cestari, L. Mendes, Thermal properties and morphology of high-density polyethylene filled with coffee dregs, J. Therm. Anal. Calorim. 114 (1) (2013) 1–4.

[18] S. Cestari, et al., Properties of recycled high density polyethylene and coffee dregs composites, Polímeros Ciência e Tecnologia 23 (6) (2013) 733–737.

[19] L. Carreira, et al., Organically-expanded graphite/octadecylamine: structural, thermal and relaxation evaluation, MSA 04 (05) (2013) 281–286.

[20] F. Ramos, L. Mendes, Recycled high-density polyethylene/gypsum composites: evaluation of the microscopic, thermal, flammability, and mechanical properties, Green Chem. Lett. Rev. 7 (2) (2014) 199–208.
[21] F. Ramos, L. Mendes, S. Cestari, Organically modified concrete waste with oleic acid, J. Therm. Anal. Calorim. 119 (3) (2015) 1895–1904.
[22] H. Pires, et al., nZnO as barrier to UV radiation on rPET/PC nanocomposites, J. Nanosci. Nanotechnol. 16 (2016) 9987–9996, *in press.*
[23] A. Carvalho, et al., Nanocomposites of recycled polycarbonate/nano-zinc oxide (rPC/nZnO): Effect of nanofiller and gamma-radiation on the properties and barrier against ultraviolet light, J. Nanosci. Nanotechnol. (2016), *in press.*
[24] S. Cestari, et al., Upcycling polymers and natural fibers waste—properties of a potential building material, Recycling 1 (2) (2016) 205–218.
[25] K. Lozano, S. Yang, Q. Zeng, Rheological analysis of vapor-grown carbon nanofiber-reinforced polyethylene composites, J. Appl. Polym. Sci. 93 (3) (2004), 1500-1500.
[26] L. Mendes, R. Rodrigues, E. Silva, Thermal, structural and morphological assessment of PVP/HA composites, J. Therm. Anal. Calorim. 101 (3) (2010) 899–905.
[27] L. Mendes, G. Ribeiro, R. Marques, *In situ* hydroxyapatite synthesis: influence of collagen on its structural and morphological characteristic, MSA 03 (08) (2012) 580–586.
[28] L. Mendes, et al., Zirconium phosphate organically intercalated/exfoliated with long chain amine, J. Therm. Anal. Calorim. 118 (3) (2014) 1461–1469.
[29] L. Mendes, D. Silva, A. Lino, Linear low-density polyethylene and zirconium phosphate nanocomposites: evidence from thermal, thermo-mechanical, morphological and low-field nuclear magnetic resonance techniques, J. Nanosci. Nanotechnol. 12 (12) (2012) 8867–8873.
[30] A. Lino, et al., High density polyethylene and zirconium phosphate nanocomposites, Polímeros 25 (5) (2015) 477–482.
[31] D. Freitas, et al., Polyamide-6/organointercalated lamellar zirconium phosphate nanocomposites: molecular mobility, crystallography and thermo-mechanical evaluation, J. Nanosci. Nanotechnol. (2016), *in press.*

Synthesis of conducting polymer/carbon material composites and their application in electrical energy storage

8

Atsushi Gabe[1], *María José Mostazo-López*[1], *David Salinas-Torres*[2], *Emilia Morallón*[1] *and Diego Cazorla-Amorós*[1]
[1]University of Alicante, Alicante, Spain, [2]University of Liège, Liège, Belgium

Chapter Outline

8.1 Introduction

Conducting polymer-based materials have received much attention because of their variety of applications such as sensors/biosensors, advanced transistors, optical limiting devices photodiodes, electromagnetic absorption, metal corrosion protection, energy storage, etc. This chapter focuses on the preparation of conducting polymer/carbon material (CP/CM) composites and their application in electrochemical energy storage, as this is one of the most studied fields of application of these materials because of the increasing concern about energy issues. The synthesis methods of CP/CM composites, as well as their properties and applications, have attracted

Hybrid Polymer Composite Materials: Processing. DOI: http://dx.doi.org/10.1016/B978-0-08-100789-1.00008-3

the interest of many researchers, as it is demonstrated by the increasing number of publications dealing with this topic. This review is divided into four sections, where the main synthesis methods of CP/CM composites, as well as their properties and application in the energy storage fields, are addressed. We will pay special attention to the preparation of advanced carbon materials from the CPs and their composites.

The importance of carbon materials is well known, not only because of their diversity of morphologies (activated carbon-AC, carbon black-CB, carbon nanotubes-CNT, carbon fibers-CF, activated carbon fibers-ACF, graphene-G, fullerene, graphene oxide-GO, reduced graphene oxide-RGO, carbon xerogels, etc), but also because of the properties derived from their intrinsic features, such as a wide range of porosity, chemical and physical stability, surface chemistry, and good electrical and mechanical properties [1–5]. As a consequence of this diversity, CMs have been claimed as very interesting candidates for numerous applications in several fields. Among them, their use in adsorption, catalysis, medical applications, sensors/biosensors, photovoltaic solar cells, energy applications (fuel cells, batteries, capacitors, etc.) should be highlighted. These properties, together with their relatively low cost, make them suitable candidates for assembling diverse composites.

Since CPs were discovered in 1963, intense investigation has been performed. MacDiarmid initiated an extensive study about CPs and their applications in sensors and energy storage, among others [6]. CPs named as "synthetic metals" are a class of polymers containing a large resonation structures with many sp^2-carbon atoms on conjugated chain structures (Fig. 8.1) that allow the delocalized transport of charge carriers [7–9]. The conductivity of CP is assigned to the delocalization of π-bonded electrons over the polymeric backbone which shows electronic properties [10].

Neutral conjugated polymers with a small conductivity (10^{-10} to 10^{-5} S cm^{-1}) can be converted into semiconductive or conductive states with conductivities of 1 to 10^4 S cm^{-1} through chemical or electrochemical redox reactions [11]. CPs are a well-known example of materials that can give pseudocapacitance properties since they present a continuous range of oxidation states with increasing electrode potential [12,13]. The pseudocapacitance arises when the CP is being charged, as it loses electrons and becomes polycation, producing anions in the solution to be inserted into the CP to maintain the electro-neutrality [6]. The CPs are typically formed either through chemical or electrochemical oxidation of the monomer [13–15]. Two oxidation reactions occur at the same time, the oxidation of the monomer and the oxidation of the polymer [16]. However, in general, pseudocapacitive materials such as CPs suffer from low stability and a short lifetime because of their degradation and aging by swelling and shrinking during redox reactions [17]. CMs are attractive scaffolds for the synthesis of CP/CM composites because of the high

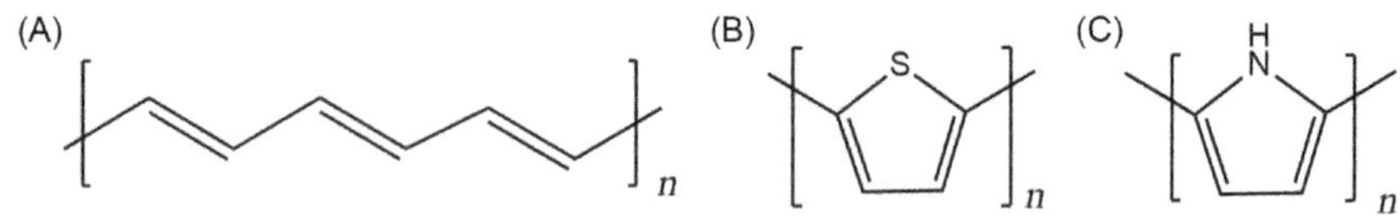

Figure 8.1 Structure of some CPs: (A) polyacetylene (PAc), (B) polythiophene (PTh), (C) polypyrrole (PPy).

mechanical strength, electrical conductivity, large surface area, flexibility, and both tunable porosity and surface chemistry, as well as easy fabrication of the electrode and facilitated electron transportation in the device [18–21].

CP/CM composites show synergistic effects, combining advantages of both materials [18,22,23], being suitable for numerous applications, such as energy storage materials (supercapacitors and batteries), energy conversion materials (fuel cells, PEMFC, solar cells, dye-sensitized solar cells), and sensors [17]. Currently, the research in this field is focused on tuning the micro-/nano-structure of the hybrid material, improving the interface between CPs and CMs, and controlling the chemical structure of the CP for achieving high electrochemical activities. Moreover, the studies on the CP/CM composites with different chemical compositions, morphology, and phase structure are crucial in attaining fundamental understanding of the interface between CP and CM, which in turn provides the basis for future improvements of the device performance [18].

8.2 Methods of synthesis of Conducting Polymer/Carbon Material composites

Regarding the routes of synthesis of CPs or CP/CM composites, the chemical and electrochemical methods are the most commonly used. Generally, the chemical method leads to powder nanomaterials and it can be easily scaled up, while the amount of CP obtained from electrochemical method is limited by the electrode surface where the polymerization happens. Consequently, CPs synthesized by electrochemical routes are usually deposited on the electrode surface as a film. Apart from these two methods, there are other ways for the synthesis of CP/CM composites (i.e., physical mixing, layer-by-layer (LBL) assembly, electrodeposition, electrospinning, and chemical grafting methods) [17,24–26].

8.2.1 Chemical polymerization method

The chemical synthesis of CPs simply consists in the oxidation of a monomer in solution by means of oxidizing agents such as $(NH_4)_2S_2O_8$, $K_2Cr_2O_7$, $FeCl_3$, $KMnO_4$, etc. In this case, the bulk production is not limited as long as there are enough monomeric units and oxidant agent in the solution. For this reason, it is preferred in industry for providing a cheap and efficient route to get large amount of polymer [27]. However, the polymer obtained shows a low level of homogeneity in comparison with other methods [6], displaying lower electrical conductivity. This drawback can get worse when the polymerization occurs inside the porosity of CMs because the monomer oxidation, and the subsequent reaction between two radical cations, can be hindered due to diffusion problems, giving ramified polymers chains. For this reason, the synthesis of CP/CM composites requires a previous monomer adsorption to facilitate a suitable diffusion of oxidizing agent to avoid diffusion problems, obtaining higher level of homogeneity. In this regard, several CMs have been studied for their combination with different CPs. Table 8.1 shows examples of the combination and some properties of CP/CM composites synthesized by chemical polymerization.

Table 8.1 Examples of conducting polymer/carbon material composites synthesized by chemical method

CM	CP	Oxidant	Thickness/nm	σ/S cm^{-1}	C/F g^{-1}(three electrode cell)	Reference
SWCNT	PANI	$(NH_4)_2S_2O_8$	10	0.000126	–	[28]
SWCNT	PEDOT	$FeCl_3$	2–6	45.2	–	[29]
MWCNT	PANI	$(NH_4)_2S_2O_8$	45–60	150	424 (1 M H_2SO_4)	[30]
MWCNT	PANI	$K_2Cr_2O_7$	65	–	360[a] (1 M H_2SO_4)	[31]
MWCNT	PPy	$FeCl_3$	–	–	200[a] (1 M H_2SO_4)	[31]
MWCNT	PEDOT	$Fe(ClO_4)_3$/ $FeCl_3$	–	–	100[a] (1 M $TEABF_4$ in acetonitrile)	[31]
MWCNT	PPy	$FeCl_3$	–	–	506 (1 M H_2SO_4)	[32]
MWCNT	PANI	$K_2Cr_2O_7$	65–70		670 (1 M H_2SO_4)	[32]
MWCNT	PEDOT	$Fe(ClO_4)_3$/ $FeCl_3$	–	–	160 (1 M $TEABF_4$ in acetonitrile)	[15]
MWCNT	PEDOT	$Fe(OTs)_3$	50–400	2–200	–	[33]
AC	PANI	$(NH_4)_2S_2O_8$	–	–	316 (1 M H_2SO_4)	[14]
AC	PANI	$(NH_4)_2S_2O_8$	–	–	273 (1 M H_2SO_4)	[34]
AC	PANI	$(NH_4)_2S_2O_8$	–	0.21	101 (1 M NaCl)	[35]
AC	PANI	$(NH_4)_2S_2O_8$	–	7–16	–	[36]
AC	PANI	V_2O_5	–	–	238 (1 M H_2SO_4)	[37]
GO	PPy	$FeCl_3$	–	–	165 (6 M KOH)	[38]

Vapor grown–graphene	PPy	$FeCl_3$	–	–	394 (6 M KOH)	[38]
RGO	PPy	$FeCl_3$	–	–	150 (6 M KOH)	[38]
NH_2-G	PPy	$FeCl_3$	–	–	225 (6 M KOH)	[38]
ACF	PANI	$(NH_4)_2S_2O_8$	0.5	–	233 (1 M H_2SO_4)	[39]
CB	PPy	H_2O_2	–	–	–	[40]
ZIF derived-Carbon	PANI	$(NH_4)_2S_2O_8$	–	–	1100 (1 M H_2SO_4)	[41]
RGO	PANI	$(NH_4)_2S_2O_8$	>50 μm	–	385 (1 M H_2SO_4)	[42]
Carbon cloth	PANI-RuO_2	$(NH_4)_2S_2O_8$	300	–	710 (1 M H_2SO_4)	[43]
CF	PANI	$(NH_4)_2S_2O_8$	84	–	180 (1 M H_2SO_4)	[44]
			280			
			1320			

[a]Results from 2-electrode cell, but values expressed per active material (3-electrode cell).

In the following subsections, some of the most representative combinations of the CP/CM composites will be briefly described to address the state-of-the-art in this issue.

8.2.1.1 CP/CNT composites

CNTs can be considered as a very interesting support material because they allow homogeneous distribution of CPs, provided that an adequate dispersion of the CNT can be achieved. Additionally, while pure CPs show poor mechanical properties, CNTs can preserve the CP as active material even when the volume change produced by intercalation and depletion of ions during doping and de-doping process occurs [31]. For that reason, CNTs have been widely investigated as support to prepare CP/CNT composites for supercapacitors using PPy, PANI, PEDOT, etc. [6,15,31,32,45,46]. For instance, Li and coworkers obtained SWCNT/PANI, which showed a PANI thickness of 10 nm for SWCNTs with a diameter of 1–2 nm, evidencing a good interaction due to the formation of charge–transfer complex instead of Van der Waals interactions [28]. In addition, an improvement of electrical conductivity as the amount of SWCNTs increased was observed, which was explained as a consequence of connective bridges between different domains of conducting PANI.

Meng et al. [30] prepared a paper-like substance (buckypaper of 25 μm) using PANI as CP and MWCNT, obtaining a composite thinner and more flexible than those obtained by conventional CNT/PANI composites. This MWCNT/PANI composite had 75 wt% of PANI, and the thickness was of 30 μm. From SEM characterization, a uniform and compact PANI film over the CNTs was observed. The development of these flexible MWCNT/PANI composites suggests a remarkable advancement for potential applications.

Another application where CP/CNT have been used is in the preparation of components in dye-sensitized solar cell (DSSC). CP/CM composites also exhibit good electrocatalytic reduction activity and they can be suitable as electrodes in DSSC [29,47]. The incorporation of a small amount of MWCNT in PEDOT exhibited an improvement in efficiency. MWCNT/PEDOT composite was synthesized by chemical polymerization using CNTs as cores, leading to composites with a high conductivity (2.10^{-2} to 2.10^{-4} S m^{-1}) [33].

8.2.1.2 CP/graphene composites

Graphene (G), as unique 2-D crystal lattice with remarkable properties, offers exceptional opportunities to address the ever-increasing global energy demands. Recent years have witnessed considerable advances in graphene-based materials and important breakthroughs in energy applications field [17]. A CP/G composite has been claimed as one of the most promising materials to be used as electrode in high-performance hybrid supercapacitors. Lai and coworkers synthesized N-doped G–PPy electrode with a high specific capacitance (393 F g^{-1}). This is a promising result, which is larger than those obtained from NH_2-graphene/PPy (225 F g^{-1}) or

GO/PPy (165 F g^{-1}) [38]. Other G/PPy composites have also shown improved capacitive performance compared to their individual components [38,48]. In addition, the unique electrical, mechanical, optical, chemical, and structural properties of G/CP composites have generated an interest in the sensor field [25].

8.2.1.3 CP/activated carbon composites

While CNTs or Gs have been widely investigated, the polymerization of CPs on activated carbons (AC) by chemical method is not very much studied. Ryu and coworkers carried out one of the first works, which consisted of aniline monomer deposition on AC and subsequent polymerization in acid solution [34]. It was reported that the PANI deposition could be controlled by an AC/PANI ratio. The AC/PANI increased the specific capacitance around 41% compared to the AC. Zhou and coworkers prepared a similar composite, but they concluded that AC/PANI ratio does not affect the uniformity of PANI distributions [37]. Recently, Yan and coworkers used an in situ polymerized AC/PANI composite as electrode material in capacitive deionization. Even though the conductivity of AC/PANI is low (0.21 S cm^{-1}), it was four times higher than the starting AC. Zengin and coworkers reported a value of conductivity of 16 S cm^{-1} for AC/PANI composite [36].

Bleda-Martínez and coworkers studied the effect of surface chemistry on the preparation of AC/PANI electrodes [14]. Several AC/PANI composites were prepared by a chemical method and a mechanical mixing method. The AC was obtained from anthracite by KOH-chemical activation and it was heat-treated to remove oxygen groups. Finally, both CMs were used to synthesize AC/PANI with different loadings. It was observed that heat-treated AC showed a lower capacitance (125 F g^{-1}) than the untreated AC (228 F g^{-1}), which was related to the larger amount of surface oxygen groups and higher porosity of the pristine AC [49] and that important differences among the composites occurred due to the different surface chemistry of the CMs, as explained below.

Cyclic voltammograms for AC/PANI composites are shown in Fig. 8.2A and B for materials obtained from pristine AC and thermally treated AC, respectively. In the case of composites obtained using pristine AC (Fig. 8.2A), the polymer coating produces a decrease of electrical double layer (228 to 182 F g^{-1}) and also reduces the redox peak at 0.5–0.7 V related to oxygen functionalities. These evidences confirm that the formation of PANI diminishes the apparent surface area and consequently the electrical double layer, and suggest that the PANI coating is formed over the CM and that there is consumption of some electroactive oxygen groups. In contrast, the electrical double layer of the samples prepared with thermally treated AC does not decrease with the polymer coating and there is no evidence of redox processes related to oxygen groups neither in the carbon material nor the composites (Fig. 8.2B). In fact, the capacitance of the composites increases with the polymer coating (from 125 to 148 F g^{-1}). The decrease in capacitance observed for the pristine AC when chemical polymerization of aniline is done, suggests that aniline reacts with oxygen functional groups during the formation of the polymer coating and, if these surface groups are on the entrance of the porosity, the polymerization produces

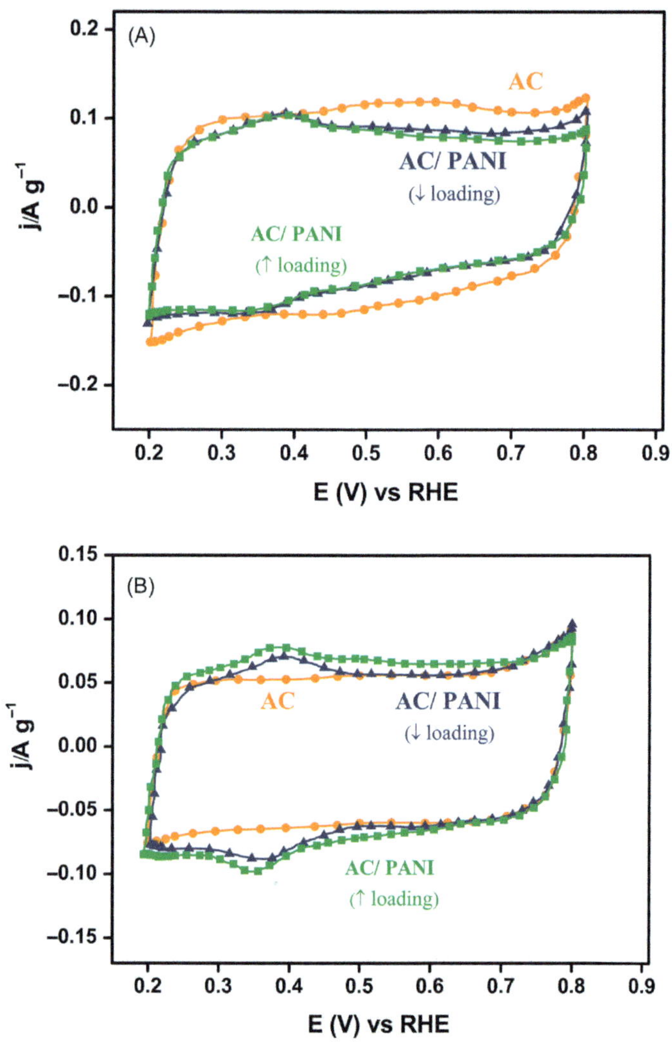

Figure 8.2 Cyclic voltammograms for electrodes prepared by chemical polymerization over (A) pristine activated carbon and (B) thermally treated activated carbon. $v = 0.5$ mV/s. 1 M H_2SO_4.

some blockage of this porosity. In the case of absence of surface oxygen groups, aniline polymerizes homogeneously along the entire AC surface, avoiding the reduction of its apparent surface area. This is an important result to be taken into account when preparing composites by chemical polymerization. Then, it is recommended to avoid high oxygen content in the carbon material and to perform the monomer adsorption isotherm in the carbon material to select the most adequate conditions for the loading of the monomer inside the porosity prior to chemical oxidation [50].

8.2.2 Electrochemical polymerization method

Regarding the electrochemical method CPs synthesis, it commonly involves the oxidation of monomers in a solution by applying a constant potential or by cyclic voltammetry in a determined potential range. In this case, the process is influenced by many parameters such as solvent, electrolyte, amount of monomer, or pH. In addition, each polymer has a different potential, current, or temperature at which the polymerization takes place giving a CP with high level of homogeneity. Therefore, the polymerization reaction can be controlled easily by electrochemical techniques, and the amount of product can also be controlled by the integrated charges used for electrosynthesis [21]. In this regard, an electrochemical method has advantages compared to the chemical route. For instance, it is a simple method and shows a high reproducibility. Also, the electrochemical polymerization has higher selectivity and sensitivity and it can be performed at room temperature and atmospheric pressure with small volumes [51]. Furthermore, it allows the tailoring of the morphology, which influences directly on properties of the nanomaterials. For this reason, electrochemical method is an effective strategy for preparing CP nanomaterials [21]. In fact, many electrochemical techniques have been used to functionalize CMs with CPs such as the potentiostatic step method, the potentiodynamic method, and multiple potentiostatic steps, among others.

Table 8.2 shows some examples of CP/CM composites synthesized by electrochemical polymerizations. In the following subsections, they will be briefly described.

8.2.2.1 CP/CNT composites

CPs over CNTs by electrochemical methods have been reported in several papers [52–54,56,59–61]. PPy and PANI are the most successful conducting polymers to get a coating on CNTs.

Huang et al. prepared SWCNT/PANI composites with different ratios by using the potentiodynamic step method [52]. It was observed that conductivity increases rapidly with the amount of PANI. This result reveals that there is a strong interaction between SWCNT and PANI, which is able to facilitate the electron transfer between polymer chains and CNT. Liu and coworkers synthesized PANI/SWCNT hybrid films by an electrochemical polymerization and degradation process. High specific capacitance was obtained after 90 cycles (502 F g^{-1}) and after electrodegradation, the capacitance value was 707 F g^{-1} due to increased polycrystalline PANI regions [58]. Jun Ge and coworkers prepared PANI/SWCNT hybrid material for their use as electrode, obtaining ultrathin and homogeneous films [57]. The ultrathin, optically homogeneous, transparent, electrically conducting films have a large specific capacitance (220 F g^{-1}) for a SWCNT density of only 10.0 mg/cm^2 and 59 wt% PANI, showing its potential for transparent and flexible energy storage devices.

Regarding CNT/PPy composites, Chen and coworkers prepared films over Pt/Ti [59]. It was revealed that the CNT/PPy composite showed better electrochemical

Table 8.2 Examples of conducting polymer/carbon material composites synthesized by electrochemical method

CM	CP	Technique	Thickness/nm	σ/S m^{-1}	C/F g^{-1} (three electrode cell)	Reference
SWCNT	PANI	CV; −0.2 to 1.2 V/SCE	–	0.0012–0.0156	–	[52]
SWCNT	PANI	CV; 0–1.6 Ag/AgCl	40–50	–	–	[53]
SWCNT	PANI	Potentiostatic deposition 0.75 V/SCE	–	–	485 (1 M H_2SO_4)	[54]
SWCNT	PPy	Galvanostatic 2 mA cm^{-2}	–	–	202 (1 M KCl)	[55]
SWCNT (modified)	PPy	Galvanostatic 2 mA cm^{-2}	–	–	223 (1 M KCl)	[55]
SWCNT	PANI	CV; −0.2 to 0.7 V/SCE	–	–	–	[56]
SWCNT	PANI	Galvanostatic 0.75 V/SCE	15	–	55[a] (1 M H_2SO_4)	[57]
SWCNT	PANI	CV; −0.2 to 0.75 V	–	–	707 (0.5 M H_2SO_4)	[58]
CNT (well-aligned)	PPy	CV; 0–0.8 V/SCE	90	–	0.7 C cm^{-2} (0.1 M $LiClO_4$)	[59]
MWCNT	PPy	Potentiostatic deposition 0.70 V/SCE	–	–	192 (0.5 M KCl)	[23]
MWCNT	PEDOT	Galvanostatic	–	–	150 (1 M $TEABF_4$/AN)	[15]
MWCNT	PEDOT	Galvanostatic 0–1.5 V/SCE	200	–	–	[60]
MWCNT (oxidized)	PPy	Galvanostatic	–	–	180 (1 M H_2SO_4)	[61]

CF	PANI	Potentiostic deposition 0.894 V/Ag/AgCl	150 nm	–	80 (0.5 M Na_2SO_4)	[62]
AC	PANI	Potentiostatic 0.3 V, 1.0 V/Ag/AgCl	–	–	350–450 (1 M HCl + 0.5 KCl)	[13]
AC	PANI	Potentiodynamic 0–0.75 V, 1.0 V/Ag/AgCl	–	–	350–450 (1 M HCl + 0.5 KCl)	[13]
AC	PANI	Multiple potentiostatic 0.3–0.75 V, 1.0 V/Ag/AgCl	–	–	350–450 (1 M HCl + 0.5 KCl)	[13]
ACF	PANI	Potentiostatic 0.3 V −1.05 V/RHE	0.5	–	233 (0.5 M H_2SO_4)	[39]
GO	PPy	CV; −0.1–1.0 V/SCE	40–50	–	–	[53]
GO	PPy	Potentiostatic 0.88 V/Ag/AgCl	–	–	–	[63]
GO	PEDOT	Potentiostatic 1.05, 0.92 V/Ag/AgCl	–	–	–	[63]
Carbon paper	PPy	CV; 0–1.2 V/SCE	7.3	–	399 (2 M KOH)	[64]
Carbon Paper	PANI	Galvanostatic 60 mA cm^{-2}	–	–	531 (1 M H_2SO_4)	[65]
Graphite (pencil drawing)	PANI	0.8 V			356 mF cm^{-2} (1 M H_2SO_4)	[66]

CV: cyclic voltammetry.
[a]Obtained from two-electrode cell.

performance than that obtained in absence of CNTs because of the large surface area, making this material specially interesting as electrode for rechargeable batteries. Wang and coworkers obtained SWCNTs/PPy composites by using a galvanostatic method [55], obtaining a SWCNT/PPy film with a lower resistance than that obtained by pure PPy.

PEDOT/CNT has been rarely reported due to their high cost. Lota and coworkers used a galvanostatic method for the deposition of a PEDOT layer over CNT in organic medium, giving a capacitance value of 150 F g^{-1} [15].

8.2.2.2 CP/G composites

PANI/graphene composites thin film was prepared by in situ electrochemical method, obtaining a chemical sensor with high sensitivity to detect hydrazine [67]. A simple method was implemented to create flexible, uniform PPy/G composite films to be used as supercapacitor electrodes [68]. These PPy/G electrodes showed a high performance ($\sim$33 Wh kg^{-1} and $\sim$1184 W kg^{-1}) and maintained the inherent flexibility of graphene films.

8.2.2.3 CP/AC composites

PANI/AC can be prepared using three different electrochemical techniques, which are detailed anon using as an example the results published by Bleda–Martínez and coworkers.

Potentiostatic step method

The potentiostatic step method was performed from the lower potential of 0.3 V to two upper potentials (0.75 and 1.0 V) that were applied until a total electrical charge of 2 C was reached. The PANI polymerization was carried out in presence and absence of AC (i.e., with the bare graphite). Two zones can be distinguished: the initial decrease in current, related to the contribution of the double layer charge, and the nucleation process. Actually, the current increase was caused by the one-dimensional chain growth by a direct monomer-unit incorporation into the existing PANI film [69]. The polymerization at 0.75 V was slower than that at 1 V. At both upper potentials, the polymerization over the AC takes shorter time than the polymerization directly over the bare graphite due to the high surface area of the AC.

Potentiodynamic method

The potentiodynamic polymerization was performed by cyclic voltammetry (30 cycles) using two different potential ranges: from 0 up to 1 V or 0.75 V. Fig. 8.4A and B shows the last 30th cycle of the potentiodynamic polymerization until 1 and 0.75 V, respectively. The PANI/AC composites polymerized up to 1 V showed higher polymerization rate than that of polymerized until 0.75 V. This is because more aniline molecules were oxidized and then incorporated to a polymer chain at

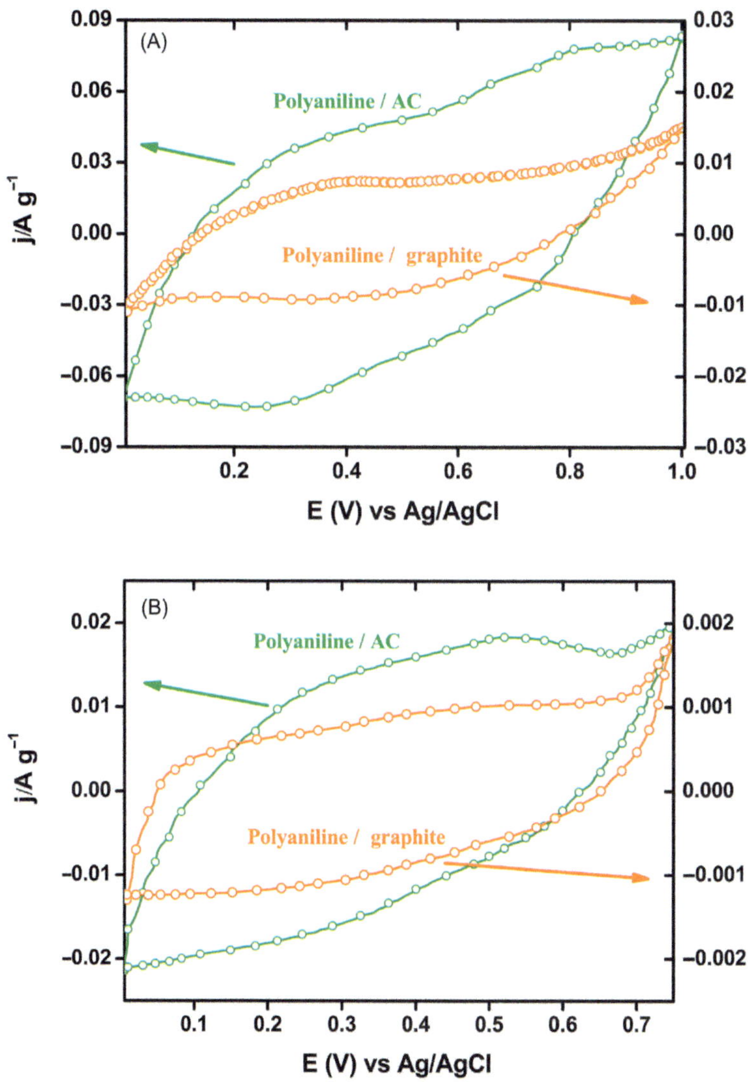

Figure 8.3 Potentiodynamic polymerization between: (A) 0 and 1 V and (B) 0 and 0.75 V. 0.15 M aniline + 1 M HCl + 0.5 M KCl solution. $V = 75$ mV/s. 30th cycle.

higher potentials. The samples with AC presented a higher current due to their higher surface area compared to the graphite (Fig. 8.3).

Multiple potentiostatic steps method

This method involved the application of 200 alternative potentiostatic pulses to the working electrode. The lower potential was 0.3 V, which was led to upper potential (0.75 or 1 V) and it was kept for 4 s. By using a polarization current shift, the potential electrode reached a lower potential for 4 s and again, it was taken to the upper potential for 4 s. These steps were repeated until the end of the experiment. It was observed that the current was higher in the case of pulses ranging from 0.3 to 1 V, indicating a larger amount of electrodeposited polyaniline due to a higher formation of radical monomer to be polymerized. From the chronopotentiometry measurements, it was confirmed that the capacitance enhancement is related to PANI redox processes, indicating the success of PANI electrodeposition.

The electrodes obtained by each electrochemical polymerization method were characterized with different techniques, such as chronopotentiometry, cyclic voltammetry, electrochemical impedance spectroscopy, and infrared spectroscopy, to know their properties as potential electrodes for supercapacitors. The results demonstrated that the composites presented a capacitance value larger than that resulting from the mixing rule using data of both individual materials. This synergistic effect is related to a higher exploitation of the polymer properties by increasing the charge transfer kinetics and a higher electron delocalization of the CP. In addition, the composite evidenced better electrical conductivity than the pristine AC, demonstrating a positive contribution of the polymer.

Regarding electropolymerization methods, the composites obtained by potentiodynamic method have higher electrical conductivity because the structure of PANI is more homogeneous in these composites. In the case of single-step potentiostatic methods, the samples evidenced larger capacitance and better performance at high current density. It was observed that samples prepared by multistep potentiostatic methods showed a performance between potentiodynamic method and single-step potentiostatic method. These results confirmed that it is necessary to select carefully the conditions of electropolymerization method depending on the requirements of the application of the desired composite.

8.2.2.4 CP/ACF composites

Salinas-Torres and coworkers prepared ACF–PANI composites by chemical and electrochemical polymerization methods [50]. Fig. 8.4 shows the cyclic voltammetry obtained for pristine ACF and both composites. The carbon material shows a quasi-rectangular shape related to the formation of pure electrical double layer while PANI/carbon composites showed several redox processes. It can be seen with the sample obtained by electrochemical polymerization that the redox peaks are more defined than in the case of the sample obtained by chemical polymerization. It was associated to secondary products when PANI is obtained by chemical polymerization, while the electrochemically prepared composite shows defined peaks associated to a larger chains with few defects [70,71]. Both composites increased the capacitance around 40% compared to pristine ACF.

The porosity of these composites were characterized by Small Angle X-ray Scattering (μSAXS). It revealed that a thin film PANI layer was deposited inside the porosity. This film showed an average thickness of 0.5 nm.

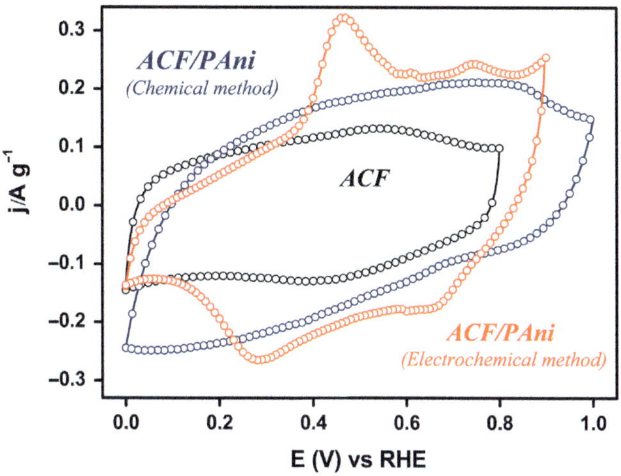

Figure 8.4 Steady state voltammogramms obtained for ACF and ACF–PANI composites. 0.5 M H_2SO_4. $V = 1$ mV s^{-1}.

8.2.3 Other synthesis methods

While most of the CP/CM composites have been prepared by chemical or electrochemical polymerization methods, there are some other ways to synthesize the CP/CM composites.

8.2.3.1 Mechanical mixing method

For obtaining AC-PANI composites by a mechanical mixing method, pure PANI was first synthesized [14]. After preparing pure PANI, composites were prepared by mixing different amounts of PANI, AC, acetylene black and binder (PVDF copolymer). It can be observed that the mechanical mixing of PANI and AC produces a decrease of the microporosity of the resulting composite, which is larger as the amount of polyaniline is increased. Furthermore, when the amount of polyaniline reaches a certain value, an important change in the voltammogram appears, being more similar to the shape of pure PANI. In addition, the capacitance values of all the composites with PANI were higher than the value displayed by pristine AC. The authors associated this change to limitations in the diffusion and migration of ions in the bulk of polyaniline and higher resistance as a consequence of the higher thickness of the layer (due to the high amount of PANI). On the other hand, Lota and coworkers also prepared the composited electrodes by mechanical mixing of PEDOT with MWCNTs. The composite gave a capacitance value of 120 F g^{-1} in acidic medium [15].

8.2.3.2 Layer-by-layer (LbL) assembly

Thin film electrodes of PANI nanofibers and functionalized MWCNTs were created by layer-by-layer (LbL) assembly for microbatteries or electrochemical capacitors.

These LbL-PANI/MWCNT films in lithium cells can reach high volumetric capacitance (~238 F cm^{-3}) and high volumetric capacity (~210 mAh cm^{-3}). In addition, rate-dependent galvanostatic tests showed that LbL-PANI/MWNT films can deliver both high power and high energy density (~220 Wh L^{-1} and ~100 kW L^{-1}) and could be promising positive electrode materials for thin film microbatteries or electrochemical capacitors [72].

8.3 Synthesis of advanced carbon materials

As it was aforementioned, the conducting polymer/carbon material composites play a key role in the manufacture of electrodes in energy storage applications [73–76]. However, conducting polymers may suffer degradation under operational conditions [77] what can be detrimental for a practical use. On the other hand, carbon materials are among the best candidates for energy storage applications and their properties can be optimized by controlling porosity, surface chemistry, and structural order. There are different strategies to synthesize advanced carbon materials [77–79], which achieve the requirements in terms of stability and energy output. One possibility is the carbonization of CP/CM composites since it may allow the introduction of heteroatoms, which are beneficial for improving electrochemical performance. This approach uses the abundant knowledge on preparation of CP/CM composites of different compositions and morphologies what is an advantage for preparing new and improved carbon materials.

The doping of carbon materials with heteroatoms (N, P, S, or B) is a subject of intense research since the presence of heteroatoms in the carbon network has an impact on the electro-donor/acceptor properties or wettability, which directly influences the final material performance. At this point, the carbonization of conducting polymers/carbon composites or the carbonization of the pure conducting polymer has become an interesting and suitable procedure because this type of polymers have heteroatoms, which could be incorporated into the carbon network during the carbonization process. In addition, the conducting polymers show many advantages as carbon precursor, such as low cost, easy synthesis and a wide variety of morphologies, even after carbonization [80–83].

The functional groups generated in the carbon matrix have an important role in the electrochemical behavior. It is well known that certain oxygen groups can improve specific capacitance by means of pseudocapacitive processes [84–86]. Regarding the effect of nitrogen groups on the electrochemical performance, the presence of nitrogen groups not only affect the specific capacitance [87,88] but also may influence the electrical conductivity [89] or electrolyte stability [77]. Heteroatoms such as sulphur, boron, or phosphorus have also attracted attention even though their effect on the capacitance enhancement has not been clearly demonstrated. However, these heteroatoms provide other properties that contribute to the better performance, such as electro-oxidation resistance [90] or a synergistic effect when they are used as a codoping agent together with nitrogen, improving the electrochemical behavior in comparison to N-doped carbon [91–93].

N-containing conducting polymers (polyaniline or polypyrrole) have been the most studied to get N-doped carbon materials because their use leads to nanostructured carbon materials with oxygen and nitrogen functionalities. Many efforts have been focused on understanding the effect of the incorporation of N-functional groups into the carbon network on the electrochemical properties of the resulting materials. For this purpose, it is necessary to know which kind of species can be formed under certain synthesis conditions. At this point, the monitoring of the evolution of the nitrogen species with the temperature is a powerful tool to produce particular nitrogen groups and allows clarifying the real effect on the material performance. In this way, Kuroki and coworkers studied the polyaniline carbonization by using solid-state NMR and it was proved that there was a correlation between the temperature and the formed nitrogen species [94]. It is established that polyaniline chains mainly consist of a combination of benzenoid diamine and quinoid dimine (Fig. 8.5), which evolve in a different way during the carbonization process.

Kuroki and coworkers observed that amine groups of the benzenoid unit lead to pyrrole and N-quaternary groups, whereas mostly pyridine species were obtained from quinoid at 600ºC. When temperature is increased up to 800ºC, the amount of

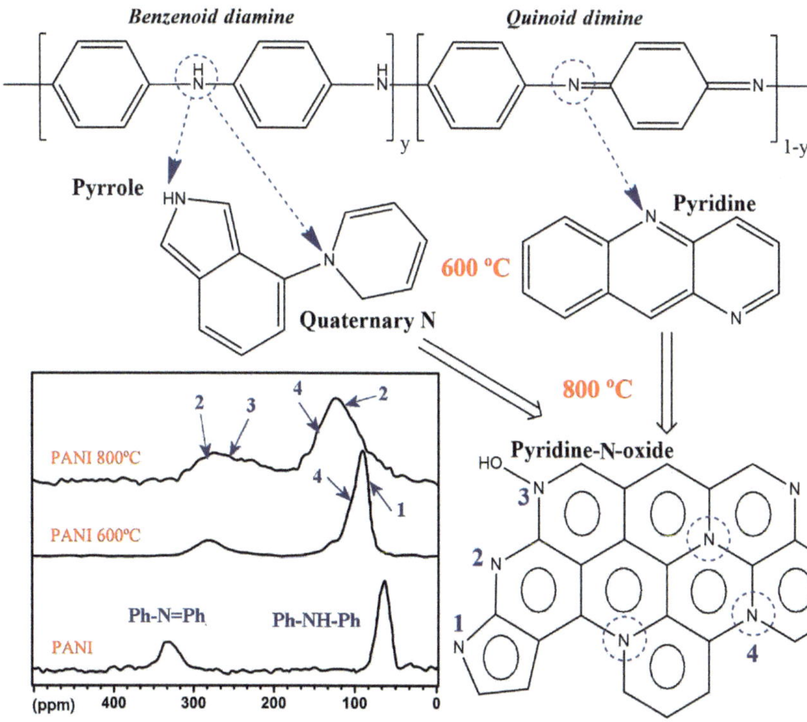

Figure 8.5 Evolution of nitrogen groups present in the polyaniline during carbonization process in inert atmosphere. NMR spectra of non-carbonized polyaniline and their carbonized at 600/800°C (inset) [adapted from [94]].

five-membered rings containing nitrogen atom diminishes, while the amount of six-membered rings increases. This transformation is reflected in the changes in RMN profiles as the temperature increases. It is noteworthy to mention that the surface chemistry of final carbon material can be tailored by controlling the temperature conditions.

Another relevant N-containing polymer is the polypyrrole. One reason to use polypyrrole instead of polyaniline to synthesize N-doped carbons, would be its higher nitrogen content as compared to polyaniline, because pyrrole monomer provides a nitrogen atom per ring, while polyaniline has a nitrogen atom for each two rings. Although the polymer composition mainly consists of neutral pyrrolic groups, XPS analysis shows that uncharged imine nitrogen atoms are produced [95]. Consequently, the polypyrrole carbonization should evolve in different nitrogen groups in comparison to polyaniline carbonization, since the initial nitrogen groups are not the same. However, the transformation does not only depend on the chosen precursor but also on the experimental conditions. For example, Qie et al. synthesized PPy nanofibers using CTAB ($(C_{16}H_{33})N(CH_3)_3$ Br) as a template and then the PPy nanofibers were heated at 650ºC under nitrogen atmosphere and activated with KOH [96]. From the nitrogen XPS of the carbonized samples, it was observed that the main contribution corresponded to pyrrole groups. Su et al. prepared PPy nanospheres using DTAB ($(C_{12}H_{25})N(CH_3)_3$ Br) as a template and they were subsequently carbonized at different temperatures under N_2 without chemical activation [97]. In this case, XPS results confirmed that the main contribution is related to quaternary N groups at 300, 600, and 800ºC. This difference could be attributed to the activation process. Both chemical and electrochemical method of PPy were reported by Ramírez-Pérez [98]. Both samples were carbonized at 500 and 800ºC and the former sample mostly shown pyrrole groups, while the latter sample exhibited a main peak related to quaternary N groups. Therefore, experimental conditions can determine the final properties (composition, texture, porosity, etc.) of the carbon material regardless of the precursor chosen. To sum up, an appropriate choice of the precursor or synthesis conditions is not only important to tailor the nitrogen groups, but also to determine the doping level. Finally, it can be said that nitrogen-containing polymer carbonization is a versatile route to get a wide range of N-doped nanostructured carbon materials.

Regarding S-containing conducting polymers, the most important is polythiophene (PTh), which has already been studied to prepare S-doped carbon materials to be used in gas storage applications [99,100]. S-doped carbon materials obtained from PTh has gained importance in their use as electrode in electrochemical applications [101]. Other S-containing conducting polymers studied in this field is poly [3,4-ethylenedioxythiophene] (PEDOT) [102–104], which is a PT derivative with a high oxygen content and remarkable electrochemical properties (see Fig. 8.6). For this reason, PEDOT has become a promising precursor to prepare advanced carbon materials.

Concerning the evolution of thiophenic sulphur group present in these conducting polymers after carbonization, it was reported that these thiophenic groups are

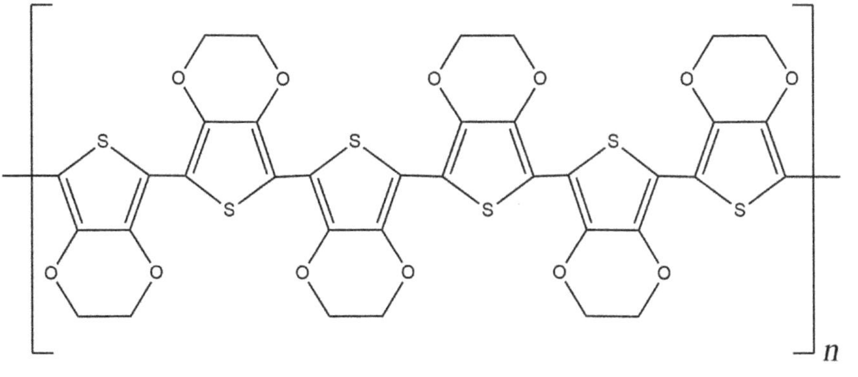

Figure 8.6 Structure of poly[3,4-ethylenedioxythiophene] (PEDOT).

mostly incorporated into graphene sheets, leading to sulphide bridges (−C−S−C−) due to the reducing conditions during the carbonization [99,100] although sulfoxides (−C−SO−C−) and sulfones (−C−SO_2−C−) groups can be formed under oxidizing atmosphere [100].

The combination between different heteroatoms (N, O, S, etc.) into the carbon network sometimes has a synergistic effect. At this point, the carbonization of conducting polymers is a suitable and simple approach to get codoped carbon materials. For example, González-Gaitán prepared functionalized CNTs by electrochemical polymerization of aminobenzene acids (ABA) and then, ABA/CNTs were carbonized at 800ºC under two different atmospheres (N_2 and 3125 ppm O_2 in N_2). It was observed that functionalized CNTs in slightly oxidizing atmosphere leads to a material with an improvement in oxygen reduction reaction compared to that obtained from inert atmosphere, pointing out that a combination of surface nitrogen and oxygen groups has a synergistic effect on electrochemical properties of functionalized CNTs [105].

However, carbon materials obtained directly from heteroatom-containing polymers could not have the adequate porosity, morphology, or structural order to be used as electrode material in energy storage applications. For that reason, carbonization of conducting polymer/carbon material composite is the best option to overcome this drawback, as carbon materials provide a wide range of porosities, morphologies and structural order. A clear example of this is the synthesis of carbon material obtained from carbonization of activated carbon fiber/PANI composite [77]. It was noticed an improvement in the electrochemical performance for carbonized samples in comparison with non-carbonized ones. In the following, a more detailed explanation of this study is presented as a representative example of this topic.

8.3.1 Carbon material based on ACF-PANI

In this work, the main objective was to prepare N-doped carbon using different loadings of polyaniline as doping agent. It was synthesized as described elsewhere [77]. Briefly, aniline adsorption inside the pores of ACF was performed and then,

Table 8.3 **C, H, and N contents (wt%) in the studied samples obtained from elemental analysis**

Sample	N (wt%)	C (wt%)	H (wt%)
ACF–PANI	4.1	93.0	1.4
ACF–PANI (600°C)	3.5	95.3	0.9
ACF–PANI (800°C)	2.7	96.5	0.5

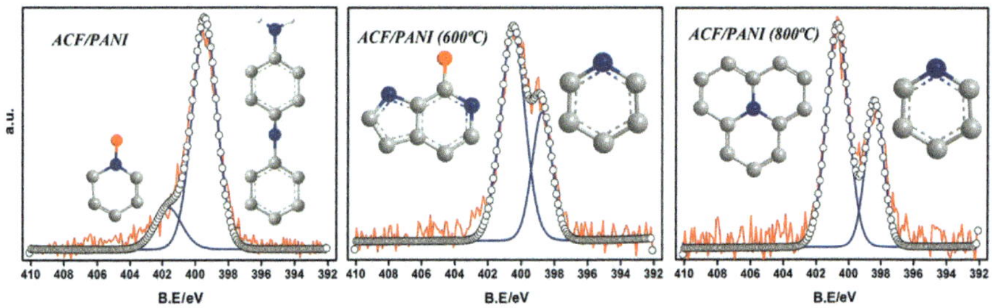

Figure 8.7 N1s XPS spectra of ACF/PANI and carbonized samples at 600/800°C (blue atom: N).

aniline was polymerized by chemical method to obtain a polyaniline thin film inside the porosity. Finally, samples were carbonized at 600/800°C under nitrogen atmosphere. The elemental composition results revealed a higher nitrogen content for nontreated samples than for heat-treated samples (Table 8.3). The amount of hydrogen and oxygen also decreased after the heat treatment due to the PANI carbonization and surface oxygen groups decomposition.

N1s XPS (Fig. 8.7) shows important changes as the temperature increases. The ACF–PANI sample (non-carbonized) has a main peak at around 399.5 eV related to neutral amine and a small peak at around 402 eV, which is attributed to oxidized N species due to PANI doping. For samples carbonized up to 600ºC two peaks appear at different binding energies compared to the ACF–PANI (i.e., at 398.7 and at 400.5 eV), which are associated to pyridine groups and positively charged nitrogen groups (pyrrole and pyridone), respectively. As temperature increases, the second peak shifts slightly to higher binding energy energies (400.7 eV), what indicates the formation of quaternary nitrogen in addition to pyrrole. The presence of pyridone should be very small in agreement with the low oxygen content calculated from elemental analysis data. These results are in agreement with the carbonization mechanism proposed for PANI [94,106].

Moreover, N_2 adsorption experiments at 77 K for ACF–PANI samples with different loadings and those carbonized at 600°C and 800°C showed Type I isotherms, which are characteristic of microporous solids. Table 8.4 contains the BET surface area and micropore volumes calculated from N_2 and CO_2 adsorption, respectively.

Table 8.4 Porous texture characterization results corresponding to all samples

Sample	A_{BET}/m^2 g^{-1}	$V_{DR}(N_2)$/cm^3 g^{-1}	$V_{DR}(CO_2)$/cm^3 g^{-1}
ACF	2300	1.06	0.51
ACF/PANI	1155	0.42	0.26
ACF/PANI (600°C)	1310	0.54	0.32
ACF/PANI (800°C)	1355	0.52	0.34

It can be seen that the adsorption capacity was significantly reduced because of the thin PANI film deposition over the surface of the fibers but adsorption capacity increases after carbonization.

In conclusion, a straightforward carbonization process of conducting polymer/carbon composite produces advanced N-doped carbon materials that have a porous texture and surface chemistry, which are useful for their use as electrodes in energy storage applications [77].

8.3.2 Strategies to transform CP or CP/carbon composites into carbon material

In the last section, a simple carbonization under nitrogen atmosphere was discussed as a method to obtain carbon materials from CPs or CP/carbon material composites. However, there are other strategies to transform CPs into valuable doped carbon materials. For instance, hydrothermal carbonization (HTC) has become an interesting route to synthesize them. Generally, this methodology consists in placing a CP-containing aqueous solution into a stainless steel autoclave with PTFE liner. This solution is then heat-treated using mild temperatures (<200°C) and under self-generated pressures [107]. However, the initial precursor is not always a CP and it could be directly a monomer [108–111], or even a carbon material derived from a CP [112]. An example of the use of a monomer as precursor was reported by Liao and coworkers [109], who prepared N-doped hollow carbon spheres (N-HCS) from a pyrrole monomer under hydrothermal conditions by using SiO_2 spheres as a hard template and subsequent removal of the silica. N-doped hollow spheres showed an average size of 370 nm and high surface area (>700 m^2 g^{-1}) attributed to the micropores, though N_2 adsorption–desorption revealed the existence of voids among spheres. The nitrogen content measured by elemental analysis was around 8.5 wt%. Taking into account the above characterization, these N-HCS can be considered as promising materials for supercapacitors. In addition, HTC synthesis has also shown to be useful when the precursor was a carbon material derived from CP. Gavrilov and coworkers [112] previously synthesized carbonized PANI as described elsewhere [113,114]. Finally, the carbonized sample was heat-treated under hydrothermal conditions using KOH solution as solvent. The carbon material based on PANI shown a high nitrogen content (> 9 wt%) and low surface area (<350 m^2 g^{-1}). From N_2 adsorption–desorption isotherms, it was observed the presence of micropores and macropores [112].

Another interesting procedure to synthesize carbon materials from conducting polymers is the microwave heating method, which consists of the interaction of matter with electromagnetic radiation. The heating lasts a short time and it takes place in a specific spot [115]. This method allows the preparation of carbon materials with different morphologies since the heating is fast and it produces the conversion of CP into nanocarbon material, retaining the morphology of precursors after carbonization. Zhang and Manohar synthesized different carbon nanostructures (nanotubes, nanospheres or nanofibers) form PPy by using MW heating [116]. These carbon materials retained their initial morphology, but the size dimensions decreased due to a dedoped process. In addition, it was observed from the HRTEM micrographs that the carbon materials had graphitic domains. This method has also been used to grow CNTs by means of ferrocene and PPy in a few seconds under air atmosphere [117], which is a great advantage over high cost conventional methods (laser ablation, arc-discharge or chemical vapor deposition). In conclusion, MW heating seems to be a promising strategy to prepare advanced carbon materials from CPs with different morphologies at low cost, as this technique does not require inert atmosphere or high temperatures.

As it can be extracted from the information compiled throughout these last sections, the carbonization of CP/CM composites or CP constitutes a powerful tool to synthesize carbon materials with different features, in terms of morphology, chemical composition, physical properties, and so forth. Some of the most representative examples reported in the literature are listed in Table 8.5.

8.4 Applications in electrical energy storage

As already mentioned, conducting polymer-based materials have been employed in different applications such as sensors, energy storage, etc. [128]. In particular, the present chapter addresses their use in energy-related devices, as these materials have been widely applied in the preparation of electrodes in batteries and supercapacitors. Inspired by the numerous studies facing with the preparation and optimization of supercapacitors conducted in our research group, a summary of the use of CP/CM composites in energy-storage devices is discussed anon.

Supercapacitors are electrical energy storage devices based on the formation of an electrical double layer generated on the interface between the electrode and the electrolyte. These devices provide larger power density than other charge storage systems, such as batteries and fuel cells, but their energy density is lower [129]. These properties depend on the components of the supercapacitor, mainly on the electrodes and electrolytes.

Conducting polymers have been proven to work as positive electrodes of asymmetric capacitors mainly due to their large capacitance and electrical

Table 8.5 **Some examples of carbon materials derived from carbonization of different CP/CM**

Precursor	Carbon material type	Synthesis/remarks	Properties	Reference
PPy Nanoparticles (NPs)	N-carbon NPs	Chem. polymerization (aq) Carbonization temperature (T_{carb}) at 1000°C in N_2	$d\sim$ 50 nm enhance ρ ~3→22 S cm^{-1} 370 S cm^{-1} at 2300	[118]
PPy/carbon nanofibers (CNFs)	N-CNFs	Chem. polymerization (aq) T_{carb} ~550–1100°C in N_2	$d\sim$ 85 nm (900°C) Micro/mesopores/macropores Good σ, graphitic domains, N (wt%) = 4–12 (pyridinic and quaternary N)	[119]
PPy nanospheres (NS)	N-CNS	Chem. Polymerization (org) T_{carb} ~900°C in N_2	Microporous spheres (100–200 nm) with small holes $S_{BET} > 450$ m^2 g^{-1} Enhance σ ~0.01→10/100 S cm^{-1} N(wt %) >3, pyridinic and quaternary N groups	[81]
PPy NTs	N-hollow CNTs	Chem. polymerization with $FeCl_3$ T_{carb} ~900°C in N_2 Chem. activation in KOH at 600°C	Capped/uncapped CNTs, diameter (inner/outer) ~50 nm/~70 nm, high S_{BET}, high nitrogen content (pyridine groups), good electrical conductivity	[80]
PPy (NFs, NSs, NTs)	Doped nanocarbons	Seeding and chem. pol. using V_2O_5 Microwave carbonization	Retain morphology and reduce size, moderate surface area, graphitic domains	[116]
PPy	N-doped CNF webs (cross-linked nanofibers)	Chem. pol. with APS Oxidative assembly template (CTAB). Chemical activation with KOH	Micropores and small mesopores Low degree of graphitizacion N(wt%) = 10 (pyrrole groups and small amount of pyridine groups)	[96]

(Continued)

Table 8.5 (Continued)

Precursor	Carbon material type	Synthesis/remarks	Properties	Reference
PPy nanospheres (PNs)	N-CNS	Chem. polymerization T_{carb} ~300–1300°C in N_2 DTAB	d~ 80–100nm Microporous and low S_{BET} High graphitizacion degree N(wt%) = 4–14 (Quaternary N)	[97]
PPy/SiO_2 core-shell	N-hollow carbon nanospheres	Hydrothermal pol. with APS using SiO_2 as template. Carbonization at 850°C in N_2	d~ 370 nm Shell thickness (~15 nm) $S_{BET} > 700\ m^2\ g^{-1}$, Amorphous structure N(wt%) >8	[109]
PANI (NF, NS, NT)	activated N-nanocarbons	Reactive-template technique (MnO_2) Chem. activ. in KOH at 800°C (without previous carbonization)	Highly porous (micro/mesopores), decrease in pyridine groups High graphitic degree Good σ	[83]
Aniline	MnO_x/N-carbons	Hydrothermal pol. with $KMnO_4$ $Mn(NO_3)_2$ at 180°C. Carbonization at 800°C in N_2	Nanoparticles (~20 nm) $S_{BET} < 250\ m^2\ g^{-1}$ PSD~15 nm	[108]
PANI nanowires	N-carbon nanowires	Chem. polymerization using CTAB as template. T_{carb} ~ 600–900°C in N_2	Morphology retained (600–800°C) ↑ Degree of graphitization Micro/mesopores High content in N(wt%) >7.	[120]
PANI (granular)	Hierarchical porous carbon (HPC)	Chem. polymerization (aq) Activation with KOH at 700°C in N_2	High S_{BET} ~2200 $m^2\ g^{-1}$ Hierarchical porous texture High content in oxygen	[121]

PANI NTs	N-activated CNTs	Chem. polymerization (aq) T_{carb} ~500°C in N_2 Activation with KOH at 700°C in N_2	d ~20–120 nm S_{BET} >3200 $m^2\ g^{-1}$ Pores >20 nm, ↓ N wt% Amorphous structure Relatively good σ	[122]
PANI NTs	N-activated CNTs	Chem. polymerization with APS T_{carb} ~700°C in Ar Activation with KOH at 700°C in Ar	140–180 nm Tubular morphology S_{BET} >3000 $m^2\ g^{-1}$, Micro/mesopores interconnected ↑ Content in oxygen, N (at.%) ~2.5 Low resistivity	[123]
PANI/doped-PANI	Nanostructured carbons (nanorods, NTs, NSs,. . .)	Chem. polymerization with APS T_{carb} ~800°C in N_2 Dopants(salicylic acids derivatives)	Wide range of diameters morphology remains micro/mesoporosity depends on the dopant ↑ Nitrogen content (mainly pyridinic and quaternary N groups).	[124]
Carbonized PANI	N-mesoporous carbon	Chem. polymerization with APS T_{carb} ~800°C in N_2 Hydrothermal treatment in KOH at 200°C	Mesoporous texture, higher N content in the surface than bulk, (pyridinic, quaternary N, pyrrolic) $\sigma = 0.38$ S cm^{-1}	[125]
PANI	N-mesoporous carbon	Chem. polymerization by nano. $CaCO_3$ template	Irregular particle morphology Low S_{BET} Interconnected pores (2–60 nm) Amorphous structure N(wt%) > 7 (pyridinic, pyrrolic, quaternary N groups) Low resistance	[126]

(Continued)

Table 8.5 (Continued)

Precursor	Carbon material type	Synthesis/remarks	Properties	Reference
Carbonized PANI	Hydrothermal carbon PANI nanostructures	Chem. Polymerization with APS Carbonization at 800°C Hydrothermal treatment in KOH at 150°C	Micro/macroporous Low S_{BET} (~300 $m^2\ g^{-1}$) N(wt%) ≈ 9 Disordered structure	[112]
PEDOT-PSS	Hierarchical porous carbon (HPC) foams	Chem. Polymerization (aq.)	Microporous/mesoporous $S_{BET} > 350\ m^2\ g^{-1}$, PSD (~6–40 nm) ↑ Graphitization degree, S (wt%) > 2.3 C-S and C-N (active sites)	([80])
PEDOT	Helical graphite	Electrochem. polCarbonization at 800°C.	Morphology retained $\sigma = 10\ S\ cm^{-1}$	[127]
Thiophene (twin monomer)	S-hollow spheres	Chem. Polymerization (H^+) using functionalized SiO_2 as template.	Hollow spheres with tailored diameter/ thickness Microporous/mesoporous $S_{BET} > 650\ m^2\ g^{-1}$ C-S (active sites)	[82]
2-Thiophenemetanol	S-doped activated carbon	Chem. polymerization with $FeCl_3$ (org.) KOH activation (600–850°C)	Mainly microporous ($S_{BET} > 2000\ m^2\ g^{-1}$), S (wt%) ≈ 2.5–8.5, particle size (10–100 μm), larger graphitic segments	[101]

conductivity [128]. However, they experience volumetric changes along cycling, reducing the lifetime of the device. A strategy for improving their performance is to develop conducting polymer/carbon material composites, since carbon can act as a support, improving the mechanical properties of the conducting polymer, while the conducting polymer contributes to the specific capacitance by providing fast pseudo-Faradaic reactions.

Frackowiak and coworkers [31,32] prepared symmetric supercapacitors using composites that combined conducting polymers (PANI, PPy and poly-(3,4-ethilene-dioxythiophene) (PEDOT)) with multiwall carbon nanotubes (MWCNTs). Carbon nanotubes play an important role in improving the mechanical properties of the conducting polymer, since they avoid detrimental mechanical changes during cycling. Also, they act as a support allowing a homogeneous distribution of the conducting polymer in the composite. Extremely high values of specific capacitance were found for both composites, providing values between 200 and 360 F g^{-1}. However, when the operation voltage was increased both symmetric capacitors experimented a loss of 40% of capacitance among cycles, evidencing the low stability of the composites. In contrast, PEDOT/MWCNTs composite showed a stable performance along cycles when working in organic medium, although the capacitance was lower (100 F g^{-1}). The composites evidenced different suitable operation and their stability was studied for the design of asymmetric supercapacitors. By using P/Py/MWCNTs as negative electrode and PANI/MWCNTs as positive electrode a capacitance value of 320 F g^{-1} was determined for every electrode working at 0.6 V.

More research to deepen into the improvement of the properties of conducting polymer/carbon composites has been developed by several researchers. Meng et al. [130] prepared PANI/MWCNTs buckypaper which evidences larger flexibility than conventional PANI/MWCNTs composites. These composites provided larger specific capacitance, lower internal resistivity, and higher stability, making them very promising in energy storage devices.

Conducting polymer/graphene composites have also evidenced an extraordinary performance as electrodes for supercapacitors. PANI/graphene composites were prepared [19,20] and they evidenced large capacitance (480 F g^{-1} at 100 mA g^{-1}), high conductivity and good stability among cycles. PANI/graphene composites were also prepared as "paper" in order to increase their flexibility by electropolymerization of PANI over graphene [131]. The composites evidenced enhanced specific capacitance (763 F g^{-1}), electrical conductivity, and stability during cycling. Similar approaches have been carried out by using polypyrrole [132], and an extraordinary capacitance of 482 F g^{-1} was determined at 0.5 A g^{-1} by using a three electrode cell configuration. The composite also evidenced good stability along cycles.

Conducting polymers/carbon fibers composites have also been used as electrodes for supercapacitors [133]. They reported an increase of 69% of capacitance related to the pseudocapacitance provided by the PANI, along with the preservation of the double layer contribution of the carbon fiber. More details about the electrochemical performance of PANI/carbon composites prepared with porous carbon on the electrochemical behavior are described in the following subsection (Table 8.6).

Table 8.6 Some examples of electrochemical properties of different composites and their measurement conditions

Electrode	Electrolyte	Voltage or potential window (V)	Current density or scan rate	Capacitance ($F\ g^{-1}$)	Cyclability	Reference
PPy/MWCNTs (20%)	1 M H_2SO_4	0.6	192 mA g^{-1}	200[a]	85% (0.6 V) 500 cycles 60% (0.8 V) 500 cycles	[31,32]
PANI/MWCNTs (20%)	1 M H_2SO_4	0.4	232 mA g^{-1}	360[a]	90% (0.4 V) 500 cycles 60% (0.6 V) 500 cycles	[31,32]
PEDOT/ MWCNTs (20%)	1 M $TEABF_4$ (acetonitrile)	1.5	225 mA g^{-1}	100[a]	Good stability	[15,31]
PANI/MWCNTs buckypaper	1 M H_2SO_4	0.8 V	200 mA g^{-1}	424[a]	89.4% (0.8 V) 1000 cycles	[130]
PANI/Graphene	2 M H_2SO_4	−0.2 to 0.8 AgCl/Ag	100 mA g^{-1}	480[b]	70%, 1.5 A g^{-1} 1000 cycles	[19,20]
PANI/Graphene paper	1 M H_2SO_4	−0.2 to 0.8 AgCl/Ag	1000 mA g^{-1}	763[b]	82%, 5 A g^{-1} 1000 cycles	[131]
PPy/Graphene	1 M H_2SO_4	− 0.2 to 0.7/SCE	500 mA g^{-1}	482[b]	95%, 50 mV s^{-1} 1000 cycles	[132]
PANI/Activated carbon fiber	1 M H_2SO_4	−0.6 to 0.6/Hg/ Hg_2SO_4	5 mV s^{-1}	222	Stable capacitance,1 A g^{-1}, 5000 cycles	[133]

[a]Capacitance measured in two electrode cell configuration and referred to one electrode. For comparison with a capacitor, it should be considered four times smaller.
[b]Capacitance measured in three electrode cell configuration.

8.4.1 *Activated carbon fiber-PANI electrodes as positive electrodes in asymmetric hybrid capacitors*

In general, conducting polymer/carbon material composites have an improved performance as electrodes for supercapacitor compared to both materials separately. There are a few parameters that play a key role on the performance of these composites depending on the carbon material. In the case of activated carbon or other porous carbons, the thickness of the film has to be adequate since specific capacitance can decrease with the loading content [34,134]. Consequently, the preparation of PANI-carbon composites with thin films of PANI inside the microporosity of a carbon material with large surface area seems to be an appropriate approach for obtaining PANI/CM composites adequate as electrodes for supercapacitors. As discussed in Section 8.2.1, the surface chemistry of the CM also plays an important role since may favor porosity-blocking upon polymerization [14].

Salinas-Torres and coworkers [39] prepared PANI/ACF composites by electrochemical and chemical polymerization and obtained composites with homogeneous distribution of PANI inside the microporosity (See Section 8.2.2). The authors used these ACF/PANI composites for preparing an asymmetric hybrid capacitor by combining them with activated carbon. The positive electrode (named as ACF/PANI) was characterized by a high surface area and pseudocapacitance contributions of PANI, whose presence on the carbon support does not reduce the ion diffusion inside the microporosity. An AC was chosen as negative electrode due to its high surface area and high overpotential for hydrogen evolution reaction [39]. For comparison, two supercapacitors were prepared using different positive electrodes: ACF and AC for obtaining ACF/AC and AC/AC supercapacitors.

Fig. 8.8 shows the galvanostatic charge–discharge cycles obtained for AC/AC, ACF/AC, and ACF-PANI/AC supercapacitors at 1.6 V of operation voltage. AC/

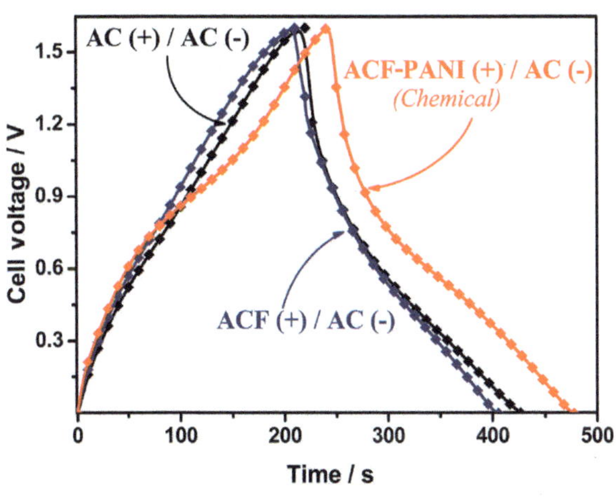

Figure 8.8 Galvanostatic charge-discharge cycles for AC/AC, ACF/AC, and ACF-PANI/AC supercapacitors at 500 mA g^{-1}. $\Delta V = 1.6$ V. 0.5 M H_2SO_4 solution.

AC and ACF/AC are composed of carbon materials and, consequently, evidence a quasi-triangular shape close to the ideal electrical double layer capacitors. In contrast to these performances, ACF-PANI/AC shows some deviations from this shape because of the pseudo-Faradaic reactions occurring on the polyaniline film. The variation of specific capacitance with increasing the current density shows that ACF-PANI/AC supercapacitor has the largest capacitance at all current densities. Also, the capacitance decreases in all cases when current density is increased as consequence of ion diffusion problems, but the decrease experimented by ACF-PANI/AC supercapacitor is around 17% lower than the diminution observed for other supercapacitors. The differences between the performance of ACs and ACFs are a consequence of the lower tortuosity of the porosity of ACFs [135]. More remarkable is the improvement observed for ACF-PANI/AC supercapacitor as consequence of the presence of polyaniline, which provides a charge by pseudocapacitive processes while avoiding diffusion problems due to its thin thickness. Also, it is responsible of the shift of oxygen evolution to more positive potentials since the oxidation of polyaniline is produce before the oxidation of the carbon material, producing an increase of the stability of the electrode, that was confirmed by a cyclability test of 1000 cycles. This test showed a capacitance retention larger than 80% in case of ACF-PANI/AC supercapacitor. These improvements result in an increase of energy and power densities of around 20–25%.

8.5 Conclusions

Porous carbons have been extensively studied as materials for different electrochemical applications including double layer capacitors because of their large surface area, chemical stability, easy processability, and long cycle life. However, in the case of double layer capacitors, two important drawbacks are the low conductivity of the highly porous carbon materials and that the energy storage occurs mainly through a double layer mechanism. This has provoked strong research on trying to improve the electrical conductivity, which may increase the power of the device, and to introduce pseudocapacitance contribution that may increase the energy stored. One interesting approach is the addition of a conducting polymer thin film. This strategy opens the possibility of preparing different composites by selecting the polymer composition and, then, generating different and new properties of the pristine carbon material. In addition, further heat treatments may produce advanced carbon materials that remain doped with heteroatoms that provide additional chemical and physical properties opening new fields of application such as in electrocatalysis. This review has focused on the preparation methods of conducting polymer/carbon material composites as well as on the carbon materials derived from them and it offers to the reader the knowledge of the great opportunities for the synthesis of tailored carbon materials by a correct selection of both the conducting polymer and carbon material and the preparation method.

Acknowledgments

The authors thank financial support from MINECO (projects CTQ2015-66080-R (MINECO/FEDER) and MAT2016-76595-R) and GV (PROMETEOII/2014/010). AG thanks the Heiwa Nakajima Foundation for the PhD Thesis grant. MJM thanks GV for VALi + d contract (ACIF/2015/374). DST thanks ULg and EC for BeIPD-Marie Curie COFUND Incoming Fellowship (No 600405).

References

[1] I. Velo-Gala, et al., Surface modifications of activated carbon by gamma irradiation, Carbon 67 (2014) 236–249.

[2] D. Cazorla-Amorós, et al., Grand challenges in carbon-based materials research, Front. Mater. (2014) 1–3.

[3] M. Inagaki, et al., New Carbons - Control of Structure and Functions, Elsevier, Amsterdam, 2000, p. 240.

[4] N. Job, et al., Carbon aerogels, cryogels and xerogels: influence of the drying method on the textural properties of porous carbon materials, Carbon 43 (12) (2005) 2481–2494.

[5] M. Knupfer, et al., Electronic properties of carbon nanostructures, Surf. Sci. Rep. 42 (1-2) (2001) 1–74.

[6] C. Peng, et al., Carbon nanotube and conducting polymer composites for supercapacitors, Progr. Nat. Sci. 18 (7) (2008) 777–788.

[7] H. Shirakawa, et al., The discovery of polyacetylene film: the dawning of an era of conducting polymers (Nobel lecture), Angew. Chem. Int. Ed. 40 (14) (2001) 2575–2580.

[8] C.M. Hangarter, et al., Hybridized conducting polymer chemiresistive nano-sensors, Nano Today 8 (1) (2013) 39–55.

[9] A.J. Heeger, et al., Semiconducting and metallic polymers: the fourth generation of polymeric materials (Nobel Lecture), Angew. Chem. Int. Ed. 40 (14) (2001) 2591–2611.

[10] M. Ates, et al., A review study of (bio)sensor systems based on conducting polymers, Mater. Sci. Eng. C Mater. Biol. Appl. 33 (4) (2013) 1853–1859.

[11] A.G. MacDiarmid, et al., "Synthetic metals": a novel role for organic polymers (Nobel lecture), Angew. Chem. Int. Ed. 40 (14) (2001) 2581–2590.

[12] P. Simon, et al., Materials for electrochemical capacitors, Nat. Mater. 7 (11) (2008) 845–854.

[13] M.J. Bleda-Martínez, et al., Electrochemical methods to enhance the capacitance in activated carbon/polyaniline composites, J. Electrochem. Soc. 155 (9) (2008) A672–A678.

[14] M.J. Bleda-Martínez, et al., Polyaniline/porous carbon electrodes by chemical polymerisation: effect of carbon surface chemistry, Electrochim. Acta 52 (15) (2007) 4962–4968.

[15] K. Lota, et al., Capacitance properties of poly(3,4-ethylenedioxythiophene)/carbon nanotubes composites, J. Phys. Chem. Solids (2004) 295–301.

[16] S. Suematsu, et al., Conducting polymer films of cross-linked structure and their QCM analysis, Electrochim. Acta 45 (22–23) (2000) 3813–3821.

[17] Y. Yang, et al., Graphene-based materials with tailored nanostructures for energy conversion and storage, Mater. Sci. Eng. R Rep. 102 (2016) 1–72.

[18] G. Yu, et al., Hybrid nanostructured materials for high-performance electrochemical capacitors, Nano Energy 2 (2) (2013) 213–234.

[19] K. Zhang, et al., Graphene/polyaniline nanofiber composites as supercapacitor electrodes, Chem. Mater. 22 (4) (2010) 1392–1401.

[20] L.L. Zhang, et al., Graphene-based materials as supercapacitor electrodes, J. Mater. Chem. 20 (29) (2010) 5983–5992.

[21] C. Li, et al., Conducting polymer nanomaterials: electrosynthesis and applications, Chem. Soc. Rev. 38 (8) (2009) 2397–2409.

[22] M. Baibarac, et al., Nanocomposites based on conducting polymers and carbon nanotubes: From fancy materials to functional applications, J. Nanosci. Nanotechnol. 6 (2) (2006) 289–302.

[23] M. Hughes, et al., Electrochemical capacitance of a nanoporous composite of carbon nanotubes and polypyrrole, Chem. Mater. 14 (4) (2002) 1610–1613.

[24] S. Zhou, et al., Graphene-wrapped polyaniline nanofibers as electrode materials for organic supercapacitors, Carbon 52 (2013) 440–450.

[25] W. Lei, et al., Conducting polymer composites with graphene for use in chemical sensors and biosensors, Microchim. Acta 181 (7–8) (2014) 707–722.

[26] Z.F. Li, et al., Fabrication of high-surface-area graphene/polyaniline nanocomposites and their application in supercapacitors, ACS Appl. Mater. Interfaces 5 (7) (2013) 2685–2691.

[27] A. Laforgue, et al., Hybrid supercapacitors based on activated carbons and conducting polymers, J. Electrochem. Soc. 148 (10) (2001) A1130–A1134.

[28] X. Li, et al., Fabrication and characterization of well-dispersed single-walled carbon nanotube/polyaniline composites, Carbon 411 (2003) 1670–1673.

[29] H.J. Shin, et al., CNT/PEDOT core/shell nanostructures as a counter electrode for dye-sensitized solar cells, Synth. Met. 161 (13–14) (2011) 1284–1288.

[30] C. Meng, et al., Flexible carbon nanotube/polyaniline paper-like films and their enhanced electrochemical properties, Electrochem. commun. 11 (1) (2009) 186–189.

[31] E. Frackowiak, et al., Supercapacitors based on conducting polymers/nanotubes composites, J. Power Sources 153 (2) (2006) 413–418.

[32] V. Khomenko, et al., Determination of the specific capacitance of conducting polymer/nanotubes composite electrodes using different cell configurations, Electrochim. Acta 50 (12) (2005) 2499–2506.

[33] K.M. Lee, et al., Effects of mesoscopic poly(3,4-ethylenedioxythiophene) films as counter electrodes for dye-sensitized solar cells, Thin Solid Films 518 (6) (2010) 1716–1721.

[34] K.S. Ryu, et al., Electrochemical capacitor with chemically polymerized conducting polymer based on activated carbon as hybrid electrodes, Synth. Met. 153 (1–3) (2005) 89–92.

[35] C. Yan, et al., Polyaniline-modified activated carbon electrodes for capacitive deionisation, Desalination 333 (2014) 101–106.

[36] H. Zengin, et al., Synthesis and characterization of polyaniline/activated carbon composites and preparation of conductive films, Mater. Chem. Phys. 120 (1) (2010) 46–53.

[37] Z.-H. Zhou, et al., In situ chemical deposition of polyaniline on activated carbon for electrochemical capacitors, Chin. J. Chem. 24 (1) (2006) 13–16.

[38] L. Lai, et al., Tuning graphene surface chemistry to prepare graphene/polypyrrole supercapacitors with improved performance, Nano Energy 1 (5) (2012) 723–731.
[39] D. Salinas-Torres, et al., Asymmetric hybrid capacitors based on activated carbon and activated carbon fibre–PANI electrodes, Electrochim. Acta 89 (2013) 326–333.
[40] W. Martínez Millán, et al., Characterization of composite materials of electroconductive polymer and cobalt as electrocatalysts for the oxygen reduction reaction, Int. J. Hydrog. Energy 34 (2) (2009) 694–702.
[41] R.R. Salunkhe, et al., Ultrahigh performance supercapacitors utilizing core–shell nanoarchitectures from a metal–organic framework-derived nanoporous carbon and a conducting polymer, Chem. Sci. (2016) 1–10.
[42] Y. Meng, et al., Hierarchical porous graphene/polyaniline composite film with superior rate performance for flexible supercapacitors, Adv. Mater. 25 (48) (2013) 6985–6990.
[43] C. Xia, et al., Highly stable supercapacitors with conducting polymer core-shell electrodes for energy storage applications, Adv. Energy Mater. 5 (8) (2015).
[44] C.P. Fonseca, et al., Influence of the polymeric coating thickness on the electrochemical performance of Carbon Fiber/PAni composites, Chem. Phys. Lett. 25 (5) (2015) 425–432.
[45] H. Zengin, et al., Carbon nanotube doped polyaniline, Adv. Mater. 14 (20) (2002) 1480–1483.
[46] E. Frackowiak, et al., Carbon materials for the electrochemical storage of energy in capacitors, Carbon 39 (2001) 937–950.
[47] K. Saranya, et al., Developments in conducting polymer based counter electrodes for dye-sensitized solar cells - an overview, Eur. Polym J. 66 (2015) 207–227.
[48] Y. Liu, et al., Ethylene glycol reduced graphene oxide/polypyrrole composite for supercapacitor, Electrochim. Acta 88 (2013) 519–525.
[49] M.J. Bleda-Martínez, et al., Chemical and electrochemical characterization of porous carbon materials, Carbon 44 (13) (2006) 2642–2651.
[50] D. Salinas-Torres, et al., Characterization of activated carbon fiber/polyaniline materials by position-resolved microbeam small-angle X-ray scattering, Carbon 50 (3) (2012) 1051–1056.
[51] D. Bélanger, et al., Electrografting: a powerful method for surface modification, Chem. Soc. Rev. 40 (7) (2011) 3995–4048.
[52] J.-E. Huang, et al., Well-dispersed single-walled carbon nanotube/polyaniline composite films, Carbon 41 (14) (2003) 2731–2736.
[53] M. Gao, et al., Aligned coaxial nanowires of carbon nanotubes sheathed with conducting polymers, Angew. Chem. 39 (20) (2000) 3664–3667.
[54] V. Gupta, et al., Polyaniline/single-wall carbon nanotube (PANI/SWCNT) composites for high performance supercapacitors, Electrochim. Acta 52 (4) (2006) 1721–1726.
[55] J. Wang, et al., Capacitance properties of single wall carbon nanotube/polypyrrole composite films, Compos. Sci. Technol. 67 (14) (2007) 2981–2985.
[56] M. Baibarac, et al., Covalent functionalization of single-walled carbon nanotubes by aniline electrochemical polymerization, Carbon 42 (15) (2004) 3143–3152.
[57] J. Ge, G. Cheng, et al., Transparent and flexible electrodes and supercapacitors using polyaniline/single-walled carbon nanotube composite thin films, Nanoscale 3 (8) (2011) 3084–3088.
[58] J. Liu, et al., A promising way to enhance the electrochemical behavior of flexible single-walled carbon nanotube/polyaniline composite films, J. Phys. Chem. C 114 (46) (2010) 19614–19620.

[59] J.H. Chen, et al., Electrochemical synthesis of polypyrrole films over each of well-aligned carbon nanotubes, Synth. Met. 125 (3) (2002) 289–294.
[60] M. Guo, et al., Fabrication of polyaniline/carbon nanotube composite modified electrode and its electrocatalytic property to the reduction of nitrite, Anal. Chim. Acta 532 (1) (2005) 71–77.
[61] E. Frackowiak, et al., Nanotubular materials as electrodes for supercapacitors, Fuel Process. Technol. 77–78 (2002) 213–219.
[62] J.D. Leary, et al., Preparation of pseudocapacitor electrodes via electrodeposition of polyaniline on nonwoven carbon fiber fabrics, J. Appl. Polym. Sci. 133 (16) (2016) 1–8.
[63] A. Österholm, et al., Electrochemical incorporation of graphene oxide into conducting polymer films, Electrochim. Acta 83 (2012) 463–470.
[64] H. Wei, et al., Electropolymerized polypyrrole nanocomposites with cobalt oxide coated on carbon paper for electrochemical energy storage, Polymer 67 (2015) 192–199.
[65] G. Wang, et al., Simple and fast synthesis of polyaniline nanofibers/carbon paper composites as supercapacitor electrodes, J. Energy Storage 7 (2016) 99–103.
[66] B. Yao, et al., Paper-based solid-state supercapacitors with pencil-drawing graphite/polyaniline networks hybrid electrodes, Nano Energy 2 (6) (2013) 1071–1078.
[67] S. Ameen, et al., Hydrazine chemical sensing by modified electrode based on in situ electrochemically synthesized polyaniline/graphene composite thin film, Sensors Actuators B Chem. 173 (2012) 177–183.
[68] A. Davies, et al., Graphene-based flexible supercapacitors: Pulse-electropolymerization of polypyrrole on free-standing graphene films, J. Phys. Chem. C 115 (35) (2011) 17612–17620.
[69] Z. Mandić, et al., The influence of counter-ions on nucleation and growth of electrochemically synthesized polyaniline film, Electrochim. Acta. 42 (9) (1997) 1389–1402.
[70] L. Duić, et al., Polymer-dimer distribution in the electrochemical synthesis of polyaniline, Electrochim. Acta. 40 (11) (1995) 1681–1688.
[71] A.A. Nekrasov, et al., Electrochemical and chemical synthesis of polyaniline on the surface of vacuum deposited polyaniline films, J. Electroanal. Chem. 412 (1–2) (1996) 133–137.
[72] M.N. Hyder, et al., Layer-by-layer assembled polyaniline nanofiber/multiwall carbon nanotube thin film electrodes for high-power and high-energy storage applications, ACS Nano 5 (11) (2011) 8552–8561.
[73] G. Wang, et al., A review of electrode materials for electrochemical supercapacitors, Chem. Soc. Rev. 41 (2) (2012) 797–828.
[74] Q. Wang, et al., Carbon materials for high volumetric performance supercapacitors: design, progress, challenges and opportunities, Energy Environ. Sci. 9 (3) (2016) 729–762.
[75] P. Xiong, et al., Recent advances on multi-component hybrid nanostructures for electrochemical capacitors, J. Power Sources 294 (2015) 31–50.
[76] J. Yan, et al., Recent advances in design and fabrication of electrochemical supercapacitors with high energy densities, Adv. Energy Mater. 4 (4) (2014).
[77] D. Salinas-Torres, et al., Improvement of carbon materials performance by nitrogen functional groups in electrochemical capacitors in organic electrolyte at severe conditions, Carbon 82 (2015) 205–213.
[78] W. Gu, et al., Review of nanostructured carbon materials for electrochemical capacitor applications: advantages and limitations of activated carbon, carbide-derived carbon,

zeolite-templated carbon, carbon aerogels, carbon nanotubes, onion-like carbon, and graphene, Wiley Interdiscip. Rev. Energy Environ. 3 (5) (2014) 424–473.
[79] H. Nishihara, et al., Templated nanocarbons for energy storage, Adv. Mater. 24 (33) (2012) 4473–4498.
[80] C. Guo, et al., N- and O-doped carbonaceous nanotubes from polypyrrole for potential application in high-performance capacitance, J. Power Sources 247 (2014) 660–666.
[81] C. Shen, et al., Facile synthesis of polypyrrole nanospheres and their carbonized products for potential application in high-performance supercapacitors, Polymer 55 (12) (2014) 2817–2824.
[82] F. Böttger-Hiller, et al., Sulphur-doped porous carbon from a thiophene-based twin monomer, Chem. Commun. 48 (85) (2012) 10568–10570.
[83] W. Chen, et al., Shape-controlled porous nanocarbons for high performance supercapacitors, J. Mater. Chem. A 2 (15) (2014) 5236–5243.
[84] M.J. Bleda-Martínez, et al., Role of surface chemistry on electric double layer capacitance of carbon materials, Carbon 43 (13) (2005) 2677–2684.
[85] T.A. Centeno, et al., The role of textural characteristics and oxygen-containing surface groups in the supercapacitor performances of activated carbons, Electrochim. Acta 52 (2) (2006) 560–566.
[86] M. Seredych, et al., Surface functional groups of carbons and the effects of their chemical character, density and accessibility to ions on electrochemical performance, Carbon 46 (11) (2008) 1475–1488.
[87] D. Hulicova-Jurcakova, et al., Combined effect of nitrogen- and oxygen-containing functional groups of microporous activated carbon on its electrochemical performance in supercapacitors, Adv. Funct. Mater. 19 (3) (2009) 438–447.
[88] O. Ornelas, et al., On the origin of the high capacitance of nitrogen-containing carbon nanotubes in acidic and alkaline electrolytes, Chem. Commun. 50 (77) (2014) 11343–11346.
[89] D. Hulicova-Jurcakova, et al., Nitrogen-enriched nonporous carbon electrodes with extraordinary supercapacitance, Adv. Funct. Mater. 19 (11) (2009) 1800–1809.
[90] R. Berenguer, et al., Enhanced electro-oxidation resistance of carbon electrodes induced by phosphorus surface groups, Carbon 95 (2015) 681–689.
[91] T. Wang, et al., Interaction between Nitrogen and sulfur in co-doped graphene and synergetic effect in supercapacitor, Sci. Rep. 5 (2015) 9591.
[92] Y. Wen, et al., Nitrogen and phosphorous co-doped graphene monolith for supercapacitors, ChemSusChem 9 (5) (2016) 513–520.
[93] C.H. Choi, et al., Binary and ternary doping of nitrogen, boron, and phosphorus into carbon for enhancing electrochemical oxygen reduction activity, ACS Nano 6 (8) (2012) 7084–7091.
[94] S. Kuroki, et al., A solid-state NMR study of the carbonization of polyaniline, Carbon 55 (2013) 160–167.
[95] A. Morozan, et al., Relationship between polypyrrole morphology and electrochemical activity towards oxygen reduction reaction, Chem. Commun. 48 (38) (2012) 4627–4629.
[96] L. Qie, et al., Nitrogen-doped porous carbon nanofiber webs as anodes for lithium ion batteries with a superhigh capacity and rate capability, Adv. Mater. 24 (15) (2012) 2047–2050.
[97] F. Su, et al., Nitrogen-containing microporous carbon nanospheres with improved capacitive properties, Energy Environ. Sci. 4 (3) (2011) 717–724.

[98] Ramírez-Pérez, A.C., 2016. Doctoral Thesis: Electrodos de materiales carbonosos dopados para aplicaciones energéticas.

[99] Y. Xia, et al., Preparation of sulfur-doped microporous carbons for the storage of hydrogen and carbon dioxide, Carbon 50 (15) (2012) 5543–5553.

[100] Sevilla, et al., Highly porous S-doped carbons, Microporous Mesoporous Mater. 158 (2012) 318–323.

[101] W. Gu, et al., Sulfur-containing activated carbons with greatly reduced content of bottle neck pores for double-layer capacitors: a case study for pseudocapacitance detection, Energy Environ. Sci. 6 (8) (2013) 2465–2476.

[102] Z. Zhao, et al., PEDOT-based composites as electrode materials for supercapacitors, Nanotechnology 27 (4) (2016) 42001.

[103] Z. Guo, et al., Sulfur, trace nitrogen and iron codoped hierarchically porous carbon foams as synergistic catalysts for oxygen reduction reaction, ACS Appl. Mater. Interfaces 6 (23) (2014) 21454–21460.

[104] D. Shao, et al., Nano-structured composite of Si/(S-doped-carbon nanowire network) as anode material for lithium-ion batteries, J. Power Sources 297 (2015) 344–350.

[105] C. González-Gaitán, et al., Functionalization of carbon nanotubes using aminobenzene acids and electrochemical methods. Electroactivity for the oxygen reduction reaction, Int. J. Hydrog. Energy 40 (34) (2015) 11242–11253.

[106] Z. Rozlívková, et al., The carbonization of granular polyaniline to produce nitrogen-containing carbon, Synth. Met. 161 (11–12) (2011) 1122–1129.

[107] W. Li, et al., Rice-like sulfur/polyaniline nanorods wrapped with reduced graphene oxide nanosheets as high-performance cathode for lithium-sulfur batteries, ChemElectroChem (2016) 999–1005.

[108] Y. Tan, et al., Facile synthesis of manganese-oxide-containing mesoporous nitrogen-doped carbon for efficient oxygen reduction, Adv. Funct. Mater. 22 (21) (2012) 4584–4591.

[109] Y. Liao, et al., Nitrogen-doped hollow carbon spheres with enhanced electrochemical capacitive properties, Mater. Res. Bull. 47 (7) (2012) 1625–1629.

[110] X. Liu, et al., Controllable fabrication of SiO2/polypyrrole core-shell particles and polypyrrole hollow spheres, Mater. Chem. Phys. 109 (1) (2008) 5–9.

[111] J. Luo, et al., Preparation of morphology-controllable polyaniline and polyaniline/graphene hydrogels for high performance binder-free supercapacitor electrodes, J. Power Sources 319 (2016) 73–81.

[112] N. Gavrilov, et al., Enhancement of electrocatalytic properties of carbonized polyaniline nanoparticles upon a hydrothermal treatment in alkaline medium, Electrochim. Acta 56 (25) (2011) 9197–9202.

[113] Trchová, et al., Evolution of polyaniline nanotubes: the oxidation of aniline in water, J. Phys. Chem. B 110 (19) (2006) 9461–9468.

[114] S. Mentus, et al., Conducting carbonized polyaniline nanotubes, Nanotechnology 20 (24) (2009) 245601.

[115] A.M. Schwenke, et al., Synthesis and modification of carbon nanomaterials utilizing microwave heating, Adv. Mater. 27 (28) (2015) 4113–4141.

[116] X. Zhang, et al., Microwave synthesis of nanocarbons from conducting polymers, Chem. Commun. 23 (2006) 2477–2479.

[117] Z. Liu, et al., Poptube approach for ultrafast carbon nanotube growth, Chem. Commun. 47 (35) (2011) 9912–9914.

[118] X.G. Li, et al., Efficient and scalable synthesis of pure polypyrrole nanoparticles applicable for advanced nanocomposites and carbon nanoparticles, J. Phys. Chem. C 114 (45) (2010) 19244–19255.
[119] L. Chen, et al., Synthesis of nitrogen-doped porous carbon nano fi bers as an efficient electrode material for supercapacitors, ACS Nano 8 (2012) 7092–7102.
[120] D.S. Yuan, et al., Nitrogen-enriched carbon nanowires from the direct carbonization of polyaniline nanowires and its electrochemical properties, Electrochem. Commun. 13 (3) (2011) 242–246.
[121] Z. Zhang, et al., Nitrogen- and oxygen-containing hierarchical porous carbon frameworks for high-performance supercapacitors, Electrochim. Acta 134 (2014) 471–477.
[122] T. Zhu, et al., Hierarchical porous and N-doped carbon nanotubes derived from polyaniline for electrode materials in supercapacitors, J. Mater. Chem. A 2 (31) (2014) 12545–12551.
[123] H. Liu, et al., Effects of nitrogen- and oxygen-containing functional groups of activated carbon nanotubes on the electrochemical performance in supercapacitors, J. Power Sources 285 (2015) 303–309.
[124] N. Gavrilov, et al., High-performance charge storage by N-containing nanostructured carbon derived from polyaniline, Carbon 50 (10) (2012) 3915–3927.
[125] M. Vujković, et al., Superior capacitive and electrocatalytic properties of carbonized nanostructured polyaniline upon a low-temperature hydrothermal treatment, Carbon 64 (2013) 472–486.
[126] H. Liu, et al., Nitrogen-rich mesoporous carbon as anode material for high-performance sodium-ion batteries, ACS Appl. Mater. Interfaces 7 (49) (2015) 27124–27130.
[127] S. Matsushita, et al., Morphology-controlled carbonaceous and graphitic materials prepared from conjugated polymers as precursors through solid-state carbonization, Synth. Met. 216 (2015) 103–112.
[128] G.A. Snook, et al., Conducting-polymer-based supercapacitor devices and electrodes, J. Power Sources 196 (1) (2011) 1–12.
[129] B.E. Conway, Electrochemical Supercapacitors: Scientific Fundamentals and Technological Applications, Kluwer-Plenum Press, New York, 1999.
[130] C. Meng, et al., Highly flexible and all-solid-state paperlike polymer supercapacitors, Nano. Lett. 10 (10) (2010) 4025–4031.
[131] H.P. Cong, et al., Flexible graphene-polyaniline composite paper for high-performance supercapacitor, Energy Environ. Sci. 6 (4) (2013) 1185–1191.
[132] D. Zhang, et al., Enhanced capacitance and rate capability of graphene/polypyrrole composite as electrode material for supercapacitors, J. Power Sources 196 (14) (2011) 5990–5996.
[133] G.J. Wilson, et al., Enhanced capacitance textile fibres for supercapacitors via an interfacial molecular templating process, Synth. Met. 160 (7–8) (2010) 655–663.
[134] Y.-R. Lin, et al., A novel method for carbon modification with minute polyaniline deposition to enhance the capacitance of porous carbon electrodes, Carbon 41 (14) (2003) 2865–2871.
[135] M.J. Bleda-Martínez, et al., Kinetics of double-layer formation: influence of porous structure and pore size distribution, Energy Fuels 24 (6) (2010) 3378–3384.

Electrochemical behaviour of graphene and carbon nanotubes based hybrid polymer composites

9

Saptarshi Dhibar
Materials Science Centre, Indian Institute of Technology Kharagpur, Kharagpur, India

Chapter Outline

9.1 Introduction

Nowadays, global energy use is rising drastically because of gradually increasing energy demand in developed nations and quickly escalating demand in emerging economies. Achieving this rising energy requirement while avoiding resource depletion and long-lasting harm to the environment needs the advancement of high-performance, inexpensive, and an environmentally benign energy storage system. The development of improved technologies intended for the production and storage of electrical energy are very important for improving the way that society utilizes energy. For that reason, significant research efforts are in progress in numerous

Hybrid Polymer Composite Materials: Processing. DOI: http://dx.doi.org/10.1016/B978-0-08-100789-1.00009-5

locations around the world to develop supercapacitors and advanced batteries for electrochemical storage devices and develop fuel cell for the direct production of electricity from chemical energy. These innovative technologies are taking a major role in mobile and automotive applications. The realization of high-performance energy storage and conversion devices will probably need the development of innovative materials. Carbon-based materials have attracted significant attention in electrochemical applications only because of their processability, abundance, stability, and comparatively environmentally friendly characteristics.

Presently, electrochemical supercapacitors have attracted considerable interest, mostly because of their long cycle life, high power density, and bridging function for the power/energy gap among traditional dielectric capacitors (which have high power output) and fuel cell/batteries (which have high energy storage) [1–2]. The first electrochemical supercapacitors patent was filed in the year 1957. Though not until the 1990s did electrochemical supercapacitors technology commence to draw some attention, in field of hybrid electric vehicles [3]. It was establish that the major function of electrochemical supercapacitors could be to improve the battery or fuel cell in a hybrid electric vehicle to supply the required power for acceleration, with a supplementary function being to recapture break energy. Additional progresses have led to the recognition that the electrochemical supercapacitors can play a vital role in complementing fuel cells or batteries in their energy storage purposes by giving back-up power supplies to protect against power disruptions. For that reason, the US Department of Energy has designated electrochemical supercapacitors as significant as batteries for future energy storage systems [4]. Several governments and enterprises have also provided money as well as time into exploring, researching, and developing telectrochemical supercapacitors technologies.

Current years have yielded major improvements in the practical and theoretical research and progress of electrochemical supercapacitors, as confirmed by a huge number of research articles and technical reports [5–10]. Simultaneously, electrochemical supercapacitors have some disadvantages: low energy density and high production costs. These have been recognized as main challenges for the continuation of electrochemical supercapacitors technology. The development of new materials for electrochemical supercapacitors is one of the most intensive approaches to overcome the problem of low energy density. In this chapter we've highlighted some present and past research on graphene, carbon nanotubes, and conducting polymer-based hybrid polymer nanocomposites for electrochemical supercapacitors.

9.1.1 Supercapacitors

Supercapacitors, also termed as an ultracapacitor, is an electrochemical storage device that has better capacity than that of conventional physical capacitors, and its charging/discharging rate capability is comparatively much improved than that of the primary/secondary batteries. Supercapacitors are eco-friendly, of high safety, and can be worked in extensive temperature ranges comprising an enormous large cycling life. Due to this reason, they exhibit superior applications in the field of transportation, electronics, aviation, and communications. Supercapacitors bridge the gap between batteries and capacitors to form fast-charging energy-storage devices of intermediate specific energy. Supercapacitors have the lesser energy density as compared to the batteries, but they can transfer the stored energy in a very small period of time [6,11]. Without losing energy storage capability, supercapacitors can work for a greater period of time than batteries, repeating millions of number of cycles. In batteries, the electrical energy is produced by the conversion of chemical energy via redox reactions at the anode and cathode, whereas in supercapacitors, the energy is stored either by the rapid reversible faradic transition redox reactions that takes place with the active materials or by the accumulation of ionic charges on the electrode/electrolyte interfaces [12]. The main advantages of supercapacitors are that they can provide high power capability (60–120 s is typical), long cycle life ($>10^5$), and outstanding reversibility (90–95% of higher). Generally, they show 20–200 times greater capacitance per unit volume or mass as compared to conventional capacitors [13–14]. For that reason, a number of applications now utilize supercapacitors or are efficiently considering them, including digital communication devices, electric vehicles, mobile phones, digital cameras, electric hybrid vehicles, pulse laser techniques, electric tools, storage of the energy produced by solar cells, and uninterruptible power supplies for computers [15–16]. One of the most important characteristics of such vehicles is their capability to give the required energy for the vehicle's range during the vehicle is stopped or start-up. The aim is to accumulate that energy as efficiently as possible, with the intention that it can be utilized in accelerating the vehicles at its next travel [17].

A capacitor is a positive electrical device that stores energy as a charge in the electrical field between two conducting plates called electrodes. Generally, capacitors can release the stored charge comparatively fast leading to high power. However, they cannot store much energy. On the other hand, conventional capacitors, also termed as electrostatic capacitors, consist of two conducting electrodes separated by an insulating layer knows as dielectric. While an external voltage is applied, charges stored on the surface of the two electrodes are isolated by an insulating dielectric layers, hence, producing an electric field. The consequent electric field allows the device to store energy. The capacitor can be described by a parameter called capacitance C, which is described as the ratio of the charge Q to the applied voltage V:

$$C = \frac{Q}{V} \tag{9.1}$$

In the case of conventional capacitor, C is directly proportional to the surface area A of each electrode and inversely proportional to the distance D between the electrodes:

$$C = \varepsilon_0 \varepsilon_r \frac{A}{D} \tag{9.2}$$

The first two factors on the right-hand side of the above equation is a constant of proportionality where ε_0 is the dielectric constant (or "permittivity") of the free space, and ε_r is the dielectric constant of the resulting materials between the electrodes.

Supercapacitors are mainly the electrochemical energy storage devices. The structure of a supercapacitor is similar to that of a battery. The supercapacitor contains of two porous electrodes by a current collector on each electrode immersed in an electrolyte separated with a dielectric porous separator. When voltage is applied across the current collectors, the potential of the negative electrode attracts the positive ions, while the positive electrode attracts the negative ions in the electrolyte. During discharging the charge assembled at both the electrode surfaces produces energy (Fig. 9.1). The constituents parts of supercapacitors, comprising the current collector, the separator, the electrolyte, in addition to the electrodes, are the most important components affecting the overall performance of a device that has to be considered when fabricating a high-performance supercapacitor device.

In the supercapacitors, the energy stored is directly proportional to the capacitance:

$$E = \frac{1}{2}CV^2 \tag{9.3}$$

The electrodes are generally constructed of high surface area A and thinner dielectrics that decreases the distance D among the electrodes. These lead to

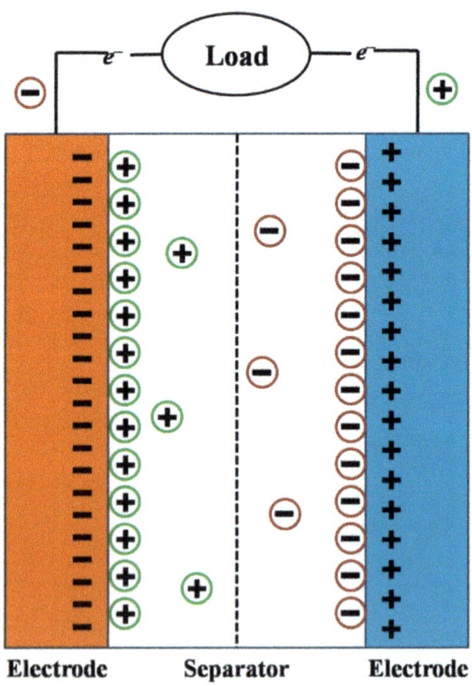

Figure 9.1 Schematic diagram of a supercapacitor device [18].

increase in both capacitance as well as energy of the supercapacitor as compared to conventional capacitors.

Based on the charge storage mechanism, supercapacitors are classified into three categories: electrochemical double layer capacitor (EDLC), pseudocapacitor, and hybrid capacitors. The EDLCs, mainly the carbon materials, are where the capacitance arises from the charge separation at the electrode/electrolyte interface. By contrast, the faradic pseudocapacitor materials, such as different metal oxides and conducting polymers, not only store energy similar to an EDLCs but also exhibit electrochemical faradic reactions among electrode materials and ions in the appropriate potential window. The hybrid capacitors are mainly the combination of EDLCs and pseudocapacitors. A graphical taxonomy of the different classes and subclasses of electrochemical supercapacitors is presented in Fig. 9.2.

9.1.1.1 Electrochemical double layer capacitors (EDLCs)

Electrochemical double layer capacitors (EDLCs) are usually built up from an electrolyte, a separator, and two carbon-based electrodes. EDLCs accumulate charges nonfaradically or electrostatically like conventional capacitors, and there is no transport of charge between electrodes and electrolytes. To store energy, they utilize an electrochemical double layer of charge. When voltage is applied, the charge accumulates on the surfaces of the electrode. Following the usual attraction of different charges, ions in the electrolyte solution diffuse across the separator into the pores of the electrode of reverse charge. However, the electrodes are engineered to prevent the recombination of the ions. Thus, a double layer of charge is produced at every electrode. These double layers coupled with a decrease in the distance among

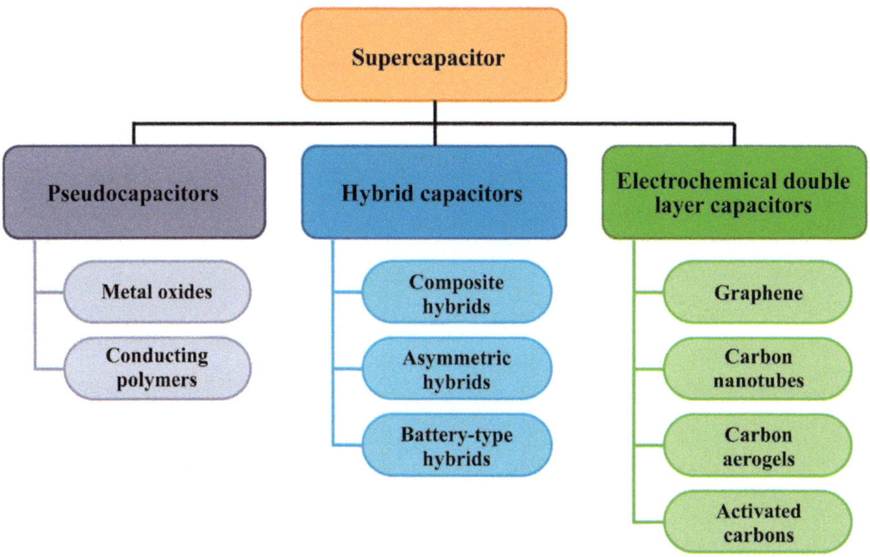

Figure 9.2 Taxonomy of supercapacitors.

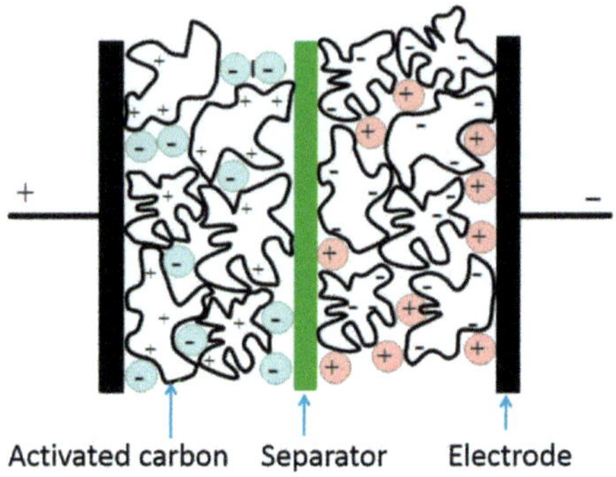

Figure 9.3 Schematic diagram of mechanism of EDLC.

the electrodes enhance the surface area, allowing EDLCs to achieve improved energy density as compared to conventional capacitors. As there is no charge transfer that takes place between electrolyte and electrode, so there are no such chemical or composition changes related with nonFaradic processes. For that reason, the charge storage in EDLCs is reversible, which allows them to achieve enormously high cycling stabilities. The schematic diagram electrochemical double layer is presented in Fig. 9.3.

If the two-electrode surfaces can be termed as E_{s1} and E_{s2}, an anion as A^-, a cation as C^+, and the electrode/electrode interface as //, then the electrochemical process for charging and discharging can be expressed by the Eqs. (9.4–9.7) [19–20].

On one electrode (say, a positive one):

$$E_{s1} + A^- \xrightarrow{\text{charging}} E_{s1}^{+}//A^- + e^- \tag{9.4}$$

$$E_{s1}^{+}//A^- + e^- \xrightarrow{\text{discharging}} E_{s1} + A^- \tag{9.5}$$

On the other electrode (say, a negative one):

$$E_{s2} + C^+ + e^- \xrightarrow{\text{charging}} E_{s2}^{-}//C^+ \tag{9.6}$$

$$E_{s2}^{-}//C^+ \xrightarrow{\text{discharging}} E_{s2} + C^+ + e^- \tag{9.7}$$

The overall charging and discharging process can be expressed by Eq. (9.8) and (9.9) as follows:

$$E_{s1} + E_{s2} + A^- + C^+ \xrightarrow{\text{charging}} E_{s1}^{+}//A^- + E_{s2}^{-}//C^+ \tag{9.8}$$

$$E_{s1}^{+}// + A^- + E_{s2}^{-}//C^+ \xrightarrow{\text{discharging}} E_{s1} + E_{s2} + A^- + C^+ \tag{9.9}$$

9.1.1.2 Pseudocapacitors

Pseudocapacitors or Faradaic supercapacitors are dissimilar from electrostatic or EDLCs. When a potential is applied to a pseudocapacitor, quick and reversible redox reactions (Faradaic reactions) take place on the electrode materials and involve the passage of charge across the double layer. This is comparable to the charging and discharging processes that take place in the case of batteries, resulting in a Faradaic current passing through the supercapacitor cell. The materials that undergo such redox reactions include various metal oxides (RuO_2, MnO_2, and Co_3O_4) and the conducting polymers (polyaniline, polypyrrole, and polythiophene) [12,21–22]. Mainly three types of Faradaic processes take place at pseudocapacitor electrodes: the reversible electrochemical doping–dedoping in conducting polymer–based electrodes, reversible adsorption (such as, adsorption of hydrogen on the surface of gold or platinum), and the redox reactions of the transition metal oxides [12]. It has been exhibited that these Faradaic electrochemical processes not only broaden the working voltage but also enhance the specific capacitance of the supercapacitors. As the electrochemical processes take place both on the surface and in the bulk close to the surface of the solid electrode, the pseudocapacitors demonstrate extreme greater capacitance values as well as energy density than that of an EDLCs. It was already reported by Conway et al. that the capacitance of a pseudocapacitors can be 10–100 times greater than the electrostatic capacitance of an EDLCs [23] though a pseudocapacitor's electrode material normally suffers from comparatively lower power density than that of an EDLCs since Faradaic processes are usually slower than nonFaradaic processes. In addition, since redox reactions take place at the electrode, a pseudocapacitor is often short of stability during cycling, as with batteries.

9.1.1.3 Hybrid supercapacitor

Hybrid capacitors attempt to expand the relative benefits and reduce the relative drawbacks of pseudocapacitors and EDLCs to improve performance characteristics. The hybrid capacitors have both Faradaic and non-Faradaic processes to store charge; therefore, they attain superior energy as well as power densities in comparison with EDLCs without sacrificing affordability and cycling permanence that have limited attainment of pseudocapacitors. Based on their electrode configuration, hybrid capacitors are classified as composite, asymmetric, and battery-type electrodes. Composite electrodes merge carbon-based materials with either conducting polymers or metal oxide materials and comprise both a physical and chemical charge storage mechanism jointly in a single electrode. The carbon-based materials facilitate a capacitive double layer of charge and additionally provide a high surface area backbone that increases the contact between the deposited pseudocapacitive materials and electrolytes. The pseudocapacitive materials are able to further increase the capacitance of the composite electrode during Faradaic reactions. Asymmetric hybrids combine a pseudocapacitor electrode with a EDLCs electrode by coupling Faradaic and nonFaradaic processes. In particular, the combining of an activated carbon negative electrode with a conducting polymer positive electrode has attracted significant attention. In an advanced hybrid electrochemical capacitor, the potential range at the cathode is broadened to the total potential window of the activated carbon, particularly from 1.5 to 4.5 V versus Li/Li^+, which is the wider potential range than that of the conventional electrochemical capacitors, in which the potential range is from 0.8 to 2.7 V versus Li/Li^+ [24].

9.1.2 Supercapacitor electrode materials

9.1.2.1 Carbon materials

Carbon materials are considered as the most promising electrode materials for the electrochemical supercapacitors industrial applications due to their benefits of high surface area, superior electrical conductivity, high temperature stability, inexpensiveness, excellent chemical stability, easy processing, abundance, and nontoxicity. Carbon-based electrochemical supercapacitors have similar charging mechanisms like conventional capacitors, which employ EDLC capacitance at the electrode–electrolyte interface. As a result, EDLC capacitance mainly depended on the specific surface area of the electrodes. It is normally believed that the higher the specific surface area, the higher the specific capacitance. However, the specific capacitance of electrochemical supercapacitors does not enhance with the increase of a specific surface area. Apart from specific surface areas, the pore size is a further significant parameter that needs to be seriously considered for the alternative of electrochemical supercapacitors electrode materials.

Another type of carbon material is the activated carbon, the most extensively employed electrode materials in EDLCs because of their large surface area, inexpensiveness, and comparatively good electrical property. Generally, activated carbons are produced by either chemical activation or physical activation or a mixture of two with various carbonaceous precursors like coal, nutshells, and wood. It utilizes a complex porous structure with micropores having a diameter of <2 nm, mesopores having a diameter from 2–50 nm, and macropores having a diameter of >50 nm to attain their high surface area. Carbon aerogels are also a carbon materials prepared with the pyrolysis of organic areogels. They are highly porous materials and synthesized by the poly-condensation of resorcinol and formaldehyde, by a sol-gel process, and subsequent pyrolysis. They have very good electrical conductivity, high porosity, high density as well as high surface area ranging among 400–1000 m^2/g. Carbon aerogels are used to fabricate supercapacitors due to their high surface area.

Carbon nanotubes

In 1991, the discovery of carbon nanotubes (CNT) opened up a new era in materials science [25]. These are the allotropes of carbon having a cylindrical nanostructure and it has the length-to-diameter ratio up to 132000000:1, significantly better than any other material. These are at least 100 times stronger than steel, but only one-sixth as heavy. For that reason nanotube fibers could strengthen almost any material. Because of its unique physical properties, such as electrical conductivity and outstanding mechanical and chemical stability, CNTs, including both multiwalled carbon nanotubes (MWCNTs) and single-walled carbon nanotubes (SWCNTs), have been recognized as promising electrode materials for electrochemical energy storage and conversion [26,27]. The SWCNTs are just like a regular straw. It has only one layer or wall. On the other hand MWCNTs are a collection of nested tubes of endlessly rising diameter. They can range from one outer and one inner tube, known as double-walled nanotube to as many as 100 tubes or more. CNTs have the superior electrical conductivity of around 5000 S/cm, thermal conductivity of 6000 W/mK, high thermal stability (stable up to 2800°C), and excellent mechanical stability (tensile strength 45 billion Pascals) [28]. The superior electrical conductivity, readily accessed surface area, easy electrolyte accessibility, and high porosity make CNTs as the choice of a high power electrode materials for electrochemical supercapacitors.

Graphene

In the present time graphene (Gr) has appeared as a rapidly growing star in the field of materials science. Graphene is the strongest and thinnest known material in the world. It has a two-dimensional (2-D) crystal, composed of monolayer of sp^2 bonded carbon arranged in a honeycombed network comprising six-membered rings [29]. It is one atom thick and can be easily obtained from graphite. Gr demonstrates superior electrical, morphological as well as mechanical properties such as excellent carrier mobility, high spring constant, and outstanding Young's modulus value. Its large surface area makes Gr an attractive selection of materials for the use of in various electronic devices. It was first discovered in 2004 by Andre Geim and Konstantin Novoselov, who acquired Gr sheets by using cohesive tape to continually split graphite crystal into growingly thinner pieces until individual atomic planes were achieved [30]. This discovery was recognized by Nobel Prize in Physics for 2010 and led to an explosion of interest in the study of Gr with respect to its many interesting properties, opening up new research areas in the field of condensed-matter physics, materials science, and has led to an investigation for extensive and diversified technological applications. The carbon atom at the edge of Gr sheet has distinctive chemical reactivity, and it has also the maximum ratio of edge carbons as compared to CNTs. Gr is a zero-gap semiconductor, since its conduction and valance bands meet at the Dirac points. It has an enormously high electron mobility value at room temperature, the value is in excess of 15,000 $cm^2\ V^{-1}\ s^{-1}$ [31]. It is also reported that Gr sheets have a resistivity value of around 10−6 Ω cm. The breaking strength of Gr is more than 100 times higher than hypothetical steel films of the same thickness having Young's modulus of 1 TPa. For that reason Gr is one of the strongest materials ever tested [32]. Because of these excellent properties Gr has been broadly investigated as unique component for several applications such as in automotive, aerospace, bio-sensors, electronic, solar cell as well as in energy storage devices [33−34]. Presently, Gr has moved from research laboratory to market in order to satisfy a huge demand of advanced materials in today's world. Due to its excellent aspect ratio, superior surface area, outstanding electrical and electrochemical properties, Gr has been considered as a potential candidate for the fabrication of supercapacitor electrode materials.

9.1.2.2 Metals oxide

Early studies in supercapacitors have mainly focused on carbon materials because of their easy accessibility, comparatively excellent electrical conductivity, and large surface area even though these carbon materials suffer from lower specific capacitance. Therefore, researchers have transferred their attention from pure carbon materials to pseudocapacitor materials, like metal oxides and conducting polymers. The specific capacitances of the metal oxides and conducting polymers are three to seven times higher than pure carbon materials. Transition metal oxides are considered to be the best candidates for supercapacitors electrode materials. Among numerous types of transition metal oxides, RuO_2 was the first considered as supercapacitor electrode. The capacitance of RuO_2 is achieved through the insertion and removal, or intercalation, of proton into its amorphous structure. The specific capacitance of RuO_2 exceeds compared to conducting polymers and carbon-based materials in its hydrous form [35]. Furthermore, the electrochemical series resistance of hydrous RuO_2 is lower as compared to other electrode materials. As a result, RuO_2 pseudocapacitors are perhaps capable of reaching better energy as well as power densities as compared to similar EDLCs and conducting polymer pseudocapacitors. Even with this potential, because of its high cost, the success of RuO_2 has been limited. Apart from RuO_2, the MnO_2 was also studied as electrode materials for supercapacitor because of its low-cost, good environmental stability, eco-friendly synthetic process, ideal capacitive behavior, and improved electrochemical properties [36]. More metal oxides that have been investigated for supercapacitors electrode materials are CoO_3, IrO_2, FeO, TiO_2, SnO_2, MoO, V_2O_5, and Fe_2O_4. It has been reported that the uniform dispersion of these metal oxides on the carbonaceous materials enhances the electrochemical properties and also considered as superior electrode materials for supercapacitors [37–38]. Usually, the transition metal oxides have comparatively low power density and poor cycling stability. The low power density can be attributed to the poor electrical conductivity of metal oxides, which restricts the electron transfer rate. Furthermore, during the charging and discharging processes, the damage to the morphology caused by the swelling and shrinkage of the electrode materials leads to the lack of charging stability. To resolve these problems, the two main methods involve combining metal oxides with conductive materials and developing porous nanostructures: the former can accelerate the reaction kinetics and the latter can buffer the stress from the swelling and shrinkage of the electrodes and supply more ion adsorption or active sites for the charge transfer reactions, as well as reduce the diffusion and transfer pathways of the electrolytes ions [39].

9.1.2.3 Conducting polymer

Over the last few decades, a new class of organic polymers has been synthesized with the capacity to conduct electric current, known as conducing polymers. The general aspect of most electrically conducing polymers is the occurrence of an extended π-conjugation system with single and double bond alteration along the polymer backbone. They are semiconducting materials in nature having low charge carrier mobility [40]. Apart from electrical properties, the distinctive chemical, optical as well as magnetic properties of the conducting polymers have led them to a wide range of technological applications in electromagnetic interference shielding, rechargeable batteries, corrosion protection coatings, sensors, electrodes, and microwave adsorption [41]. Amongst different conducting polymers, polyaniline (PANI), polypyrrole (PPy), polythiophene (PTh), and its derivatives are extensively investigated because of their easy synthesis, high electrical conductivity, exclusive electrochemical properties, relatively cheap, and good environmental stability [42]. Conducting polymers are regarded as promising electrode materials for supercapacitor mainly because of two reasons: (1) they provide a high specific capacitance: not only surface, but also bulk materials are involved in charge storage mechanism, and (2) in the charge state they have high conductivity. As a result, devices with low ESR are feasible. Conducting polymers cost lower than other electrode materials, and their plastic properties readily permit their manufacture as thin film. The storage in conducting polymers is due to a Faradaic process that takes place at the electrode materials. During oxidation the ions are transferred to the polymer chain, while at reduction the ions are released back into the solution. In conducting polymer film, the charging takes place throughout the bulk volume of the thin film not just on the surface; for that reason they reached high levels of specific capacitance.

Polyaniline (PANI) is a semi flexible rod polymer family and has been known for over a century since the synthesis of so-called aniline blocks. PANI is one of the most studied conducting polymers over the past 50 years. The electrical conductivity of the PANI is comparatively very high at 10 S/cm. PANI is one of the smart polymers among the all conducting polymers and organic semiconductors because of its attractive processing properties. It can be easily synthesized by chemical or electrochemical oxidation of aniline in presence of oxidants like ammonium persulfate in acidic medium. It is mainly available in three different oxidations states: (1) leucoemeraldine—white/color and colorless, (2) perinigraniline—blue/violet, (3) emeraldine—green for emeraldine salt, blue for emeraldine base. Among all the forms, the leucoemeraldine slat form of PANI is most electrically conducting. Because of the existence of charges and conformation of the different oxidation states, PANI has been considered as a promising candidate for the fabrication of electrode materials for supercapacitors. The general structure of PANI is shown in Fig. 9.4A.

In 2000 The Nobel Prize in Chemistry was awarded for work on conducting polymer including polypyrrole (PPy). PPy is synthesized generally by chemical or electropolymerization techniques. In chemical polymerization methods, monomer is polymerized in a solution (aqueous or mild acid solution) by using an oxidant, while for electropolymerization methods, pyrrole monomer is polymerized in a conductive substrate by applying an external potential. PPy is an insulator and its oxidized derivatives are a good conductor of electricity. The electrical conductivity of the PPy is relatively very high in the range from 2 to 100 S/cm. It has some significant properties such as good thermal stability, easy synthesis procedure, good environmental stability, and superior electrical conductivity all of which make PPy a

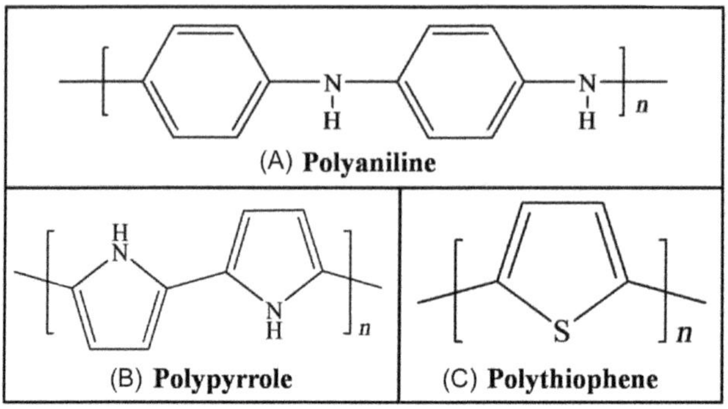

Figure 9.4 The structures of (A) polyaniline, (B) polypyrrole, and (C) polythiophene.

potential candidate for supercapacitor electrode materials. It can also be utilized for the manufacture of chemical sensors, artificial muscles, and is a potential candidate for drug delivery. The structure of PPy is shown in Fig. 9.4B.

Polythiophenes (PTs) have much in common with PPy. Similar to PPy, PTs can also be synthesized by using a chemical oxidant. This reaction should be carried out in a nonaqueous medium as the limited solubility of PTs. Mainly, the direct polymerization of thiophene monomer has been successfully used to prepare PTs using $FeCl_3$ in the presence of chloroform solvent that results in the molecular weight between 3000 and 30,0000. The electrical conductivity of the PT and its derivatives are comparatively very high but the mechanical properties of the electrochemically prepared PTs is very poor. Electrochemical switching is not so efficient in aqueous medium. For that reason various researchers have utilized the ability of attached functional groups such as polyether to improve the hydrophilicity and enhance the electrochemical behavior in an aqueous solution [43]. PT and its derivatives are considered as electrode materials for supercapacitors because of its excellent electrical properties, environmental stability as well as high-room-temperature electrical conductivities. The structure of PT is shown in Fig. 9.4C.

9.1.3 Electrolyte

The electrolyte plays a vital role in the capacitive performance, the safety, and the lifetime of a supercapacitor. Generally, it is made up of a solvent and dissolved chemical that dissociates into positive cation and negative anions, producing the electrolyte electrically conductive. If the electrolyte contains more ions then its conductivity is very high. In supercapacitors, the electrolytes are the electrically conductive link between the two electrodes. A superior electrolyte for supercapacitors should have the properties such as, wide voltage range, low resistivity, low-cost, low toxicity, and high electrochemical stability. Generally, three types of electrolytes are traditionally used for the investigation of electrochemical properties of the electrode materials: (1) aqueous electrolyte, (2) organic electrolytes, and (3) ionic liquids.

1. *Aqueous electrolyte*: Various aqueous electrolytes like KCl, KOH, K_2SO_4, H_2SO_4, NH_4Cl, and $LiClO_4$ have been used in supercapacitor electrolytes. These types of electrolytes have some benefits like low-cost, high ionic concentration, and low ionic radius. Small voltage window is the main disadvantage of these electrolytes.
2. *Organic electrolytes*: Different organic electrolytes such as propylene carbonate, acetonitrile, tetrahydrofuran, tetraehtylammonium tertafluroborate, diethyl carbonate, triethylmethylammonium terafluroborate, and tetraethylphosponium tetrafluoborate were used in supercapacitor electrolytes. These types of electrolytes have many disadvantages such as high cost, high resistivity, and hazardous purification process.
3. *Ionic liquids*: Various types of ionic liquids such as imidazolium pyrrolidinium, aliphatic quaternary ammonium salts with anions like tetrafluroborate, trifloromethanesulfonate, bis (trifluoromethanesulfonyl)imide, bis(fluorosulfonyl)imide, and hexafluorophosphate were used in supercapacitor electrolytes. They have some superior properties such as high thermal and electrical stability, high conductivity, low flammability, and a wide voltage range (2 to 6 V). But the major drawback of these electrolytes is that they are much more expensive.

9.2 Carbon nanotubes based hybrid nanocomposites for supercapacitors

9.2.1 Multi-walled carbon nanotubes based

Over the past few years many researchers around the world have developed multi-walled carbon nanotubes (MWCNTs)—based hybrid nanocomposites for supercapacitor applications. Conducting polymers doped with transition metal ions in presence of CNTs has attracted much research attention. Dhibar et al. synthesized $CuCl_2$ doped PANI at different doping levels (1, 2, 3, and 4 wt%) by in situ polymerization techniques where ammonium persulfate (APS) used as oxidant and obtained superior electrochemical properties for PANI at 2 wt% doping level [44]. After that, they prepared PANI at Cu^{2+} ion (2 wt%) doped PANI in presence of MWCNTs and checked the electrochemical properties [45]. From the morphological analysis they obtained that the MWCNTs surfaces are uniformly coated by $CuCl_2$ doped PANI. The electrochemical characterizations were carried out by a there-electrode system where 1 M KCl solution used as an electrolyte, and they obtained the highest specific capacitance of 724 F/g at 10 mV/s scan rate. The PANI C2 CNT nanocomposites also achieved 91% of capacitance retention after 1000 cycles. This group also synthesized the $CuCl_2$ doped PPy nanocomposites in presence of MWCNTs and obtained the highest specific capacitance of 312 F/g at 10 mV/s scan rate and capacitance retention of 90% after 1000 cycles [46]. Shi et al. prepared PPy coated MWCNTs nanocomposite by chemical polymerization techniques where $(CTA)_2S_2O_8$ used as oxidant [47]. From the morphological analysis it was confirmed that the MWCNTs are uniformly coated by PPy. The electrochemical characterizations were done in three-electrode system where 0.5 M Na_2SO_4 solution used as electrolyte and obtained the highest specific capacitance of 4.1 F/cm at a 2 mV/s scan rate and superior cyclic stability also achieved by the nanocomposite. The MWCNTs/PANI/MnO_2 ternary coaxial nanostructure was prepared by Li et al. via a simple wet chemical procedure [48]. They studied the electrochemical characterization with a three-electrode system in 0.5 M Na_2SO_4 solution and obtained the highest specific capacitance of 330 F/g at 5 mV/s scan rate. The nanostructure also retains their 77% of capacitance after 1000 consecutive cycles. Zhang et al. synthesized the PTh/MWCNTs nanocomposite by in situ polymerization techniques in an oil-in-ionic liquid microemulsion [49]. From the morphological analysis they obtained that the MWCNTs were uniformly coated by PTh having a thickness of 2–3 nm. They studied the electrochemical analysis with a three-electrode system in 1 M Na_2SO_4 electrolyte and achieved the highest specific capacitance of 216 F/g at 1 A/g current density and also good cycle stability. Recently, Singu et al. prepared a MWCNTs/PANI nanocomposite via in situ emulsion polymerization techniques [50]. A maximum specific capacitance of 360 F/g was obtained for the nanocomposites, at a discharge current density of 0.4 A/g in 1 M H_2SO_4 electrolyte solution, with 98.3% capacitance retention over 5000 charge/discharge cycles. A new approach has been developed by Chen et al. for the fabrication of PPy coated MWCNTs, based on the use of indigo carmine dye as a dispersant for MWCNTs and anionic dopant for PPy polymerization and obtained highest specific capacitance of 2.55 F/cm^2 at a scan rate of 100 mV/s [51]. They also obtained that the nanocomposite showed good capacitance retention and good cycling stability at high mass loading. Shivkkumar et al. synthesized PANI nanofibers/MWCNTs nanocomposite by interfacial polymerization techniques [52]. They studied the electrochemical characterization in a two-electrode system with the help of 1 M H_2SO_4 electrolyte and obtained the highest specific capacitance of 606 F/g at 1 A/g current density. Researchers are now developing the MWCNTs and conducting polymer–based flexible supercapacitor electrodes. Lin et al. prepared PANI/MWCNTs-based flexible and transparent electrodes by an easy electrodeposition process [53]. The electrochemical characterization was carried out by a two-electrode system in H_3PO_4-PVA gel electrolyte and obtained the highest specific capacitance of 233 F/g at a current density of 1 A/g. Wang et al. fabricated PANI/MWCNTs-based flexible supercapacitor where the PANI nanowires having a diameter of 8 nm, and an aspect ratio over 100 are deposited inside the hollow channels of MWCNTs by an in situ electropolymerization method [54]. The electrochemical characterization of the composite film was measured by a three-electrode setup where 1 M KOH used as electrolyte and obtained the highest specific capacitance of 296 F/g at a current density of 1.6 A/g with perfect cycling life (95% of capacitance retention after 2000 cycles). Dhibar et al. synthesized silver nanoparticles decorated PANI/MWCNTs nanocomposite by in situ polymerization methods using APS as oxidizing agent in the presence of DBSA and $AgNO_3$ and investigated as electrode materials for supercapacitors [55]. From the morphological analysis it was obtained that the MWCNTs surfaces are uniformly coated by PANI and between the coated MWCNTs there is the presence of uniformly distributed Ag nanoparticles. The electrochemical characterizations were carried out in three-electrode system where 1 M KCl used as electrolyte. The Ag-PANI/MWCNTs nanocomposite showed the highest specific capacitance of 528 F/g at 5 mV/s scan rate, highest energy density of 187.73 Wh/kg at 5 mV/s scan rate, better power density of

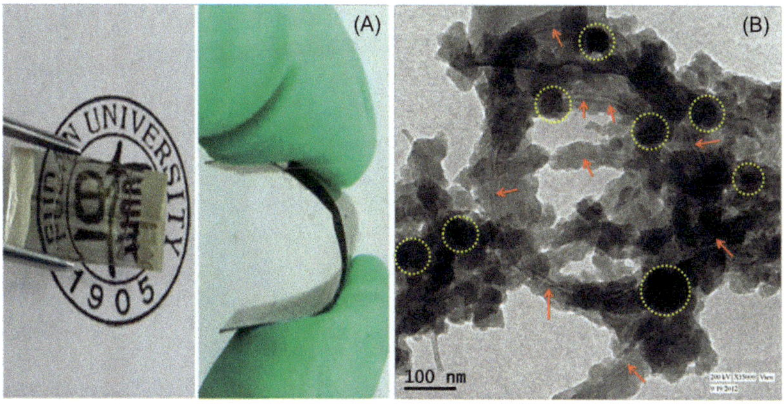

Figure 9.5 (A) Photograph of transparent and flexible supercapacitor [Lin *et al. Sci. Rep.* 2013 (Ref. 53)]; (B) HRTEM image of Ag-PANI/MWCNTs [Dhibar *et al. Ind. Eng. Chem. Res.* 2014 (Ref. 55)].

4185 W/kg at 200 mV/s scan rate, and also maintained 94% of their initial capacitance over 1000 cycles, respectively. The nanocomposite also attends maximum electrical conductivity of 4.24 S/cm and better thermal stability up to 350°C. (Figure 9.5).

9.2.2 Single-walled carbon nanotubes based

Apart from MWCNTs, the SWCNTs can also play a major part in the fabrication of supercapacitor electrode materials. Wang et al. synthesized the $NiCo_2O_4$-SWCNTs nanocomposite by controlled hydrolysis reaction in an ethanol–water mixed solvent [56]. The electrochemical tests were carried out by three-electrode systems in 2 M KOH solution and the nanocomposite showed the highest specific capacitance of 1642 F/g at a 0.5 A/g current density and also achieved excellent cycling stability of 94.1% retention after 2000 cycles. Niu et al. reported the "skeleton/skin" strategy for the preparation of free-standing, thin and flexible SWCNT/PANI hybrid films by a simple in situ electrochemical polymerization method [57]. The electrochemical tests were done with the help of a three-electrode cell in 0.5 M H_2SO_4 and 0.5 M Na_2SO_4 electrolyte. The SWCNT/PANI hybrid film showed the highest specific capacitance of 236 F/g in 30 seconds PANI deposition time, high energy density of 131 Wh/kg, and a power density of 62.5 kW/kg, and retains their 85% capacitance after 1000 cycles, respectively. Gupta et al. synthesized PANI/SWCNTs nanocomposite by in situ potentiostatic deposition of PANI onto SWCNTs [58]. A three-electrode cell was used for electrochemical tests and 1 M H_2SO_4 used as electrolyte. The nanocomposite showed the highest specific capacitance of 462 F/g at a 10 mA/cm^2 and also retains their 94% of capacitance after 1500 cycles. Dhibar et al. synthesized $MnCl_2$ doped

PANI/SWCNTs by in situ polymerization techniques and investigated as electrode materials for supercapacitors [59]. The electrochemical characterizations were carried out by a two-electrode system and in 1 M aqueous KCl electrolyte. The SWCNTs surfaces are uniformly coated by Mn-doped PANI, confirmed from the morphological analysis. The Mn-PANI/SWCNT nanocomposite showed the highest specific capacitance of 546 F/g at 0.5 A/g current density, better energy density of 194.13 Wh/kg at 0.5 A/g current density, highest power density of 2571.38 W/kg at 3 A/g current density, and the nanocomposite retained about 88.13% of their original capacitance after 1000 cycles, respectively. A high electrical conductivity of 9.64 S/cm and superior thermal stability was also obtained for the Mn-PANI/SWCNT nanocomposite. A facile and in situ method was developed to synthesized SWCNT@MnO_2/PPy hybrid films as supercapacitor electrode without binder at room temperature was reported by Liang et al. [60]. They studied the electrochemical characterization with a three-electrode system in 1 M Na_2SO_4 electrolyte and achieved the highest specific capacitance of 351 F/g based on the total weight of the electrodes with energy density of 39.7 Wh/kg and power density of 10 kW/kg and losses of the capacitance of 5.6% after 10,000 cycles. Apart from these types of supercapacitor electrodes, the SWCNTs-based flexible supercapacitors were also developed by different research groups. Ge et al. developed the transparent and flexible supercapacitors electrodes using a PANI/SWCNT composite thin film where polyethylene terephthalate (PET) was used as a flexible and transparent substrate [61]. This flexible and transparent electrode showed the specific capacitance of 55 F/g at a current density of 2.6 A/g. Wang et al. fabricated PANI/SWCNT/cloth for flexible supercapacitors [62]. Here a nonwoven wiper cloth was used for flexible substrate. This flexible supercapacitor achieved the specific capacitance of 410 F/g and also retains 90% of its original capacitance after 3000 charge/discharge cycles. Thin and flexible supercapacitor based on SWCNT/PANI was developed by Souza et al. [63]. In this case a very thin and flexible all-solid device was manufactured using two electrodes (PET covered with SWCNT/PANI nanocomposite separated by a H_2SO_4-PVA gel electrolyte). This flexible and thin supercapacitor electrode showed the specific capacitance of 76.7 F/cm^3. Ge et al. fabricated SWCNTs wrapped around the cellulose fibers as the conductive skin, while ultrathin ($\sim$50 nm) and ultralong (tens of microns) PANI nanoribbons were synthesized in situ between microporous cellulose fibers and interpenetrated within the SWCNTs network [64]. They had done all the electrochemical tests with a three-electrode setup in 1 M H_2SO_4 electrolyte. The SWCNTs/PANI nanoribbons showed the good volumetric capacitance of 40.5 F/cm^3, and areal capacitance of 0.33 F/cm^3, and retained 79% of the initial capacitance after 1000 cycles, respectively (Figure 9.6).

9.2.3 CNT–metal oxide supercapacitors

Carbon nanotubes and metal oxide–based supercapacitors have been the subject of great research interest in the past few years. Various researchers around the world have developed CNT and metal oxide–based electrode materials for

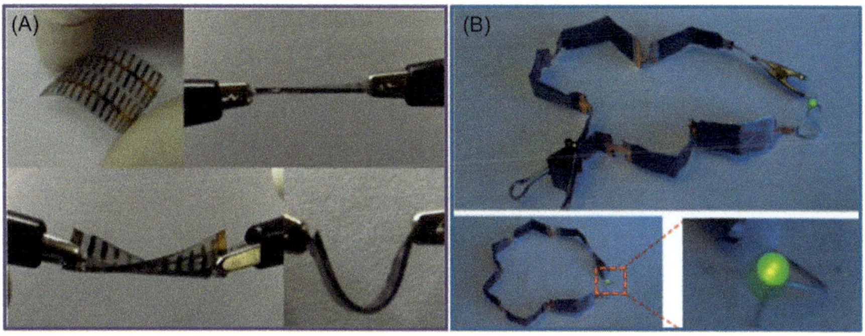

Figure 9.6 (A) Digital images of the SWNT/PANI nanocomposite deposited over PET and the all-solid device straighten, twisted and folded [Souza et al. *J. Power Sources*, 2014, (Ref. 63)]; (B) Picture of six solid-state supercapacitors based on SWCNT/PANI nanoribbon paper electrodes after folding 500 time were connected in series to light a green LED. The clamp represented the folding line. Insets: picture of released supercapacitors after folding and the enlarged picture of the lighting green LED [Ge et al. *Nano Today* 2015 (Ref. 64)].

supercapacitors. Wang et al. prepared a Co_3O_4@MWCNT nanocable by a simple hydrothermal procedure [65]. From the morphological analysis they confirmed that the core MWCNTs was uniformly coated with Co_3O_4. The electrochemical cauterizations were carried out in a three-electrode cell in aqueous 0.5 M KOH solution and achieved the highest specific capacitance of 590 F/g at the current density 15 A/g, and there is no capacitance decay after 2000 full cycles with high Coulombic efficiency (almost 100%). Jiang et al. fabricated nickel (Ni)-functionalized CNT forest grown directly on a silicon substrate using a Fe/Al/Mo stacking layer that functioned as both the catalyst material and subsequently a conductive current collecting layer in pseudocapacitor applications [66]. They studied the electrochemical characterization using a typical three-electrode setup in 0.1 M KOH aqueous solution. The Ni-functionalized CNT forest showed the specific capacitance of 1.26 F/cm^3 and retained 94.2% of its initial capacity after 10,000 cycles. The fabrication of PANI/MWCNT hybrid in presence of MnO_2 was reported by Yang et al. [67]. At first they wrapped the CNTs in MnO_2 shell, and then PANI covered the CNTs to replace MnO_2 as the polymerization reaction proceeded. They studied the electrochemical tests in a there-electrode cell in 1 M H_2SO_4 electrolyte and achieved the highest specific capacitance of 764 F/g at a 0.25 A/g current density and the nanocomposite also retained 89% of their initial capacitance after 1000 cycles. Li et al. fabricated Ni-manganese oxide/MWCNTs/carbon fiber paper (CFP) nanocomposite with 3-D porous structure by an electrochemical deposition process [68]. They measured the electrochemical tests with the help of a three-electrode system and in 0.1 M Na_2SO_4 aqueous electrolyte and obtained the highest specific capacitance value of 961.5 F/g at 10 mV/s scan rate and the nanocomposite remained 89.32% of the initial capacitance after 1000 cycles. The CNT@PPy@MnO_2 core-double shell sponge was fabricated by Wu et al. with the help of a chemical vapor deposition (CVD) followed by hydrothermal processes [69]. From the morphological analysis they confirmed that PPy layer coated on the surface of individual CNTs and the surface become rough in CNT@PPy@MnO_2 sponge due to the outside MnO_2 layer. Here the electrochemical characterizations were measure in a three-electrode configuration in 2 M KCl aqueous electrolyte. The CNT@PPy@MnO_2 sponge achieved the highest specific capacitance of 325 F/g at a scan rate of 2 mV/s and retained 90.2% of their initial capacitance after 1000 charge–discharge cycles. Adekunle et al. prepared a MWCNT/metal (Ni, Co, Fe) oxide and investigated as electrode materials for supercapacitors [70]. They performed the electrochemical characterization in a three-electrode setup in two different electrolytes, 1 M H_2SO_4 and in 1 M Na_2SO_4 and achieved the highest specific capacitance for MWCNT-NiO based electrode, which is 2119 F/g in 1 M H_2SO_4 electrolyte. They also reported that this electrode exhibited high electrochemical reproducibility with no significant changes over 1000 cyclic voltammetry cycles. Guan et al. fabricated a hierarchical structure consisting of iron oxide nanoparticles decorated on 3-D ultrathin graphite foam–CNT forest substrate as GF–CNT@Fe_2O_3 [71]. From the morphological analysis they confirmed that the CNT covered with tiny nanoparticles having ~10 nm in size. They studied the electrochemical behavior using a three-electrode system with 2 M KCl as the electrolyte and obtained the highest specific capacitance of ~ 470.5 mF/cm^2 and ~ 95.4% of the capacitance was retained after 5000 cycles of charge–discharge. More recently, the decoration of Fe_2O_3 nanoparticles onto the surface of the MWCNTs for the making of Fe_2O_3/MWCNTs thin films with the help of combining dip-coating, and successive ionic adsorption and reaction methods were discussed by Raut et al. [72]. From the morphological analysis they confirmed that the hematite Fe_2O_3 nanoparticles with a particle size of less than 10 nm were uniformly coated onto the surface of the MWCNTs. They studied the electrochemical characterization in a three-electrode setup in 1 M Na_2SO_3 electrolyte and achieved the highest specific capacitance value for Fe_2O_3/MWCNTs of 431 F/g at 5 mV/s scan rate, high energy density of 38 Wh/kg, high power density of 800 W/kg and also retained 65% capacitance after 500 cycles. Boukhalfa et al. reported the atomic layer uniform deposition of smooth nanostructured vanadium oxide coating on the surface of the MWCNTs and investigated as electrode materials for supercapacitor [73]. They analyzed the electrochemical characterization with a three-electrode system in aqueous KCl electrolyte and obtained the specific capacitance of ~1550 F/g at 1 A/g current density and the nanocomposite also retained 92% of their original capacitance after 5000 cycles. Recently, Ko et al. fabricated a MnO_2/MWCNTs-based flexible network as binder-free electrodes and a PVA/H_2SO_4 electrolytes for the formation of a stretchable solid-state supercapacitor [74]. They studied the electrochemical characterization with a three-electrode system and achieved the highest specific capacitance of 324 F/g at 0.5 A/g current density, maximum energy density of 7.2 Wh/kg, better power density 3.3 W/kg, and retained ~100% of capacitance retention after 5000 cycles (Fig. 9.7).

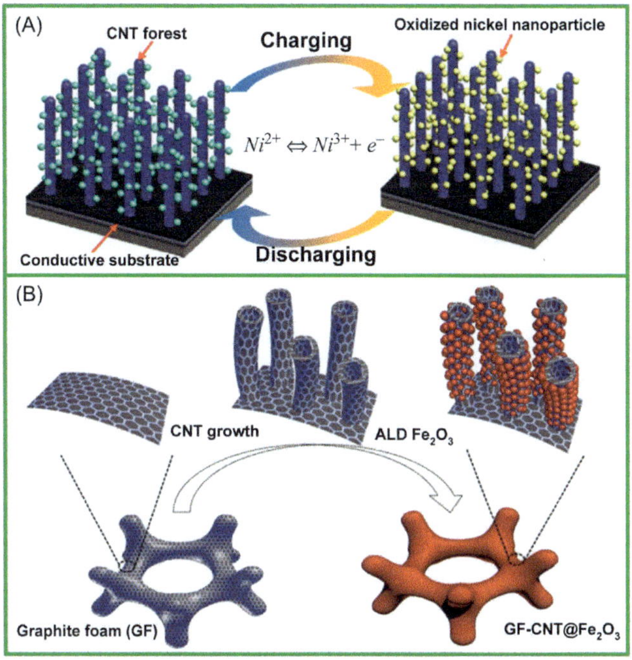

Figure 9.7 (A) Schematic of the pseudocapacitor using CNT forests functionalized with oxidized nickel nanoparticles as the electrode [Jiang et al. *Nano Lett.* 2013 (Ref. 66)], (B) Growth procedure of GF–CNT@Fe_2O_3 starting from graphite foam [Guan et al. *ACS Nano* 2015 (Ref. 71)].

9.3 Graphene-based hybrid nanocomposites for supercapacitors

9.3.1 Graphene polymer hybrid

Over the past few years supercapacitor electrode materials based on Gr and conducting polymers have attracted great interest of research. Different nanocomposites based on Gr and conducting polymers such as PANI, PPy, and PT are studied by different groups. More recently, Ouyang et al. fabricated rGO/PANI beads by a freeze-casting method and were investigated as an electrode materials supercapacitor [75]. The uniform coating over rGO was confirmed by morphological analysis. They have done the electrochemical cauterization with a three-electrode system in 2 M KCl aqueous electrolyte and achieved the highest specific capacitance of 669 F/g at 63.5 wt% of PANI loading. Zhang et al. prepared GOPPy–F nanocomposite by in situ polymerization techniques and studied the electrochemical testes with a three-electrode setup in 2 M H_2SO_4 aqueous electrolyte [76]. The GOPPy–F nanocomposite exhibited the excellent electrocapacitive performances with a high specific capacitance of 500 F/g and retained 70% of original capacitance after 1000 cycles. The conducting polymers such as PEDOT, PANI, and PPy were directly coated on the RGO via in situ polymerization techniques reported by Zhao et al. [77]. The uniform coating of PEDOT, PANI, and PPy over RGO was confirmed by morphological analysis. They studied the electrochemical characterization by a three-electrode system and achieved the better specific capacitance for RGO-PANI and which was 361 F/g at a 0.3 current density and also retained 82% capacitance retention after 1000 cycles. Li et al. prepared Gr/PANI nanocomposite by reducing GO in the presence of different contents of PANI nanoparticles [78]. They evaluated the electrochemical performances with a three-electrode cell system in 1 M H_2SO_4 electrolyte and obtained the highest specific capacitance of 257 F/g at 0.1 A/g current density, and the nanocomposite still remained about 98% of initial capacitance after 1000 cycles. Presently, the Gr/PPy hybrid aerogels with 3-D hierarchical structure was fabricated by He et al. [79]. The electrochemical tests were performed using 1 M KOH as electrolyte solution in a three-electrode system and reached the highest specific capacitance of 418 F/g at a current density of 0.5 A/g and the nanocomposite maintained 74% of capacitance after 1000 cycles. Li et al. synthesized GO/PPy nanowire composite materials by in situ chemical polymerization method and investigated as electrode materials for supercapacitors [80]. From the morphological study they confirmed that PPy nanowires with 40 nm in diameter were uniformly dispersed on the surface of the GO nanosheets. They tested the electrochemical performances in a three-electrode setup in 1 M KCl aqueous solution and achieved the better specific capacitance of 728 F/g at a 0.5 A/g current density and also the nanocomposite retained their 93% initial capacitance after 1000 cycles. Recently, Chen et al. prepared rGO/PANI composites through an effective in situ one-pot synthesis route that includes the reduction of GO by aniline under weak alkali condition via hydrothermal method and then followed by in situ polymerization of aniline [81]. They carried out the electrochemical characterizations in three-electrode system in 1 M H_2SO_4 electrolyte. They achieved the maximum specific capacitance of 524.4 F/g with a mass ratio of aniline to GO 10:1 at a current density of 0.5 A/g for the rGO/PANI nanocomposite and specific capacity retention rate of 81.1% after 2000 cycles at 100 mV/s scan rate. Further, Luo et al. prepared PANI nanofiber/nitrogen-doped Gr composite hydrogel by integration polymerization of aniline followed by a hydrothermal process and studied the electrochemical behavior [82]. A three-electrode cell system was used for electrochemical study in 1 M H_2SO_4 aqueous electrolyte, and the nanocomposite hydrogel attend the highest specific capacitance of 610 F/g at 1 A/g current density and retained 94.4% of initial capacitance after 1000 charge/discharge cycles. The molybdenum disulfide/RGO@PANI nanocomposite prepared through a two-stage synthetic method including hydrothermal and polymerized reaction reported by Li et al. [83]. The electrochemical characterizations of MoS_2/RGO@PANI nanocomposite were carried out with the help of a three-electrode system in 1 M H_2SO_4 electrolyte and achieved the highest specific capacitance of 1224 F/g at a 1 A/g current density and retained 82.5% of initial capacitance after 3000 loops at 10 A/g current density (Fig. 9.8).

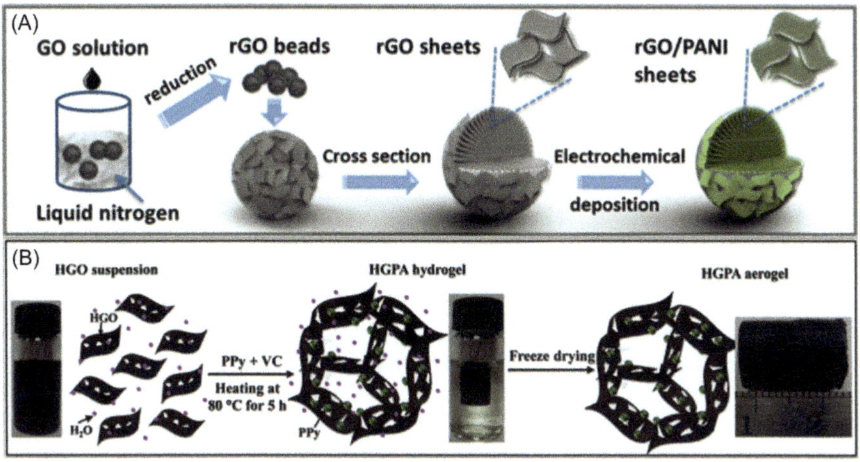

Figure 9.8 (A) Illustration of the freeze-casting process to make rGO beads and subsequent electrodeposition of rGO/PANI beads [Ouyang et al. *ACS Appl. Mater. Interfaces* 2016 (ref. 75)]; (B) the formation schematic illustration of Gr/PPy hybrid aerogels and the corresponding digital photograph [He et al. *J. Power Sources* 2016 (ref. 79)].

9.3.2 Modified graphene based supercapacitors

Chemical modifications are needed to separate Gr sheets without destroying too much of the sp^2 structure. Recently, Li et al. synthesized a PPy-Gr sheet nanocomposite after making a reduced Gr react with tertazine derivatives through inverse electron demand Dials–Alder reaction and investigated as electrode materials for supercapacitors [84]. They studied the electrochemical characterization with a three-electrode configuration in 0.1 M $TBAPF_6$ electrolyte and obtained the highest specific capacitance of 326 F/g at a 0.5 A/g current density and decreased 11% of capacitance after 500 cycles and a total loss of 38.5% of capacitance after the 5000 cycles. Wang et al. fabricated sulfonated a Gr/MnO_2/PANI nanocomposite synthesized via a dilute in situ polymerization technique using $KMnO_4$ as oxidant under neutral condition [85]. From the morphological analysis they confirmed that MnO_2/PANI nanoparticles having diameter of 5–10 nm deposit onto sulfonated Gr nanosheets. The electrochemical study was performed with the help of a three-electrode system in 1 M Na_2SO_4 electrolyte and achieved the highest specific capacitance of 276 F/g at 1 A/g current density and retained 88.3% of initial capacitance after 3000 cycles. Bora et al. synthesized a sulfonated Gr and PPy nanocomposite by interfacial polymerization procedure and investigated it as electrode material for supercapacitors [86]. They performed the electrochemical characterization with a three-electrode system in 1 M H_2SO_4 electrolyte and obtained the highest specific capacitance for the sulfonated Gr/PPy nanocomposite of 360 F/g at a 1 A/g current density and retained 90% of initial capacitance after 500 charging/discharging cycles. Xinming et al. synthesized hydrogen bonded/PANI (HbG/PANI) nanocomposite by electrochemical polymerization technique for electrochromo–supercapacitor application [87]. From the morphological analysis it was confirmed that a thin film uniformly covered the surface of HbG. They studied the electrochemical performance with a three-electrode system in 1 M KOH electrolyte and obtained the highest specific capacitance of 598 F/g at 1 A/g current density and maintained 96.5% of initial capacitance after 5000 cycles. Iamprasertkun et al. fabricated nitrogen-doped RGO aerogel coated on carboxyl-modified carbon fiber by hydrothermal reduction of GO with hydrazine via a freezing-dye method [88]. The electrochemical characteristics were carried out in three electrode system in 1 M H_2SO_4 and 0.5 Na_2SO_4 aqueous electrolytes and obtained the high specific capacitance of 764.53 F/g at 1 A/g current density, high specific power of 6525.56 W/kg, high specific energy of 245 Wh/kg, and a capacity retention of 86% over 3000 charge–discharge cycles, respectively. An innovative approach was applied to produce an interconnected and oxygen-functionalized porous RGO electrode (PRGO-O electrode) by a facile chemical treatment route ZnO-template and an alkaline treatment with a NaOH etchant was reported by Hwang et al. [89]. The highest specific capacitance of 337.2 F/g was obtained for the nanocomposite in a three-electrode cell in 1 M H_2SO_4 electrolyte. With the help of a two-electrode cell the nanocomposite achieved the highest specific capacitance of 322.1 F/g with high energy density of 38.8 Wh/kg. Rajagopalan et al. fabricated surfactant-treated Gr/PANI nanocomposites with the MnO_2 template-aided oxidative polymerization of aniline on the surfactant-treated Gr sheets [90]. The electrochemical performance of the Gr/PANI nanocomposite was studied in a three-electrode system in 2 M H_2SO_4 electrolyte and obtained the highest specific capacitance value of 436 F/g at a 1 A/g current density and retained 91.7% of capacitance after 500 cycles. Lu et al. fabricated dye-functionalized Gr/PANI nanocomposite for supercapacitor application [91]. They performed the electrochemical performances of electrodes in a three-electrode system using 1 M H_2SO_4 as an electrolyte and achieved a high specific capacitance of 579 F/g at a current density at a 0.5 A/g current density with good cycling stability and rate capacity. Lai et al. synthesized four different surface-functionalized Gr such as, GO, RGO, N_2 doped Gr and primary amine modified Gr in presence of PANI by in situ polymerization techniques as the electrode materials for supercapacitors [92]. They studied the electrochemical performance with a three-electrode cell using 0.5 M H_2SO_4 as the electrolyte and achieved the highest specific capacitance of 500 F/g for amine modified Gr/PANI nanocomposite and also obtained good cyclability with no loss of capacitance over 680 cycles. Du et al. fabricated PANI-modified oriented Gr hydrogel films as the free-standing electrode for flexible solid-state supercapacitors [93]. The flexible solid-state supercapacitor exhibited high specific capacitance of 530 F/g at the current density of 0.5 A/g and remained 80% of its original capacitance value up to 10,000 charge–discharge cycles at the current density of 10 A/g (Fig. 9.9).

9.3.3 Graphene–metal oxide supercapacitors

Graphene and metal oxide–based supercapacitors have attracted much more research attention over the past few years. Lee et al. prepared a Gr/Mn_3O_4

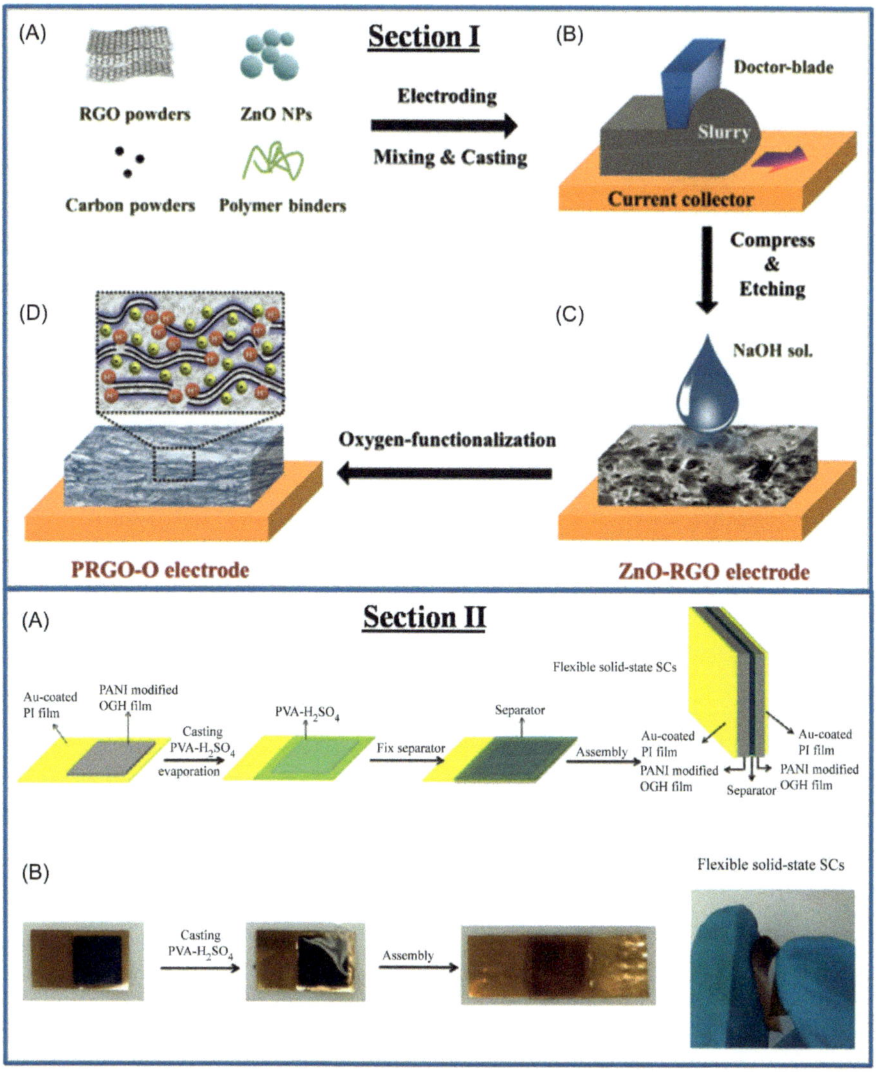

Figure 9.9 (Section I) Schematic illustration of the fabrication process of the PRGO-O electrode [Hwang et al. *Crabon* 2016 (Ref. 89)]; (Section II) Schematic diagram of the assembly process of flexible solid-state supercapacitors and photographs of the fabrication process of flexible solid-state supercapacitors based on the PANI-modified oriented Gr hydrogel film. [Du et al. *ACS Appl. Mater. Interfaces* 2015 (Ref. 93)].

composite by a simple hydrothermal process using ethylene glycol as a reducing agent and investigated as electrode materials for supercapacitors [94]. From the morphological analysis they confirmed that Mn_3O_4 nanorod of 100 nm to 1 μm length were well dispersed on Gr sheets. They studied the electrochemical characterization by a beaker-type three-electrode cell in 1 M Na_2SO_4 electrolyte and achieved the highest specific capacitance for the composite of 121 F/g at a 0.5 A/g current density and retained almost 100% of capacitance after 10,000 cycles. Yu et al. developed a "conductive wrapping" method to greatly improved the supercapacitor performance of Gr/MnO_2-based nanostructured electrodes [95]. They performed the electrochemical testes with a three-electrode configuration in 0.5 M Na_2SO_4 electrolyte solution and obtained the highest specific capacitance of ~380 F/g at 0.1 mA/cm^2 current density and an exhibited excellent cycling performance with >95% capacitance retention over 3000 cycles. Fan et al. synthesized Fe_3O_4@carbon nanosheets composite using ammonium ferric citrate as the Fe_3O_4/carbon precursor and GO as the structure-directing agent under hydrothermal process [96]. With the help of a three-electrode electrochemical setup and KOH/PVA gel electrolyte they performed the electrochemical characterization and achieved the excellent specific capacitance of 586 F/g at 0.5 A/g current density, largest energy density of 18.3 Wh/kg at power density of 351 W/kg, and remained 70.8% of its initial capacitance after 5000 cycles, respectively. Wang et al. fabricated Co_3O_4 nanosheets arrays on 3-D porous Gr/nickel foam through a hydrothermal synthesis procedure [97]. The electrochemical performance of the abovementioned nanocomposite was studied via a three-electrode glass cell in 6 M KOH electrolyte and achieved the highest specific capacitance value of 3533 F/g at a current density of 1 A/g, and the nanocomposite retained 94% of initial capacitance after 2000 cycles. The exfoliation of bulk-layered materials into 2-D nanosheets dramatically enhances their surface area and opens up novel properties. For that reason, Mendoza–Sánchez et al. synthesized 2-D Gr-MnO_2 nanosheets by a novel single-step liquid-phase coexfoliation method [98]. They tested the electrochemical characterization in a 0.5 M K_2SO_4 aqueous electrolyte and achieved a geometric capacitance of 80 mF/cm^2 (single electrode), volumetric capacitances of 300 F/cm^3 (single electrode) and 40.9 F/cm^3 (symmetric device), and volumetric energy and power of 5.5 mWh/cm^3 and 2193.1 mW/cm^3 (symmetric device), respectively. Luo et al. successfully developed an electrostatic spray deposition (ESD) route to fabricate the $NiCo_2O_4$-rGO nanocomposite supported on the nickel foam substrates with excellent electrochemical performance [99]. The electrochemical tests of the nanocomposite were conducted using three-electrode system in 2 M KOH electrolyte and exhibited highest specific capacitance of 777.1 F/g at a current density of 5 A/g and about 99.3% of the capacitance retained after 3000 cycles. The chemically anchored metal oxide nanoparticles onto Gr nanosheets of the resultant composites—SnO_2/GNs, MnO_2/GNs, and RuO_2/GNs—coated over conductive carbon fabric substrates for supercapacitor electrode was reported by Rakhi et al. [100]. They performed the electrochemical characterization in two symmetric electrodes, separated by a thin polymer separator in a 30 wt% of KOH aqueous electrolyte. They obtained the specific capacitance values for SnO_2/GNs (195 F/g), MnO_2/GNs (235 F/g), and RuO_2/GNs (365 F/g) at scan rate of a 20 mV/s, respectively. They also achieved capacitance loss of SnO_2/GNs (78%), MnO_2/GNs (73%), and RuO_2/GNs (90%) after 6000 consecutive cycles. With the help of a simple hydrothermal synthesis procedure, Co_3O_4 nanowires were in situ synthesized on 3-D Gr foam grown by CVD techniques for supercapacitor electrode materials was reported by Dong et al. [101]. The electrochemical behavior were carried out in a three-electrode system in 0.1 M NaOH electrolyte and exhibited highest specific capacitance of 1100 F/g at a 10 A/g current density with excellent cycling stability. Recently, Wu et al. prepared MnO_2/PANI/RGO by in situ polymerization techniques and were investigated as electrode materials for supercapacitors [102]. Initially, PANI coated on RGO to form PANI/RGO hybrid nanosheets and after that nanoflaky MnO_2 in situ grown on PANI/RGO hybrid nanosheets to obtained sandwich-like nanosheets MnO_2/PANI/RGO. They studied the electrochemical characterization with a three-electrode system in 1 M KOH electrolyte and achieved the highest specific capacitance of 1090.2 F/g at a 0.5 A/g current density and the hybrid nanosheets also maintain 82.3% of initial capacitance after 5000 consecutive cycles at 0.5 A/g. Due to the large requirement of portable consumer electronics in the present era, much effort was devoted to the development of lightweight, flexible and even wearable technologies. For that reason, Sankar et al. synthesized a $CoFe_2O_4$/rGO/PANI composite by in situ chemical oxidative polymerization for flexible supercapacitor electrode [103]. They performed the electrochemical behavior in a three-electrode system in 1 M KOH electrolyte and obtained the highest specific capacitance of 8.59 F/m at 1 mV/s scan rate with excellent cycling stability (Fig. 9.10).

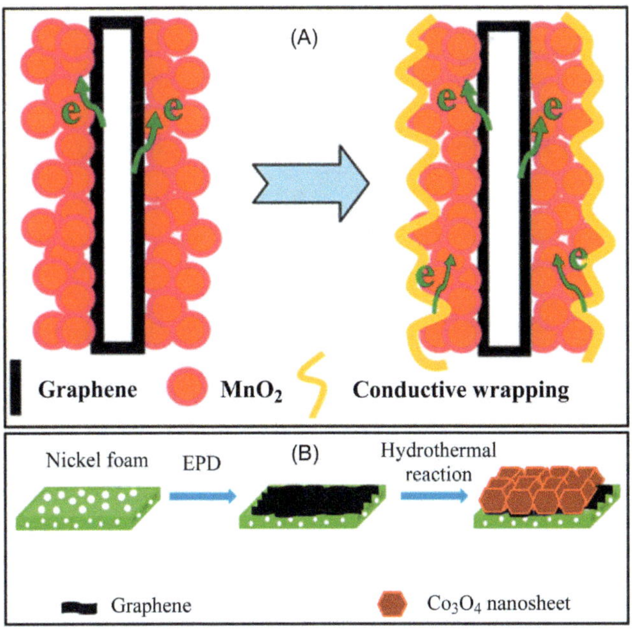

Figure 9.10 (A) Schematic illustration showing conductive wrapping of Gr/MnO_2 to introduce an additional electron transport path (in a discharge cycle) [Yu et al. *Nano Lett.* 2011 (Ref. 95)]; (B) Schematic illustration for the fabrication of thin Co_3O_4 nanosheet arrays on 3-D porous Gr/Ni foam [Wang et al. *Electrochim. Acta* 2016 (Ref. 97)].

9.3.4 Asymmetric supercapacitors

The asymmetric hybrid capacitors are the combination of both Faradaic and non-Faradaic methods by coupling a pseudocapacitor electrode with an EDLC electrode. Where the positive electrode was based on a real pseudocapacitive metal oxide electrode but not a composite electrode, the negative electrode was based on a EDLCs-activated carbon electrode. The development of such an asymmetric supercapacitor has been of great research interest in the past two or three years. A simple, one-step, two-electrode electrochemical composite depositing for fabrication amorphous MnO/Gr-CNT hybrid was reported by Wang et al. [104]. They constructed the asymmetric supercapacitor consisting of a commercial-activated carbon negative electrode and an a-MnO_x/rGO-CNT as positive electrode. They studied the electrochemical behavior with a three-electrode system, and the ternary a-MnO_x/rGO-CNT hybrid electrode exhibited a high specific capacitance of 440 F/g at 5 mV/s scan rate and high capacitance retention of 60% at 1000 mV/s scan rate. Wang et al. developed a high-energy density and power density supercapacitor using an asymmetric configuration of Fe_2O_3 and MnO_2 nanoparticles incorporated into 3-D macroporous Gr film electrodes [105]. To build the aqueous asymmetric supercapacitor devices, the 3-D Fe_2O_3/m-RGO and 3-D MnO_2/m-RGO films were employed as the anode and cathode, respectively. The electrochemical performances were carried out in a three-electrode system in 2 M LiCl electrolytes and at scan rate of 5 mV/s, a maximum specific capacitance of 382 F/g was obtained for the 3-D MnO_2/m-RGO film, while that of the 3-D Fe_2O_3/m-RGO film was 260 F/g, having a capacitance retention of 83 and 90% after 5000 cycles, respectively. Du et al. fabricated an asymmetric supercapacitor of $Cu_2O/CuMoO_4$ nanosheets directly grown on Ni foam synthesized by a facile hydrothermal method [106]. The electrochemical performances of the asymmetric supercapacitor were explored in a two-electrode mode in which $Cu_2O/CuMoO_4$ acted as the positive electrode while the activated carbon electrode was used as the negative electrode. All the electrochemical measurements were performed in 2 M KOH solution at room temperature and obtained highest specific capacitance of 4264 F/g, highest areal capacitance of 9.38 F/cm^2 at a current density of 1 A/g, better energy density of 75.1 Wh/kg, superior power density of 420 W/kg, and remains around 86.6% of capacitance after 3000 cycles at a current density of 5 A/g, respectively. Shen et al. synthesized a sulfonated Gr nanosheets/carboxylated MWCNT/PANI (sGNS/cMWCNT/PANI) nanocomposite through an interfacial polymerization method [107]. They fabricated the asymmetric supercapacitor using sGNS/cMWCNT/PANI as the positive electrode and activated porous Gr (aGNS) as the negative electrode. They performed the electrochemical behavior in a two-electrode cell in 1 M aqueous H_2SO_4 electrolyte and exhibited a preferable specific capacitance of 107 F/g at 1 A/g current density, high energy density of 20.5 Wh/kg at a power density of 25 kW/kg, and the asymmetric supercapacitor device also exhibited a superior long cycle life with 91% retention of the initial specific capacitance after 5000 cycles, respectively. The fabrication of an asymmetric supercapacitor based on Gr/CNT-PANI as a cathode and Gr/CNT as an anode was reported by Cheng et al. [108].

From the morphological analysis they confirmed the vertically aligned PANI coating on the Gr surfaces. The electrochemical characterization was carried out in a two-electrode system in 1 M KCl aqueous electrolyte and obtained the specific capacitance of 271 F/g at a current density of 0.3 A/g, highest energy density of 188 Wh/kg, maximum power density of 200 W/kg, and capacitance retention of 82% after 1000 cycles, respectively. A low-cost high-performance solid-state flexible asymmetric supercapacitor with α-MnO_2 nanowires and amorphous Fe_2O_3 nanotubes grown on flexible carbon fabric was fabricated by Yang et al. through facile scalable wet chemical methods, where the conductive carbon cloth not only served as a lightweight and flexible current collector but also acted as an excellent scaffold for active materials [109]. They fabricated the solid-state flexible asymmetric supercapacitor with low-cost MnO_2 nanowires as a positive electrode and Fe_2O_3 nanotubes as a negative electrode. The solid-state flexible asymmetric supercapacitor device achieved a high specific and volumetric capacitance of 91.3 F/g and 1.5 F/cm^3 at a current density of 2 mA/cm^2, a high energy density of 0.55 mWh/cm^3, high power density of 139.1 mW/cm^3, and also exhibited 84% retention of the initial capacitance after 5000 cycles. Salunkhe et al. prepared a CNT-$Ni(OH)_2$ composite by a simple one-step and inexpensive chemical coprecipitation method and fabricated the asymmetric supercapacitor using CNT/$Ni(OH)_2$ composite as positive electrode and rGO as negative electrode [110]. The above-mentioned asymmetric supercapacitor achieved the specific capacitance of 78 F/g at a 2 A/g current density and a high energy density of 35 Wh/kg at a power density of 1.8 kW/kg (Fig. 9.11).

9.4 Graphene and carbon nanotubes based ternary nanocomposites

Over the past few years the ternary nanocomposite based on Gr and CNT have attracted much more research attention for the next generation of electrode materials for supercapacitors. The Gr nanosheets/CNT/PANI ternary nanocomposite synthesized by in situ polymerization techniques for supercapacitors was reported by Yan et al. [111]. From the morphological analysis they confirmed that the PANI particles homogeneously coated on the surface of the CNTs as well as on Gr nanosheets. They studied the electrochemical characterization with the help of a three-electrode setup in 6 M KOH aqueous electrolyte at room temperature and achieved the highest specific capacitance of 1035 F/g at a 1 mV/s scan rate and also retained their 94% of initial capacitance after 1000 cycles. Ning et al. synthesized 3-D hybrid materials composed of 2-D fish scale-like PANI nanosheet arrays on GO sheets and CNTs with a one-step process using a simplified template-free polymerization method [112]. They performed all electrochemical experiments in 1 M H_2SO_4 using a three-electrode system and obtained the highest specific capacitance of 589 F/g at a 0.2 A/g current density and also the hybrid nanocomposite retained 81% of initial capacitance after 1000 cycles. Sun et al. synthesized Gr nanosheets/acid-treated MWCNTs-poly(1,5−diaminoanthraquinone) ternary nanocomposite

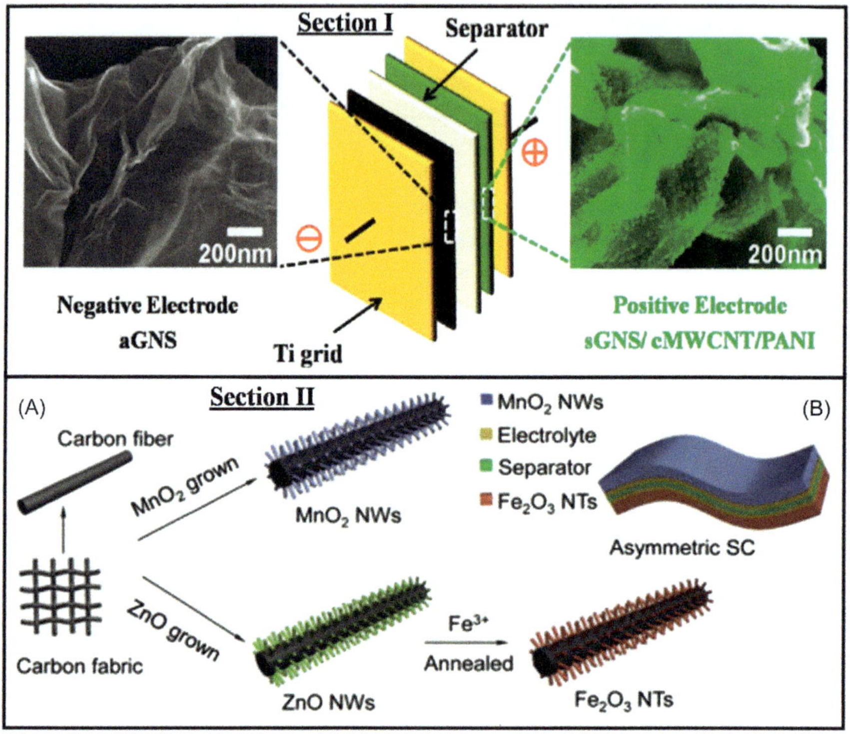

Figure 9.11 (Section I) Schematic illustration of the fabricated asymmetric supercapacitor device based on a sGNS/cWMCNT/PANI composite as the positive electrode and aGNS as the negative electrode in 1 M H_2SO_4 Electrolyte [Shen et al. *ACS Appl. Mater. Interfaces* 2013 (Ref. 107)]; (Section II) (A) Schematic diagram illustrated the synthesis procedure of MnO_2 nanowires and Fe_2O_3 nanotubes on carbon cloth. (B) Schematic sketch illustrating the designed asymmetric supercapacitor device [Yang et al. *Nano Lett.* 2014 (ref. 109)].

using $Ce(SO_4)_2$ as an oxidant and camphor sulphonic acid as a dopant [113]. They used both three- and two-electrode cell configurations to measure the electrochemical performances in 1 M organic electrolyte (tetraethylammonium tetrafluoroborate–acetonitrile) and achieved the highest specific capacitance of 80.8 F/g at 0.5 A/g current density, high energy density of 86.4 Wh/kg at a power density of 0.73 kW/kg and also maintained 97% of capacitance after 10,000 cycles, respectively. Recently, Wang et al. prepared a GO/PPy/MWCNTs ternary nanocomposite by in situ polymerization procedure and investigated it as electrode materials for supercapacitors [114]. They performed the electrochemical measurements by a three-electrode system in 1 M $NaNO_3$ aqueous electrolyte and achieved better specific capacitance of 406.7 F/g at a current density of 0.5 A/g and also retained 92% of initial capacitance after 1000 cycles. Similarly, Lu et al. fabricated Gr/PPy/CNT ternary nanocomposite by in situ polymerization method and studied the electrochemical behavior [115]. From the morphological analysis they confirmed that the

PPy particles are uniformly wrapped around the CNTs as well as on Gr. All the electrochemical measurements were conducted using a three-electrode system in 1 M KCl electrolyte and obtained the highest specific capacitance of 361 F/g at a 0.2 A/g current density and also retained 96% of capacitance after 2000 cycles. Dhibar et al. synthesized a ternary nanocomposite based on Gr, SWCNTs and poly (3-methylthiophene) [Gr-SWCNTs-PMT] by a simple in situ chemical oxidative polymerization technique using $FeCl_3$ as the oxidant and investigated as potential electrode materials for supercapacitors [116]. From the morphological analysis they confirmed that the uniform coating of PMT on both SWCNTs and Gr surfaces and these coated SWCNTs and Gr were interconnected with each other and formed a sandwich structure and also there is the formation of a bridge between the coated Gr with the coated SWCNTs. They performed the electrochemical characterization with the help of a three-electrode system in 1 M KCl electrolyte and achieved the highest specific capacitance of 551 F/g at a current density of 0.5 A/g for the ternary nanocomposite. The Gr-SWCNTs-PMT ternary nanocomposite achieve the highest energy density of 48.97 Wh/kg at a 0.5 A/g current density as well as highest power density of 1579.35 W/kg at a current density of 3 A/g. The Gr-SWCNTs-PMT ternary nanocomposite still retained 93% of its initial capacitance after 1000 cycles. The ternary nanocomposite of MnO_2/GNS/CNTs was fabricated through a facile chemical method involving in situ growth of MnO_2 particles on the surface of a GO/CNT hybrid followed by the chemical reduction of GO, as reported by Ramezani et al. [117]. They carried out the electrochemical measurements in a standard three-electrode cell using 1 M Na_2SO_4 aqueous solution as electrolyte and the ternary nanocomposite obtained the highest specific capacitance of 367 F/g at 20 mV/s scan rate and retained 88% of initial capacitance after 3000 cycles. Further, Lu et al. prepared a GO/PANI/CNT precursor with a flow-directed assembly from a complex dispersion of GO and PANI/CNT, followed by reoxidation and redoping of the reduced PANI in the composite to restore the conducting PANI structure [118]. They conducted all the electrochemical measurements using a three-electrode system in 1 M HCl electrolyte and obtained the ternary nanocomposite highest specific capacitance of 569 F/g at a current density of 0.1 A/g and also maintained 96% of capacitance retention after 5000 continuous charge/discharge cycles (Fig. 9.12).

9.5 Modern applications of supercapacitors

Supercapacitors are generally used where a huge amount of power is required for a comparatively short time and also where an extreme number of charge/discharge cycles or a longer lifetime is required. At the beginning, supercapacitors were used as start-up capacitors for large engines in submarines and tanks. They started to come into sight on diesel trucks and railroad locomotive when the cost decreased. In the year 2000 supercapacitors gained more attention in the electric car industry because of their capability to charge more quickly than batteries. That makes them

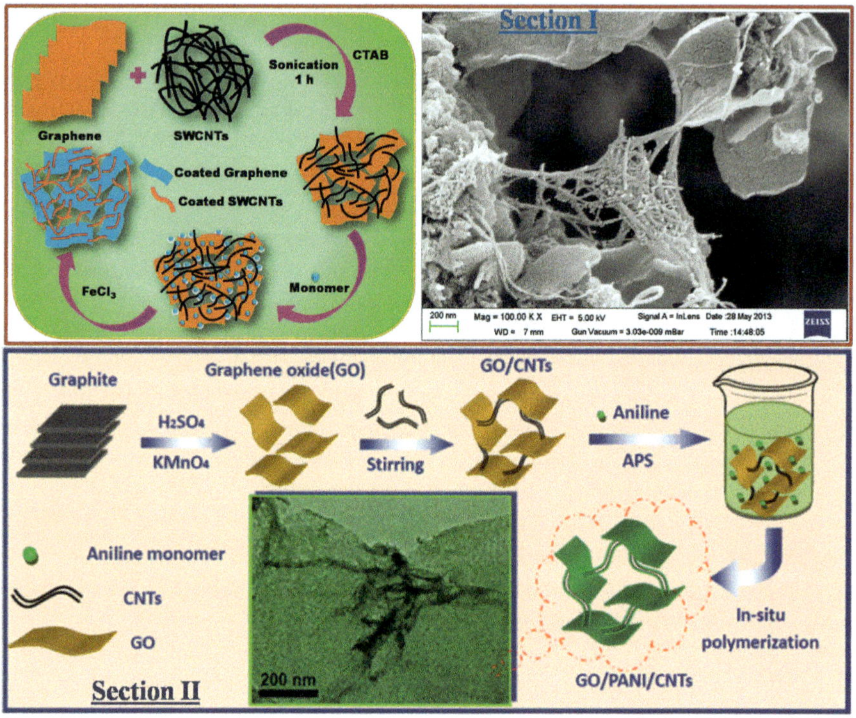

Figure 9.12 (Section I) Schematic representation of the ternary nanocomposite synthesis process and FESEM image of Gr-SWCNTs-PMT ternary nanocomposite [Dhibar et al. *Ind. Eng. Chem. Res.* 2014 (ref. 116)]; (Section II) Illustrations of the fabrication of PANI–GO–CNT ternary hybrid materials by oxidative polymerization. Fish scale-like PANI nanosheet arrays on the GO sheets and CNT [Ning et al. *Carbon* 2013 (Ref.112)].

typically appropriate for regenerative braking applications. China is experimenting with a new form of electric bus (capabus) that uses onboard supercapacitors. The bus recharges rapidly whenever it arrives at bus stop (under so-called electric umbrellas) and fully charged in the terminus. From the year 2003 to 2008 Mannheim Stadtbahn in Mannheim, Germany, operated a light-rail vehicle (LRV) with the help of supercapacitors for regenerative breaking. The supercapacitor permitted the LRVs to operate in the area of Heidelberg without the overhead wires. In August 2012, the CSR Zhouzhou Electric Locomotive Corporation of China presented a prototype two-car light metro train prepared with a roof-mounted supercapacitor offering regenerative braking and the capability to work without overhead wires at the time of charging at stations. The Toyota racing car utilized a hybrid drivetrain with supercapacitors. The energy storage devices made the Toyota and Audi hybrids the fastest race cars. In 2007, an elastic screwdriver was manufactured with the help of supercapacitor units. Supercapacitor can also drive low power equipment like photographic flash, PC card, portable media players, flashlights, and automated meter-reading equipment.

Figure 9.13 Different modern applications of supercapacitors.

Research and development has brought rapid enhancement in cost and superiority of the supercapacitors. This significant technology still has many miles to go in terms of technical assurance and practical applicability. There are four major markets where supercapacitors can be used in near future: consumer, industrial, stationary, and transport energy storage power management. The different modern applications of supercapacitors are depicted in Fig. 9.13.

9.6 Summary

The electrochemical properties of the graphene and CNT-based hybrid polymer nanocomposites has been discussed in this chapter, which has mainly highlighted the electrochemical characteristics of CNTs-based hybrid nanocomposites (both MWCNTs- and SWCNTs-based); graphene-based hybrid polymer nanocomposites includes graphene polymer hybrids, modified Gr based supercapacitors, Gr-metal oxide supercapacitors, and asymmetric supercapacitors, etc., and Gr- and CNT-based ternary nanocomposites, respectively. The modern applications of supercapacitors were also discussed in this chapter. It can be observed that the incorporation of CNTs (both MWCNTs and SWCNTs), metal oxides, Gr, modified Gr, and GO in the conducting polymers enhances the electrochemical behavior dramatically. This is because of the superior properties of CNTs and Gr. Different attractive morphologies also affect to improve the electrochemical behavior of the electrode materials. Presently, the flexible, stretchable, as well as transparent supercapacitor can take a major part in the research for the development of electrode materials for supercapacitors. Researchers now want to develop relatively very-low-cost as well as green electrode materials for supercapacitors.

References

[1] C. Largeot, C. Portet, J. Chmiola, P. Taberna, Y. Gogotsi, P. Simon, Relation between the ion size and pore size for an electric double-layer capacitor, J. Am. Chem. Soc. 130 (2008) 2730.

[2] S. Kandalkar, D. Dhawale, C. Kim, C. Lokhande, Chemical synthesis of cobalt oxide thin film electrode for supercapacitor application, Synth. Met. 160 (2010) 1299.

[3] R. Kotz, M. Carlen, Principles and applications of electrochemical capacitors, Electrochim. Acta. 45 (2000) 2483.

[4] P. Simon, Y. Gogotsi, Materials for electrochemical capacitors, Nat. Mater. 7 (2008) 845.

[5] E. Frackowiak, Carbon materials for supercapacitor application, Phys. Chem. Chem. Phys. 9 (2007) 1774.

[6] L.L. Zhang, X.S. Zhao, Carbon-based materials as supercapacitor electrodes, Chem. Soc. Rev. 38 (2009) 2520.

[7] M. Inagaki, H. Konno, O. Tanaike, Carbon materials for electrochemical capacitors, J. Power Sources 195 (2010) 7880.

[8] S.W. Zhang, G.Z. Chen, Manganese oxide based materials for supercapacitors, Energy Mater. 3 (2008) 186.

[9] E. Frackowiak, F. Beguin, Carbon materials for the electrochemical storage of energy in capacitors, Carbon N. Y. 39 (2001) 937.

[10] P. Simon, A. Burke, Nanostructured carbons: double-layer capacitance and more, Interface 17 (2008) 38.

[11] P. Simon, Y. Gogotsi, Capacitive energy storage in nanostructured carbon-electrolyte systems, Acc. Chem. Res. 46 (2012) 1094.

[12] B.E. Conway, Electrochemical Supercapacitors, Scientific Fundamental and Technological Applications, *Kluwer Academic/Plenum Publishers*, New York, 1999.

[13] Y. Zhang, Y. Gui, X. Wu, H. Feng, A. Zhang, L. Wang, et al., Preparation of nanostructures NiO and their electrochemical capacitive behaviors, Int. J. Hydrog. Energy. 34 (2009) 2467.

[14] C. Chen, D. Zhao, X. Wang, Influence of addition of tantalum oxide on electrochemical capacitor performance of molybdenum nitride, Mater Chem Phys. 97 (2006) 156.

[15] C.-Z. Yuan, B. Gao, X.-G. Zhang, Electrochemical capacitance of nio/ru0.35v0.65o2 asymmetric electrochemical capacitor, J Power Sources 173 (2007) 606.

[16] J.R. Miller, Electrochemical capacitor thermal management issues at high-rate cycling, Electrochim. Acta. 52 (2006) 1703.

[17] A.K. Cuentas Gallegos, M.E. Rincon, Carbon nanofiber and PEDOT-PSS bilayer systems as electrodes for symmetric and asymmetric electrochemical capacitor Cells, J Power Sources. 162 (2006) 743.

[18] L.L. Zhang, R. Zhou, X.S. Zhao, Graphene-based materials as supercapacitor electrodes, J. Mater. Chem. 20 (2010) 5983.

[19] J.P. Zheng, J. Huang, T.R. Jow, The limitations of energy density for electrochemical capacitors, J. Electrochem. Soc. 144 (1997) 2026.

[20] G. Wang, L. Zhang, J. Zhang, A review of electrode materials for electrochemical supercapacitors, Chem. Soc. Rev. 41 (2012) 797.

[21] W. Sugimoto, H. Iwata, Y. Murakami, Y. Takasu, Electrochemical Capacitor Behavior of Layered Ruthenic Acid Hydrate, J. Electrochem. Soc. 151 (2004) A1181.

[22] X. Dong, W. Shen, J. Gu, L. Xiong, Y. Zhu, H. Li, et al., MnO_2-embedded-in-mesoporous-carbon-wall structure for use as electrochemical capacitors, J. Phys. Chem. B. 110 (2006) 6015.

[23] B.E. Conway, V. Birss, J. Wojtowicz, The role and utilization of pseudocapacitance for energy storage by supercapacitors, J. Power Sources 66 (1997) 1.

[24] Y. Zhang, H. Feng, X. Wu, L. Wang, A. Zhang, T. Xia, et al., Progress of electrochemical capacitor electrode materials: a review, Int. J. Hydrog. Energy. 34 (2009) 4889.

[25] S. Iijima, Helical microtubules of graphite carbon, Nature. 354 (1991) 56.

[26] D.N. Futaba, K. Hata, T. Yamada, T. Hiraoka, Y. Hayamizu, Y. Kakudate, et al., Shape-engineerable and highly densely packed single-walled carbon nanotubes and their application as supercapacitor electrodes, Nat. Mater. 5 (2006) 987.

[27] R.H. Baughman, A.A. Zakhidov, W.A. de Heer, Carbon nanotubes-the route toward applications, Science 297 (2002) 787.

[28] S. Berber, Y.K. Kwon, D. Tománek, Unusually high thermal conductivity of carbon nanotubes, Phys. Rev. Lett. 84 (2000) 4613.

[29] D. Li, R.B. Kaner, Graphene-based materials, Science 320 (2008) 1170.

[30] K. Novoselov, A. Geim, S. Morozov, D. Jiang, Y. Zhang, S. Dubonos, et al., Electric field effect in atomically thin carbon films, Science 306 (2004) 666.

[31] K. Geim, K.S. Novoselov, The rise of graphene, Nature Mater. 6 (2007) 183.

[32] C. Lee, X. Wei, J.W. Kysar, J. Hone, Measurement of the elastic properties and intrinsic strength of monolayer graphene, Science 321 (2008) 385.

[33] M. Pumera, Graphene-based nanomaterials for energy storage, Energy Environ. Sci. 4 (2011) 668.

[34] X. Huang, Z. Yin, S. Xu, X. Qi, Q. He, Q. Zhang, et al., Graphene-based materials: synthesis, characterization, properties, and applications, Small. 7 (2011) 1876.

[35] J.P. Zheng, P.J. Cygan, T.R. Jaw, Hydrous ruthenium oxide as an electrode material for electrochemical capacitors, J. Electrochem. Soc. 142 (1995) 2699.

[36] J. Wei, I. Zhitomirsky, Manganese oxide films for electrochemical supercapacitors, J. Mater. Process. Tech. 186 (2007) 356.

[37] J.H. Park, J.M. Ko, O.O. Park, Carbon nanotube/RuO_2 nanocomposite electrodes for supercapacitors, J. Electrochem. Soc. 150 (2003) A864.

[38] I.H. Kim, J.H. Kim, Y.H. Lee, K.B. Kim, Synthesis and characterization of electrochemically prepared ruthenium oxide on carbon nanotube film substrate for supercapacitor applications, J. Electrochem. Soc. 152 (2005) A2170.

[39] F. Shi, L. Li, X. Wang, C. Gu, J. Tu, Metal oxide/hydroxide-based materials for supercapacitors, RSC Adv. 4 (2014) 41910.

[40] A.B. Kaiser, Systematic Conductivity Behavior in Conducting Polymers: Effects of Hererogeneous Disorder, Adv. Mater. 13 (2001) 927.

[41] J.A. Osaheni, A.S. Jenekhe, H. Vanherzeele, J.S. Meth, Y. Sun, A.G. MacDiarmid, Nonlinear optical properties of polyaniline and derivatives, J. Phys. Chem. 96 (1992) 2830.

[42] E.T. Kang, K.G. Neoh, K.L. Tan, Polyaniline: a polymer with many interesting intrinsic redox states, Prog. Polym. Sci. 23 (1998) 277.

[43] P. Marque, J. Roncali, F. Garnier, Electrolyte effect on the electrochemical properties of poly(3-methylthiophene) thin films, J. Electroanal. Chem. 218 (1987) 107.
[44] S. Dhibar, S. Sahoo, C.K. Das, R. Singh, Investigations on copper chloride doped polyaniline composites as efficient electrode materials for supercapacitor applications, J. Mater. Sci: Mater. Electron. 24 (2013) 576.
[45] S. Dhibar, S. Sahoo, C.K. Das, Copper chloride-doped polyaniline/multiwalled carbon nanotubes nanocomposites: superior electrode material for supercapacitor applications, Polym. Compos. 34 (2013) 517.
[46] S. Dhibar, S. Sahoo, C.K. Das, Fabrication of transition-metal-doped polypyrrole/multiwalled carbon nanotubes nanocomposites for supercapacitor applications, J. Appl. Polym. Sci. 130 (2013) 554.
[47] K. Shi, I. Zhitomirsky, Fabrication of polypyrrole-coated carbon nanotubes using oxidant – surfactant nanocrystals for supercapacitor electrodes with high mass loading and enhanced performance, ACS Appl. Mater. Interfaces 5 (2013) 13161.
[48] Q. Li, J. Liu, J. Zou, A. Chunder, Y. Chen, L. Zhai, Synthesis and electrochemical performance of multi-walled carbon nanotube/polyaniline/MnO_2 ternary coaxial nanostructures for supercapacitors, J. Power Sources 196 (2011) 565.
[49] H. Zhang, Z. Hu, M. Li, L. Hu, S. Jiao, A high-performance supercapacitor based on a polythiophene/multiwalled carbon nanotubes composite by electropolymerization in an ionic liquid microemulsion, J. Mater. Chem. A 2 (2014) 17024.
[50] B.S. Singu, P. Srinivasan, K.R. Yoon, Emulsion polymerization method for polyaniline-multiwalled carbon nanotube nanocomposites as supercapacitor materials, J Solid State Electrochem. (2016). Available from: http://dx.doi.org/10.1007/s10008-016-3309-1.
[51] S. Chen, I. Zhitomirsky, Polypyrrole coated carbon nanotubes for supercapacitors, prepared using indigo carmine as a dispersant and dopant, Mater. Lett. 135 (2014) 47.
[52] S.R. Sivakkumar, W.J. Kim, J.-A. Choi, D.R. MacFarlane, M. Forsyth, D.-W. Kim, Electrochemical performance of polyaniline nanofibres and polyaniline/multi-walled carbon nanotube composite as an electrode material for aqueous redox supercapacitors, J. Power Sources 171 (2007) 1062.
[53] H. Lin, L. Li, J. Ren, Z. Cai, L. Qiu, Z. Yang, et al., Conducting polymer composite film incorporated with aligned carbon nanotubes for transparent, flexible and efficient supercapacitor, Sci Rep. 3 (2013) 1353.
[54] R. Wang, Q. Wu, X. Zhang, Z. Yang, L. Gao, J. Ni, et al., Flexible Supercapacitors based on a polyaniline nanowire-infilled 10 nm-diameter carbon nanotube porous membrane by in situ electrochemical polymerization, J. Mater. Chem. A 4 (2016) 12602.
[55] S. Dhibar, C.K. Das, Silver nanoparticles decorated polyaniline/multiwalled carbon nanotubes nanocomposite for high-performance supercapacitor electrode, Ind. Eng. Chem. Res. 53 (2014) 3495.
[56] X. Wang, X. Han, M. Lim, N. Singh, C.L. Gan, M. Jan, et al., Nickel cobalt oxide-single wall carbon nanotube composite material for superior cycling stability and high-performance supercapacitor application, J. Phys. Chem. C 116 (2012) 12448.
[57] Z. Niu, P. Luan, Q. Shao, H. Dong, J. Li, J. Chen, et al., A "Skeleton/Skin" strategy for preparing ultrathin free-standing single-walled carbon nanotube/polyaniline films for high performance supercapacitor electrodes, Energy Environ. Sci. 5 (2012) 8726.
[58] V. Gupta, N. Miura, Influence of the microstructure on the supercapacitive behavior of polyaniline/single-wall carbon nanotube composites, J. Power Sources 157 (2006) 616.
[59] S. Dhibar, P. Bhattacharya, G. Hatui, S. Sahoo, C.K. Das, Transition metal-doped polyaniline/single-walled carbon nanotubes nanocomposites: efficient electrode material for high performance supercapacitors, ACS Sustain. Chem. Eng. 2 (2014) 1114.

[60] K. Liang, T. Gu, Z. Cao, X. Tang, W. Hu, B. Wei, In situ synthesis of SWNTs@MnO_2/polypyrrole hybrid film as binder-free supercapacitor electrode, Nano Energy 9 (2014) 245.
[61] J. Ge, G. Cheng, L. Chen, Transparent and flexible electrodes and supercapacitors using polyaniline/single-walled carbon nanotube composite thin films, Nanoscale 3 (2011) 3084.
[62] K. Wang, P. Zhao, X. Zhou, H. Wu, Z. Wei, Flexible supercapacitors based on cloth-supported electrodes of conducting polymer nanowire array/SWCNT composites, J. Mater. Chem. 21 (2011) 16373.
[63] V.H. Rodrigues de Souza, M.M. Oliveira, A.J.G. Zarbin, Thin and flexible all-solid supercapacitor prepared from novel single wall carbon nanotubes/polyaniline thin films obtained in liquid-liquid interfaces, J. Power Sources 260 (2014) 34.
[64] D. Ge, L. Yang, L. Fan, C. Zhang, X. Xiao, Y. Gogotsi, et al., Foldable supercapacitors from triple networks of macroporous cellulose fibers, single-walled carbon nanotubes and polyaniline nanoribbons, Nano Energy 11 (2015) 568.
[65] X. Wang, M. Li, Z. Chang, Y. Yang, Y. Wu, X. Liu, Co_3O_4@MWCNT nanocable as cathode with superior electrochemical performance for supercapacitors, ACS Appl. Mater. Interfaces 7 (2015) 2280.
[66] Y. Jiang, P. Wang, X. Zang, Y. Yang, A. Kozinda, L. Lin, Uniformly embedded metal oxide nanoparticles in vertically aligned carbon nanotube forests as pseudocapacitor electrodes for enhanced energy storage, Nano. Lett. 13 (2013) 3524.
[67] F. Yang, M. Xu, S.-J. Bao, Q.-Q. Sun, MnO_2-assisted fabrication of PANI/MWCNT composite and its application as a supercapacitor, RSC Adv. 4 (2014) 33569.
[68] Y.-H. Li, Q.-Y. Li, H.-Q. Wang, Y.-G. Huang, X.-H. Zhang, Q. Wu, et al., Synthesis and electrochemical properties of nickel−manganese oxide on Mwcnts/CFP substrate as a supercapacitor electrode, Appl. Energy 153 (2015) 78.
[69] P. Li, Y. Yang, E. Shi, Q. Shen, Y. Shang, S. Wu, et al., Core-double-shell, carbon nanotube@polypyrrole@MnO_2 sponge as freestanding, compressible supercapacitor electrode, ACS Appl. Mater. Interfaces 6 (2014) 5228.
[70] A.S. Adekunle, K.I. Ozoemena, B.O. Agboola, MWCNTs/metal (Ni, Co, Fe) oxide nanocomposite as potential material for supercapacitors application in acidic and neutral media, J. Solid State Electrochem. 17 (2013) 1311.
[71] C. Guan, J. Liu, Y. Wang, L. Mao, Z. Fan, Z. Shen, et al., Iron oxide-decorated carbon for supercapacitor anodes with ultrahigh energy density and outstanding cycling stability, ACS Nano 9 (2015) 5198.
[72] S.S. Raut, B.R. Sankapal, Comparative studies on MWCNTs, Fe_2O_3 and Fe_2O_3/MWCNTs thin films towards supercapacitor application, New J. Chem. 40 (2016) 2619.
[73] S. Boukhalfa, K. Evanoff, G. Yushin, Atomic layer deposition of vanadium oxide on carbon nanotubes for high-power supercapacitor electrodes, Energy Environ. Sci. 5 (2012) 6872.
[74] W.-Y. Ko, Y.-F. Chen, K.-M. Lu, K.-J. Lin, Porous honeycomb structures formed from interconnected MnO_2 sheets on CNT-coated substrates for flexible all-solid-state supercapacitors, Sci. Rep. 6 (2016) 18887.
[75] A. Ouyang, A. Cao, S. Hu, Y. Li, R. Xu, J. Wei, et al., Polymer-coated graphene aerogel beads and supercapacitor application, ACS Appl. Mater. Interfaces 8 (2016) 11179.
[76] L.L. Zhang, S. Zhao, X.N. Tian, X.S. Zhao, Layered graphene oxide nanostructures with sandwiched conducting polymers as supercapacitor electrodes, Langmuir. 26 (2010) 17624.

[77] J. Zhang, X.S. Zhao, Conducting polymers directly coated on reduced graphene oxide sheets as high-performance supercapacitor electrodes, J. Phys. Chem. C 116 (2012) 5420.
[78] Z.-F. Li, H. Zhang, Q. Liu, L. Sun, L. Stanciu, J. Xie, Fabrication of high-surface-area graphene/polyaniline nanocomposites and their application in supercapacitors, ACS Appl. Mater. Interfaces 5 (2013) 2685.
[79] Y. He, Y. Bai, X. Yang, J. Zhang, L. Kang, H. Xu, et al., Holey graphene/polypyrrole nanoparticle hybrid aerogels with three-dimensional hierarchical porous structure for high performance supercapacitor, J. Power Sources 317 (2016) 10.
[80] J. Li, H. Xie, Y. Li, Fabrication of graphene oxide/polypyrrole nanowire composite for high performance supercapacitor electrodes, J. Power Sources 241 (2013) 388.
[81] N. Chen, Y. Ren, P. Kong, L. Tan, H. Feng, Y. Luo, In situ one-pot preparation of reduced graphene oxide/polyaniline composite for high-performance electrochemical capacitors, Appl. Surf. Sci. 392 (2017) 71.
[82] J. Luo, W. Zhong, Y. Zou, C. Xiong, W. Yang, Preparation of morphology-controllable polyaniline and polyaniline/graphene hydrogels for high performance binder-free supercapacitor electrodes, J. Power Sources 319 (2016) 73.
[83] X. Li, C. Zhang, S. Xin, Z. Yang, Y. Li, D. Zhang, et al., Facile synthesis of MoS_2/reduced graphene oxide@polyaniline for high-performance supercapacitors, ACS Appl. Mater. Interfaces 8 (2016) 21373.
[84] Y. Li, G. Louarn, P.-H. Aubert, V. Alain-Rizzo, L. Galmiche, P. Audebert, et al., Polypyrrole-modified graphene sheet nanocomposites as new efficient materials for supercapacitors, Carbon N. Y. 105 (2016) 510.
[85] G. Wang, Q. Tang, H. Bao, X. Li, G. Wang, Synthesis of hierarchical sulfonated graphene/Mno_2/polyaniline ternary composite and its improved electrochemical performance, J. Power Sources 241 (2013) 231.
[86] C. Bora, J. Sharma, S. Dolui, Polypyrrole/sulfonated graphene composite as electrode material for supercapacitor, J. Phys. Chem. C 118 (2014) 29688.
[87] W. Xinming, W. Qiguan, Z. Wenzhi, W. Yan, C. Weixing, Enhanced electrochemical performance of hydrogen-bonded graphene/polyaniline for electrochromo-supercapacitor, J. Mater. Sci. 51 (2016) 7731.
[88] P. Iamprasertkun, A. Krittayavathananon, M. Sawangphruk, N-doped reduced graphene oxide aerogel coated on carboxylmodified carbon fiber paper for high-performance ionic-liquid supercapacitors, Carbon. N. Y. 102 (2016) 455.
[89] Y.-H. Hwang, S.M. Lee, Y.J. Kim, Y.H. Kahng, K. Lee, A new approach of structural and chemical modification on graphene electrodes for high-performance supercapacitors, Carbon. N. Y. 100 (2016) 7.
[90] B. Rajagopalan, S.H. Hur, J.S. Chung, Surfactant-treated graphene covered polyaniline nanowires for supercapacitor electrode, Nanoscale Res. Lett. 10 (2015) 183.
[91] Q.-F. Lü, G. Chen, T.-T. Lin, Y. Yu, Dye-functionalized graphene/polyaniline nanocomposite as an electrode for efficient electrochemical supercapacitor, Compos. Sci. Technol. 115 (2015) 80.
[92] L. Lai, H. Yang, L. Wang, B.K. Teh, J. Zhong, H. Chou, et al., Preparation of supercapacitor electrodes through selection of graphene surface functionalities, ACS Nano 6 (2012) 5941.
[93] P. Du, H.C. Liu, C. Yi, K. Wang, X. Gong, Polyaniline-modified oriented graphene hydrogel film as the free standing electrode for flexible solid-state supercapacitors, ACS Appl. Mater. Interfaces 7 (2015) 23932.

[94] J.W. Lee, A.S. Hall, J.-D. Kim, T.E. Mallouk, A facile and template-free hydrothermal synthesis of Mn_3O_4 nanorods on graphene sheets for supercapacitor electrodes with long cycle stability, Chem. Mater. 24 (2012) 1158.
[95] G. Yu, L. Hu, N. Liu, H. Wang, M. Vosgueritchian, Y. Yang, et al., Enhancing the supercapacitor performance of graphene/MnO_2 nanostructured electrodes by conductive wrapping, Nano. Lett. 11 (2011) 4438.
[96] H. Fan, R. Niu, J. Duan, W. Liu, W. Shen, Fe_3O_4@carbon nanosheets for all-solid-state supercapacitor electrodes, ACS Appl. Mater. Interfaces 8 (2016) 19475.
[97] H. Fan, L. Quan, M. Yuan, S. Zhu, K. Wang, Y. Zhong, et al., Thin Co_3O_4 nanosheet array on 3D porous graphene/nickel foam as a binder-free electrode for high-performance supercapacitors, Electrochim. Acta. 188 (2016) 222.
[98] B. Mendoza-Sánchez, J. Coelho, A. Pokle, V. Nicolosi, A 2D graphene-manganese oxide nanosheet hybrid synthesized by a single step liquid-phase co-exfoliation method for supercapacitor applications, Electrochim. Acta. 174 (2015) 696.
[99] Y. Luo, H. Zhang, D. Guo, J. Ma, Q. Li, L. Chen, et al., Porous $NiCo_2O_4$-reduced graphene oxide (Rgo) composite with superior capacitance retention for supercapacitors, Electrochim. Acta. 132 (2014) 332.
[100] R.B. Rakhi, W. Chen, D. Cha, H.N. Alshareef, High performance supercapacitors using metal oxide anchored graphene nanosheet electrodes, J. Mater. Chem. 21 (2011) 16197.
[101] X.-C. Dong, H. Xu, X.-W. Wang, Y.-X. Huang, M.B. Chan-Park, H. Zhang, et al., 3D graphene-cobalt oxide electrode for high-performance supercapacitor and enzymeless glucose detection, ACS Nano 6 (2012) 3206.
[102] Q. Wu, M. Chen, S. Wang, X. Zhang, L. Huan, G. Diao, Preparation of sandwich-like ternary hierarchical nanosheets manganese dioxide/polyaniline/reduced graphene oxide as electrode material for supercapacitor, Chem. Eng. J. 304 (2016) 29.
[103] K.V. Sankar, R.K. Selvan, Fabrication of flexible fiber supercapacitor using covalently grafted $CoFe_2O_4$/reduced graphene oxide/polyaniline and its electrochemical performances, Electrochim. Acta. 213 (2016) 469.
[104] Y.-S. Wang, S.-M. Li, S.-T. Hsiao, W.-H. Liao, S.-Y. Yang, C.-C.M. Ma, et al., Electrochemical composite deposition of porous cactus-like manganese oxide/reduced graphene oxide–carbon nanotube hybrids for high-power asymmetric supercapacitors, J. Mater. Chem. C 3 (2015) 4987.
[105] M. Yang, K.G. Lee, S.J. Lee, S.B. Lee, Y.-K. Han, B.G. Choi, Three-dimensional expanded graphene – metal oxide film via solid state microwave irradiation for aqueous asymmetric supercapacitors, ACS Appl. Mater. Interfaces 7 (2015) 22364.
[106] D. Du, R. Lan, W. Xu, R. Beanland, H. Wang, S. Tao, Preparation of a hybrid Cu_2O/$CuMoO_4$ nanosheet electrode for high-performance asymmetric supercapacitors, J. Mater. Chem. A 4 (2016) 17749.
[107] J. Shen, C. Yang, X. Li, G. Wang, High-performance asymmetric supercapacitor based on nanoarchitectured polyaniline/graphene/carbon nanotube and activated graphene electrodes, ACS Appl. Mater. Interfaces 5 (2013) 8467.
[108] Q. Cheng, J. Tang, N. Shinya, L.-C. Qin, Polyaniline modified graphene and carbon nanotube composite electrode for asymmetric supercapacitors of high energy density, J. Power Sources 241 (2013) 423.
[109] P. Yang, Y. Ding, Z. Lin, Z. Chen, Y. Li, P. Qiang, et al., Low-cost high-performance solid-state asymmetric supercapacitors based on MnO_2 nanowires and Fe_2O_3 nanotubes, Nano. Lett. 14 (2014) 731.

[110] R.R. Salunkhe, J. Lin, V. Malgras, S.X. Dou, J.H. Kim, Y. Yamauchi, Large-scale synthesis of coaxial carbon nanotube/$Ni(OH)_2$ composites for asymmetric supercapacitor application, Nano Energy 11 (2015) 211.
[111] J. Yan, T. Wei, Z. Fan, W. Qian, M. Zhang, X. Shen, et al., Preparation of graphene nanosheet/carbon nanotube/polyaniline composite as electrode material for supercapacitors, J. Power Sources 195 (2010) 3041.
[112] G. Ning, T.L.J. Yan, C. Xu, T. Wei, Z. Fan, Three-dimensional hybrid materials of fish scale-like polyaniline nanosheet arrays on graphene oxide and carbon nanotube for high-performance ultracapacitors, Carbon N. Y. 54 (2013) 241.
[113] M. Sun, G. Wang, C. Yang, H. Jiang, C. Li, A graphene/carbon nanotube@p-conjugated polymer nanocomposite for high-performance organic supercapacitor electrodes, J. Mater. Chem. A 3 (2015) 3880.
[114] B. Wang, J. Qiu, H. Feng, E. Sakai, Preparation of graphene oxide/polypyrrole/multi-walled carbon nanotube composite and its application in supercapacitors, Electrochim. Acta. 151 (2015) 230.
[115] X. Lu, F. Zhang, H. Dou, C. Yuan, S. Yang, L. Hao, et al., Preparation and electrochemical capacitance of hierarchical graphene/polypyrrole/carbon nanotube ternary composites, Electrochim. Acta. 69 (2012) 160.
[116] S. Dhibar, P. Bhattacharya, D. Ghosh, G. Hatui, C.K. Das, Graphene – single-walled carbon nanotubes – poly(3-methylthiophene) ternary nanocomposite for supercapacitor electrode materials, Ind. Eng. Chem. Res. 53 (2014) 13030.
[117] M. Ramezani, M. Fathi, F. Mahboubi, Facile synthesis of ternary MnO_2/graphene nanosheets/carbon nanotubes composites with high rate capability for supercapacitor applications, Electrochim. Acta. 174 (2015) 345.
[118] X. Lu, H. Dou, S. Yang, L. Hao, L. Zhang, L. Shen, et al., Fabrication and electrochemical capacitance of hierarchical graphene/polyaniline/carbon nanotube ternary composite film, Electrochim. Acta. 56 (2011) 9224.

Processing of ferroelectric polymer composites

10

Muklesur Rahman[1] *and Prabir K. Mukherjee*[2]
[1]Aliah University, Kolkata, India, [2]Government College of Engineering and Textile Technology, Serampore, India

Chapter Outline

10.1 Introduction

"Ferroelectric Polymer Composites" have been considered as a substitute for conventionally widely used inorganic ferroelectric as well as for pristine ferroelectric polymers. Polymers with typical features like flexibility, softness, light weight, relatively low acoustic impedance, low thermal conductivity, etc., are widely used in our daily life and in practically all areas of technology. On the other hand, "ferroelectricity" was first discovered in Rochelle salt, an organic molecule containing tartrate ions [1–4], and ferroelectric materials with versatile properties that are useful in important applications viz., data storage, sensing, actuation, energy harvesting, and electro-optic devices [5]. From the history of development of ferroelectric materials, described briefly in next section, it is noted that, although the first discovered ferroelectric material was organic, inorganic ferroelectrics are much more compatible with technological demand and hence widely used in technologies. However, inorganic ferroelectrics have few drawbacks: they are usually heavy, brittle, toxic, and require high temperature processing, which limits their technological usefulness, and nontoxic substitutes of inorganic ferroelectrics is in demand for future technologies. In the last several years, significant advances in polymer-based ferroelectrics, mainly polyvinylidene fluoride (PVDF) and its copolymers with the ferroelectric properties comparable to inorganic materials, have been observed. These polymer-based ferroelectric materials are soft and flexible, light weight,

Hybrid Polymer Composite Materials: Processing. DOI: http://dx.doi.org/10.1016/B978-0-08-100789-1.00010-1

lead-free, and hence are promising for fulfilling the demand for environmental friendly technology, but low spontaneous polarization, slow switching time, etc., offer constraints to these pristine ferroelectric polymers. Moreover, fast-growing technology demands new material systems with diverse functional aspects. Single-phase materials, due to their structural constraints, in general, cannot meet such diverse functions. Composite technology, where a novel heterogeneous material is developed whose properties are determined by the number of different phases of the material, the volume fraction of the phases, the properties of individual phases, and the ways in which different phases are interconnected, has been used to develop functional materials compatible with technologically demanding properties [6–8]. In recent years polymer nanocomposites have drawn considerable research interest because of their considerable betterment in functional properties including thermal, mechanical, and electrical compared to those in the respective pristine polymers. Inorganic and organic nanofillers have been used on a variety of polymers to achieve polymer composites where desired. A large enhancement of the physical and mechanical properties has also been reported with the dispersion of metal or semiconductor nanoparticles within the polymer matrix. Ferroelectric polymer composites are generally prepared to enhance or modify the dielectric, ferroelectric, and other properties depending on the technological application. Different preparation methods with different materials have been developed to prepare nanocomposites. In this chapter the different preparation methods and characterization ferroelectric polymer composites with enhanced ferroelectric and dielectric functions will be discussed.

10.2 Ferroelectric materials and ferroelectric polymers

In 1917, A. M. Nicolson, J. A. Anderson, and W. G. Cady, while investigating the piezoelectric properties of Rochelle salt, an organic molecule containing tartrate ions, noted certain anomalies in dielectric behavior. However, this salt was first separated by Elie Seigmette in France during the middle of the 16th century and piezoelectricity was established by Curie's brother in 1880 [9,10]. The significant anomalies included the (1) existence of hysteresis between applied electric field (E) and polarization (P), and (2) a sudden change in the piezoelectric activity at 23°C, later recognized as the first observation of the Curie point [11,12]. Almost after three years of these observations, the physical properties of Rochelle salt were described in detail in a series of papers by J. Valasek in which the analogy between the dielectric properties of Rochelle salt and ferromagnetism was identified [1–4]. Hence, ferroelectrics are described as the polar substances of either solid (crystalline or polymeric) or liquid crystal, in which spontaneously generated electric polarization (P) can be reversed by inverting the external field (E), which usually results in a hysteresis between polarization and the electric field (P–E hysteresis loop) and have a Curie temperature T_c for a paraelectric-to-ferroelectric phase transition. Moreover, the polar crystal structure gives second-order optical nonlinearity, causing second-harmonic generation (SHG) activity and a nonlinear electro-optic effect.

The next ferroelectric substance, whose structure is not similar to that of Rochelle salt, was discovered by Busch and Scherrer in a salt, potassium dihydrogen phosphate KH_2PO_4 (KDP), in 1935 [13]. Subsequently, it was found that the primary phosphates and arsenates of alkalis and ammonium, all forming tetragonal crystals, also possess ferroelectric properties, but a Curie temperature at liquid air temperatures. The spontaneous polarization in these crystals is an order of magnitude greater than Rochelle salt, and they appear to have no lower Curie point, the ferroelectric state extending to absolute zero. But the rapid advances of the field only occurred after the development of a new class of ferroelectric in perovskite oxides such as barium titanate $BaTiO_3$, which forms in the perovskite lattice with Ba ions at the corners of the cubic unit cell, O ions at the centers, Ti ions at body centers [14], and lead zirconate titanate (PZT) [15,16]. These ferroelectrics exhibit a behavior more spectacular and interesting than that of Rochelle salt or the alkali phosphates and arsenates. Anomalies in the dielectric properties of barium titanate ceramic material were observed by Wainer and Salomon [17,18] in 1942, and it was soon established by von Hippel [11] and independently by Wul and coworkers that $BaTiO_3$ is ferroelectric [19–24]. Later on it was concluded that, from the structural viewpoint, among 32 crystalline classes 21 are without a symmetry center. And 20 of them are piezoelectric, within which 10 of them possess a unique polar axis that is spontaneously polarized [25,26]. These 10 polar classes are referred to as pyroelectric, whose spontaneous polarization varies with temperature. These materials are also called ferroelectric if this polarization is switchable with external electric field, which results in hysteresis loop between polarization and electric field. With these unparalleled electric and optical properties, ferroelectrics are fascinating for a diverse practical application in ferroelectric random access memory (FeRAM), ferroelectric field-effect transistors, data storage, sensing, actuation, energy harvesting, and electro-optic devices. Today, PZT and other perovskite oxide ferroelectrics remain the most widely used ferroelectric material, as they exhibit superior ferroelectric properties orders of magnitude higher than those of molecular crystals.

In early 1970s, it took wide attention when Kawai unambiguously discovered strong piezoelectricity in its uniaxialy drawn and poled PVDF film [27]. This newly reported piezoelectric polymers possesses *mm2* symmetry and piezoelectric constants d_{31}, d_{32}, d_{33}, d_{15}, and d_{24} specifically different from the earlier reported piezoelectric polymers with $\infty 2$ symmetry and piezoelectric constants d_{14} and d_{25} [28–30]. Moreover, a series of published research works noted that oriented and poled polymers were both piezoelectric as well as pyroelectric [31–34]. Within two years of the discovery of piezoelectricity, reports on pyroelectricity and nonlinear optical response-second-harmonic generation (SHG) in PVDF films led to the discovery of its ferroelectric properties [35]. Subsequently, many novel ferroelectric polymers have also been discovered and explored, including aromatic and aliphatic polyurea, copolymers of vinylidene cyanide (VDCN), odd-numbered polyamides, poly-L-lactic acid (PLLA), many biopolymers and synthetic polypeptides, ferroelectric liquid crystal polymers, and polymer composites of organic and inorganic piezoelectric ceramics, etc. Fig. 10.1 shows the molecular composition of some

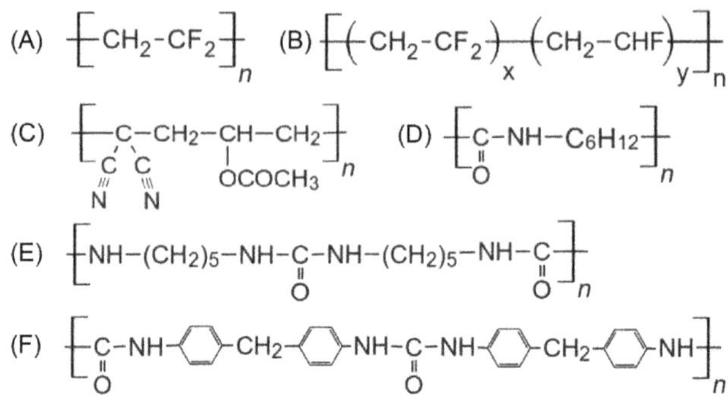

Figure 10.1 Molecular compositions of some ferroelectric polymer (A) PVDF (B) P(VDF-TrFE) (C) poly(vinylidene-cyanide-co-vinylacetate) (D) Polyamide 7 (PA-7) (E) aliphatic polyurea 5 and (F) aliphatic polyurea.

ferroelectric polymers. Among them, PVDF and its copolymers, are the most studied and widely used molecular ferroelectrics, as it has several advantageous properties including a relatively large remnant polarization, a short switching time, and a good thermal stability over the number of other organic ferroelectric materials known to exist [36]. Hence, in this chapter mainly ferroelectric polymer PVDF-based composite systems will be discussed. The monomer unit of this polymer is $-CH_2-CF_2-$, semi-crystalline in nature: half crystalline and half amorphous. The crystalline region consists of at least four crystal polymorphs named Form I (β-phase), Form II (α-phase), Form III (γ-phase), and Form IV (δ-phase) [37,38]. Form II or α-phase is the most common polymorph of PVDF having conformation structure TGTG (T-trans, G-gauche). The β-phase of the PVDF polymer has planer all-trans (TTTT) conformation, and the H and F atoms are attached in the chain in such a way that the dipole moments associated with two C−H and two C−F bonds add up and align in the direction perpendicular to the carbon backbone to give higher dipole moments per unit cell and hence forms a polar phase [39,40]. For γ-phase, the molecules adopt an intermediate conformation T_3GT_3G' and form a polar crystalline because of its parallel packing. Form IV or δ-phase polymorph comprises a parallel packing of TGTG' molecules. Among these four common polymorphs, only α -phase (Form II) is nonpolar, and the remaining three are polar. Among these three polar phases the β phase is found to have the highest ferroelectric properties, having large spontaneous polarization along the b-axis, which is parallel to C-F dipole moment, and perpendicular to the C-axis, polymer chain direction, as shown in Fig. 10.2 [41,42]. The differences in electro negativity of fluorine, carbon, and hydrogen in the all-trans (TTTT) conformation resulted in an enhancement of spontaneous polarization up to 8 $\mu C/cm^2$, on the same order of barium titanate. This opens the door for practical applications of PVDF and its copolymers, which are soft and flexible, lightweight, lead-free, processed at relatively low temperature, and biocompatible. Today, PVDF and its copolymers are the most widely used ferroelectric polymers, though a number of exciting new molecular ferroelectrics have been developed in recent years, with properties approaching those of $BaTiO_3$,

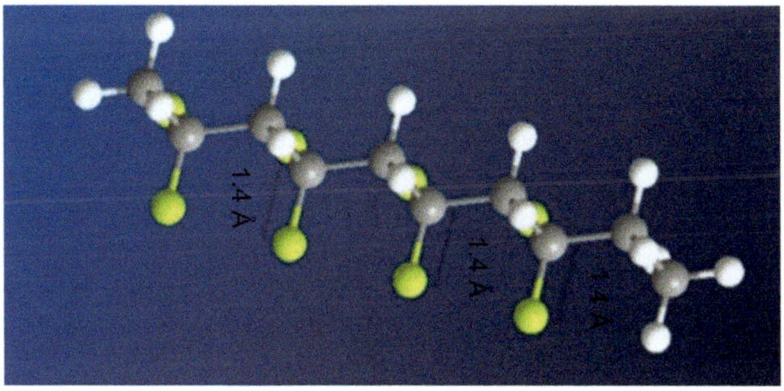

Figure 10.2 *All-trans* (TTTT) molecular conformation of ferroelectric β-phase PVDF.

which has motivated this perspective. It is the β phase with X-ray diffraction angle (2θ) 20.7°, 36.6°, and 56.9°, that attracts the researchers for its high piezoelectric and ferroelectric properties including polar hysteresis [27,35,43]. So lots of research work is going on for yielding high β-phase content in the material. However, the existence of ferroelectricity in β-phase PVDF remains a controversial issue, as many researchers doubt that the β -phase is ferroelectric owing to its low crystallinity and other mechanisms that account for its piezoelectricity, pyroelectricity, and polar hysteresis. Moreover, there is no eminent ferroelectric-to-paraelectric phase transition, no Curie –Weiss type dielectric peak occurs between the β-phase and other crystalline phases, until the β -phase melts at 170°C [44,45]. Different phases exist in PVDF depending on various processing parameters like type of solvent in solution casting, solvent evaporation temperatures, fillers in the polymer matrix, stretching load, and annealing to make stable β-phase, controlled annealing, or electrical poling of the prepared PVDF film [46–48]. Copolymerization of PVDF with TrFE (PVDF-TrFE) [49–51], tertrafluoroethylene (PVDF-TFE) [52,53], ethylene tetrafluorethulene (PVDF-ETFE) [54,55], vinylidene cynide (PVDCS) [56,57] results in increased crystallinity in comparison with the pristine PVDF and stronger polarization under an applied electric field, which is crucial for electronics applications.

10.3 Ferroelectric polymer (PVDF) composites to enhance ferroelectric phase

Considering PVDF as a ferroelectric component, the collective ferroelectric properties of individual chains of PVDF rely strongly upon the manner of assembling the polymer chains into a crystalline lattice and on the hierarchical morphological suprastructure, as well as the degree of crystallinity and crystal orientation [58]. A high b-axis orientation of the ferroelectric crystals, parallel to the direction of the applied electric field, is of prime importance for successful device performance, with the degree of crystallinity being as high as possible. There are two additional important characteristics that must be achieved efficiently: (1) ferroelectric β-crystals that, with their all-trans planar zigzag conformation (TTTT), provide superior ferroelectricity, piezoelectricity, and pyroelectricity in comparison to those of either α or γ-PVDF crystals [59]; and (2) a morphologically homogeneous and very low surface roughness. Therefore, many attempts have been made to induce the electroactive β phase in PVDF by various methods such as solution growth [60], melt-quenching [61,62], mechanical stretching [38,63], application of high pressure [64,65], addition of metal salts [66] formation of a nanocomposite [67,68], polarization via an applied field [69], and electrospinning [70] blending with polymers consisting of carbonyl group polymers like poly(methyl methracrylate) (PMMA), poly(ethyl acrylate) (PEA), or poly(vinyl acetate) (PVAc) [71–74]. The miscibility of PVDF and these carbonyl groups arise from hydrogen bonding between the double-bonded oxygen of the carbonyl group and the acidic hydrogen of CH_2-CF_2 group [75]. PVDF (high crystalline polymer) and poly(methylmethacrylate) (PMMA; amorphous polymer) are a very rare combination that exhibits compatibility in a blend [76,77]. Moreover, a method to fabricate ferroelectric PVDF films with low surface roughness has also been reported using "melt and quench" process, one of the simplest way to fabricate PVDF:PMMA and other blend films with β-crystals [71]. Due to its simplicity in operation and equipment solution blending is mostly used in the laboratories. Amorphous PMMA in a blend film effectively slows the rapid crystallization of PVDF upon quenching, giving rise to a thin and flat ferroelectric film with nanometer scale β-type PVDF crystals [78–80]. The crystal phase of PVDF is very sensitive to blending with PMMA. In Fig. 10.3, the AFM images present the solidification behavior of PVDF:PMMA thin films with different wt% of PMMA. The images show that with increasing PMMA content, the PVDF crystal growth is suppressed and the spherulite size decreases to nanocrystalline [71,81]. From the incidence reflection absorption spectroscopy (GIRAS) and XRD characterization of PVDF:PMMA, as shown in Fig. 10.4A, it has been observed that β-phase dominates in the blends comprising $10-30$ wt% PMMA. A comparative study of GIRAS data for PVDF/PMMA (80:20) blend films with cooled slowly cooled, as-quenched and as-quenched-and-annealed at 150°C on an Al substrate present that the slowly cooled sample dominantly exhibits the paraelectric α-PVDF crystalline structure—identified by the characteristic IR absorption bands at 610 and 796 cm^{-1}, whereas the as-quenched blend sample clearly shows the formation of a mixture of ferroelectric β- and γ-crystals from the representative absorption bands at 1280 and 1234 cm^{-1}, respectively. But the relative fraction of β-crystals (F(β)) in an as-quenched blend sample is about less than 50%, as calculated from the relative intensities of characteristic β- and γ-absorption peaks [71,82]. The fraction of β-phase F(β) can be calculated from FTIR spectra using the following equation [83].

$$F(\beta) = \frac{X_\beta}{X_\gamma + X_\beta} = \frac{A_\beta}{\left(\frac{K_\beta}{K_\gamma}\right)A_\gamma + A_\beta} \tag{10.1}$$

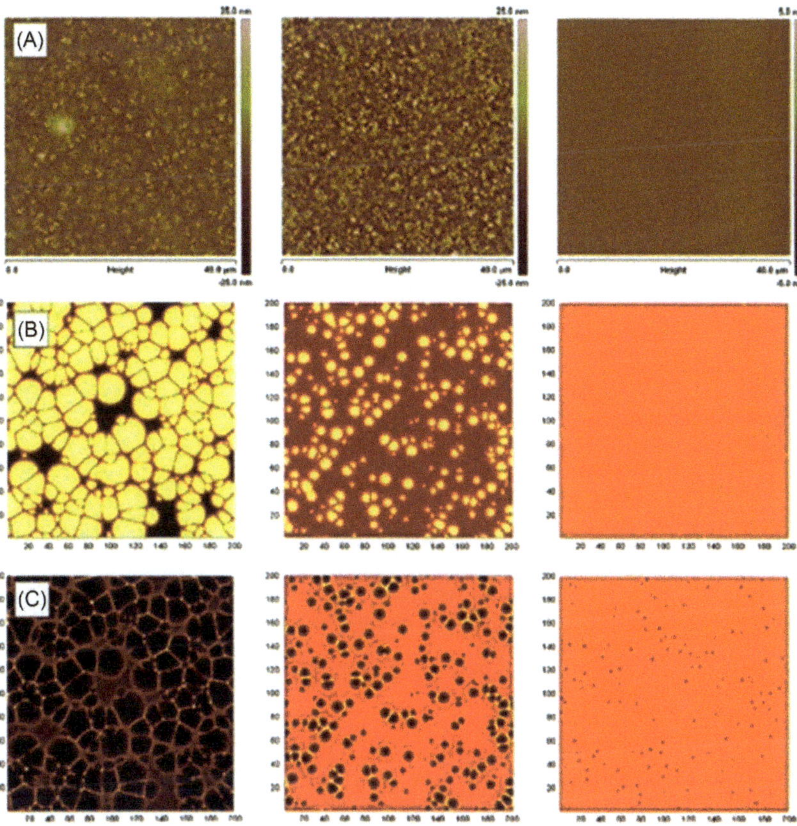

Figure 10.3 Solidification behavior of PVDF:PMMA blend thin films of compositions from left to right of 98:2, 90:10, and 70:30 (w/w): (A) AFM height profiles (40 × 40 μm); (B) PVDF crystallization (phase-field simulation on arbitrary length scale); (C) PMMA redistribution upon crystallization (phase-field simulation on arbitrary length scale). The displayed simulation results refer to PMMA with a molecular weight, Mw, of 91 kg/mol, but comparable images were obtained from 50 and 2 kg/mol.
After M. Li, N. Stingelin, J.J. Michels, M.-J. Spijkman, K. Asadi, K. Feldman, et al., Macromolecues ferroelectric phase diagram of PVDF:PMMA 45 (2012) 7477–7485.

where, X_γ and X_β represent the percentage of crystallinity, A_γ and A_β are the absorbance values of γ and β phases, respectively. K_γ and K_β are the absorption coefficients of the respective wave-numbers.

The subsequent annealing of the sample significantly enhanced the amount of β-crystallites to more than 90%. In addition, the ice quenched-and-annealed PVDF/PMMA blends exhibit very smooth film surfaces without apparent crystalline microdomains. The total fraction of β-crystals gradually decreases with the amount of PMMA in the blend film quenched-and-annealed at 150°C. Fig. 10.4B

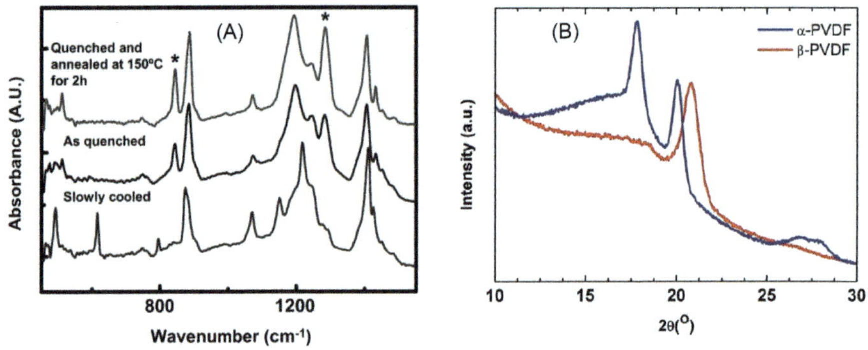

Figure 10.4 (A) GIRAS spectra of thin PVDF/PMMA (80:20) films spin-coated onto an Al substrate with various thermal treatments. The characteristic absorbance peaks at 840 and 1280 cm^{-1} indicate the presence of β-crystals, as indicated with asterisks (after Kang et al. [71]). (B) Grazing incidence X-ray diffraction scans of thin PVDF:PMMA (70:30) films produced by wire bar coating from DMF, followed by melting at 200 °C for 2 hin a vacuum. The slowly cooled blend film crystallizes in the α- phase, while the ice quenched and annealed blend film forms the β-phase of PVDF.
After M. Li, N. Stingelin, J.J. Michels, M.-J. Spijkman, K. Asadi, K. Feldman, et al., Macromolecues ferroelectric phase diagram of PVDF:PMMA 45 (2012) 7477–7485.

presents a typical example of XRD diffractographs of PVDF:PMMA (70:30) blend films to establish the formation of β-phase in PVDF:PMMA blend. The molten and slowly cooled blend film crystallizes in the α-phase characterized by strong peaks at diffraction angles 2θ of 17.6° and of 20°, assigned to the (100) and (110) reflections, respectively. The film melted, the ice quenched and subsequently annealed at 140 °C, crystallizes in the β-phase as indicated by a diffraction peak at 2θ of 20.7°, corresponding to the overlapping (110) and (200) reflections [84,85]. The reason of formation of β-phase in PVDF:PMMA blend upon quenching is that amorphous PMMA significantly hinders the α-crystallization and promotes β-phase formation. Li et al. determined the percentage of PVDF β-phase in PVDF:PMMA blends, calculated from FTIR spectra for different molecular weight of the components [81]. The percentage of β-crystallinity increases up to 80 wt% of PMMA, but decreases on further increase in the wt% of PMMA. Although the β-phase in PVDF:PMMA blends increases to some extent, but the remnant polarization monotonically decreases with PMMA content ranging from 4.8 μC cm^{-2} for a blend ratio of 90:10 to 0.25 μC cm^{-2} for a ratio of 40:60. Degree of crystallinity, calculated from the enthalpy of fusion deduced from thermal analysis, decreases with PMMA fraction, as shown in the Fig. 10.5.

To enhance the ferroelectric β-phase, polymer composites have been prepared by incorporating different materials viz., modified clay [86,87], palladium nanoparticle [88], and gold nanoparticle (Au-NPs) [68] etc.

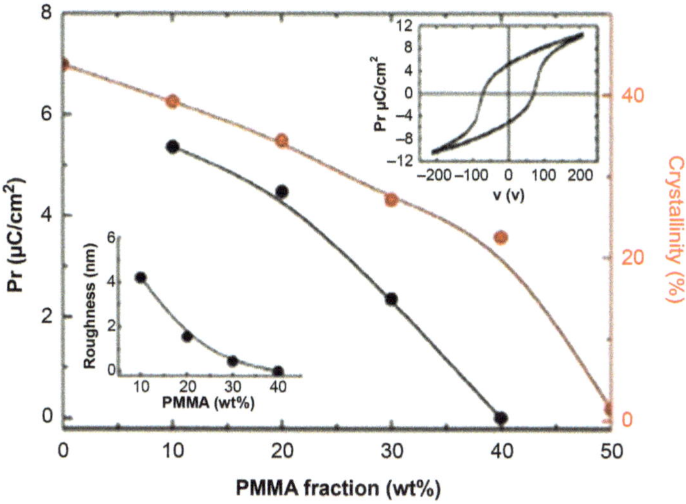

Figure 10.5 Variation of degree of crystallinity, remnant polarization and surface roughness as function of wt% of PMMA in PVDF:PMMA prepared from melt and quench process. After M. Li, N. Stingelin, J.J. Michels, M.-J. Spijkman, K. Asadi, K. Feldman, et al., Macromolecues ferroelectric phase diagram of PVDF:PMMA 45 (2012) 7477–7485.

It has been reported that in the melt-quenching method this β polymorph is not always retained. PVDF composites with carbon nanotube (CNT) processed through sonication [89–91] and eletrospining [92–98] have also demonstrated remarkable enhancement in β-phase PVDF formation. CNTs [99] are perhaps one of the most interesting new materials to emerge during the past decade with outstanding electronic and mechanical properties [100,101]. There are mainly two types of carbon nanotubes, namely, the single walled carbon nanotube (SWCNT) and the multiwalled carbon nanotube (MWCNT). In sonication method, PVDF is dissolved in dimethylacetamide (DMAc). The CNTs are also dispersed in DMAc by sonication for few minutes and then mixed with the PVDF solution. The PVDF-CNT/MWCNT mixture solution stirred and sonicated at room temperature for hours to and then gently heated at about 50°C for days for removing DMAc. The mixture dried further to obtain the PVDF-CNT composite. It has been pointed out that drawing and poling of this composite prepared from sonication method helps additional β-phase formation [102]. Electrospinning is a simple and versatile technique for fabricating ultrafine fibers with diameters ranging from several micrometers down to a few nanometers and have been used to process PVDF-CNT/MWCNT composite fibers [103–105]. Fig. 10.6 shows a typical schematic presentation of an electrospinning setup with a rotating wire-framed drum [103]. The viscous fluid of PVDF-CNT/MWCNT is prepared by dissolving PVDF in imethylformamide (DMF) and acetone (1:1) and dispersing nanotubes in carboxyl agent with acetone and finally blending the mixtures. This mixture is loaded in a syringe with a stainless steel spinneret of different diameters.

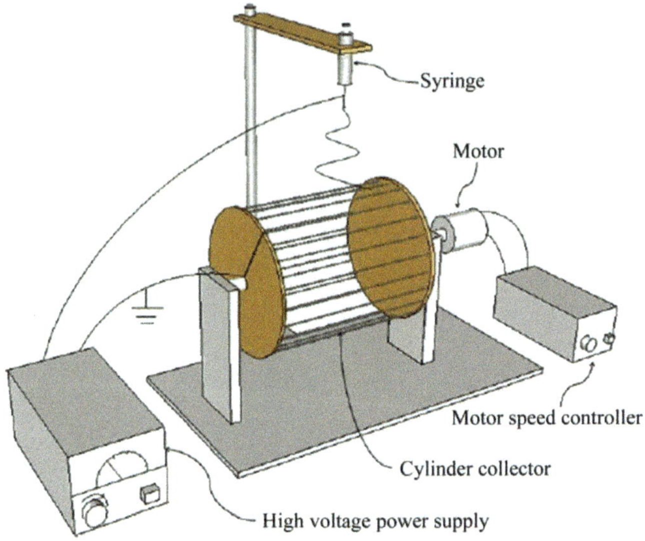

Figure 10.6 Schematic presentation of an electrospinning setup with a rotating wire-framed drum.
After S.-H. Wang, Y. Wan, B. Sun, L.-Z. Liu, W. Xu. Mechanical and electrical properties of electrospun PVDF/MWCNT ultrafine fibers using rotating collector. Nanoscale Res. Letts 9 (2014) 522–526.

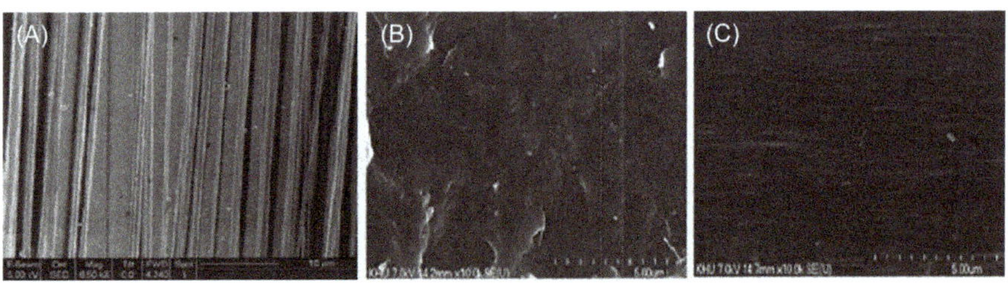

Figure 10.7 SEM images of the (A) aligned electrospun PVDF/MWCNTs (2 wt%) (after Wang et al. [103]) sonicated PVDF/MWCNTs (1 wt%) composite (B) undrawn, and (C) drawn.
After G.H. Kim, S.M. Hong, Y. Seo, Piezoelectric properties of poly(vinylidene fluoride) and carbon nanotube blends: β-phase development. Phys. Chem. Chem. Phys. 11 (2009) 10506–10512.

The spinneret is connected to the positive electrode of a high-voltage dc power supply of about kV voltage. The negative electrode of the high-voltage dc power supply was attached to the rotating drum, which acts as collector. Temperature and relative humidity of the chamber can be controlled externally. Fig. 10.7 presents comparative SEM images of PVDF/MWCNT composites processed by electrospinning and sonication methods (undrawn and drawn), respectively. The drawing of a sonicated composite blend is generally done by a programmable drawing machine in hot nitrogen atmosphere with different drawing rates. In

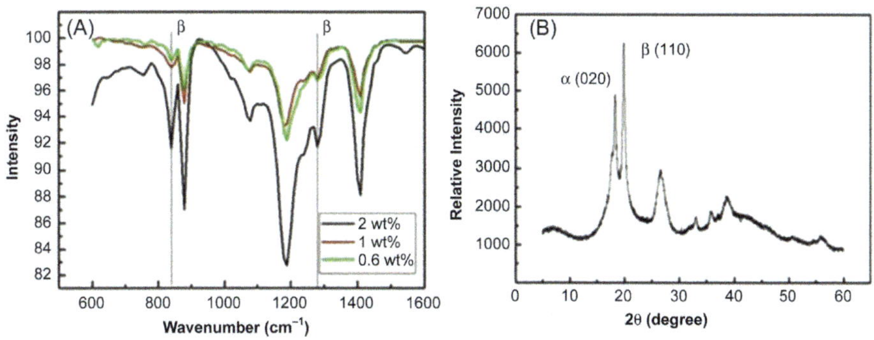

Figure 10.8 (A) FTIR spectra (0.6, 1, and 2 wt%) and (B) XRD diffraction (2 wt%) patterns of the aligned electrospun PVDF/MWCNT fibers.
After S.-H. Wang, Y. Wan, B. Sun, L.-Z. Liu, W. Xu. Mechanical and electrical properties of electrospun PVDF/MWCNT ultrafine fibers using rotating collector. Nanoscale Res. Letts 9 (2014) 522–526.

electrospun PVDF-MWCNT composites the MWCNTs are found well oriented along the fiber axis of the fibers. However, due to their inherent dispersion characteristics as well as strong electric field applied for electrospinning, most MWCNTs form agglomerates or exhibit curved or wavy conformation rather than straight. For the composites processed through sonication method the MWCNTs are uniformly dispersed. FTIR (at 837 and 1273 cm^{-1}) spectra and XRD ($2\theta = 19.9°$) pattern shown in Fig. 10.8 of electrospun PVDF-MWNCT show enhancement in of β-phase with increasing MWCNT. Such enhancement in β-phase is not only attributed to the higher electric field during electrospinning produces polarity direction of the PVDF fiber, facilitating the growth of the crystalline structure, but also to the increasing amounts of the MWCNTs in the composited fibers [106]. The PVDF-CNT composite processed from sonication and evaporation also shows considerable enhancement in ferroelectric β-phase as well as remnant polarization. In general, the peaks for β-phase in FTIR spectra appear at 840 cm^{-1} (CH_2 rocking) and 1280 cm^{-1} (CF_2 stretching). With the increase in wt% of MWCNT in composite the characterization absorption of β-phase increases while that for α-phase decreases, as MWCNT influences the crystallization of PVDF. This increase in crystallization of PVDF may be due to the alternation of kinetics of crystallization as the added MWCNTs act as nuclei for PVDF. By the electrostatic interaction of functional groups on the MWCNT with the CF_2 dipole, the PVF_2 chain becomes more straightened, forming the zigzag conformation of the β -phase, instead of the coiled α-phase [88,107]. From the calculation of crystallization using Eq. (10.1), it has been shown that the content of β −phase in the composite changes drastically with drawing, as shown in Fig. 10.9. The characteristics of the P-E hysteresis also changes with drawing and poling. For undrawn and poled composites films the increase in the β-phase content increases with CNT concentration and sudden enhancement is reported for concentration more than 1 wt%. For the composite, without drawing but poling if

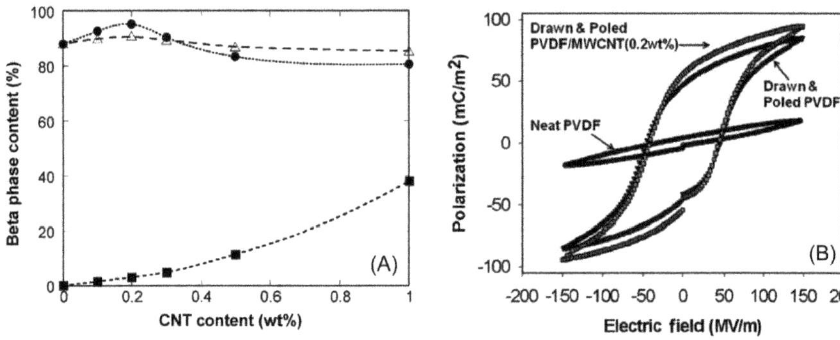

Figure 10.9 Variation of (A) β-phase content as a function of MWCNT content (wt%): (•) drawn & poled films, (Δ) drawn & unpoled films, (■) undrawn & poled films and (B) P-E hysteresis loops.
After S.-H. Wang, Y. Wan, B. Sun, L.-Z. Liu, W. Xu. Mechanical and electrical properties of electrospun PVDF/MWCNT ultrafine fibers using rotating collector. Nanoscale Res. Letts 9 (2014) 522–526.; G.H. Kim, S.M. Hong, Y. Seo, Piezoelectric properties of poly (vinylidene fluoride) and carbon nanotube blends: β-phase development. Phys. Chem. Chem. Phys. 11 (2009) 10506–10512; © 2015 Society of Plastics Engineers.

the added MWCNT amount goes over the percolation threshold, the conductivity of the composites increases. Increasing conductivity also increases the orientation of PVDF chains near the MWCNT hence enables the transformation of the α-phase into β-phase [89,108]. For drawing and poling, the largest content of β-phase found at a particular wt% of nanoparticles further increases in concentration and reduces the β-phase.

It is concluded that the maximum value of β-phase is rather referable to interfacial charge accumulation at the boundaries of the layers due to different current densities in the MWCNT and ferroelectric β-phase and nonferroelectric phases [102,109]. Hence, it is clear that for PVDF-CNT composites, the effect of poling is less significant than drawing [110]. Another possible explanation of enhancing β-phase in the composites is that the CNT surface has zigzag carbon atom that match well with the all-trans conformation of β PVDF and as a result may induce crystallization of PVDF in the β-polymorphic structure [91]. However, all these explanations are phenomenological, exact origin on the β-phase formation, there are no extensive theoretical studies reported. Certainly the addition of pristine MWCNTs as a nucleation agent accelerate crystallization of PVDF, but the percentage of nonpolar α-crystals are much more than polar β crystals. Surface modification of nanotubes has been proved as effective for obtaining homogeneous composites [91,111]. Very recently Ke et al. [112] extensively studied the effect of surface functionalization of carbon nanotubes with carboxyl (c-CNT), amino (a-CNT) and hydroxyl (h-CNT) groups as well as the pristine one (u-CNT) in β-phase formation in PVDF-CNT composite. The nonocomposites are prepared using a micro-compounder with dried PVDF pellets and MWCNT powder. Although it has been observed that u-CNT exhibits the best macrodispersion in PVDF, followed by

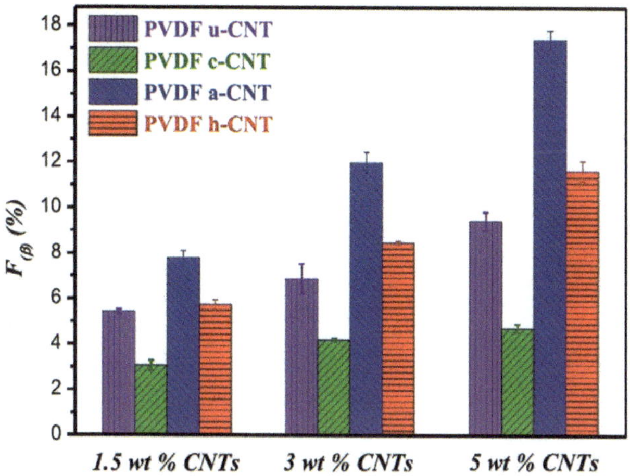

Figure 10.10 Relative percentage of β-phase F(β) in PVDF nanocomposites filled with different CNTs.
After K. Ke, P. Pötschke, D. Jehnichen, D. Fischer, V. Brigitte, Polymer achieving β-phase poly(vinylidene fluoride) from melt cooling: effect of surface functionalized carbon nanotubes 55 (2014) 611–619.

a-CNT and h-CNT, while c-CNT exhibits the worst dispersion in PVDF matrix, but calculating the fraction of β-phase $F(\beta)$ in the composite from FTIR for different concentration of functionalized nanotubes, quantitatively summarized in Fig. 10.10, illustrates that nanocomposites with amino group functionalized MWCNTs showed the highest percentage of β-phase (17.4%) formation in PVDF, followed by those with hydroxyl groups (11.6%) and unmodified MWCNTs (9.4%). The nanocomposites containing MWCNTs with carboxyl groups, which are assumed to be able to well interact with the dipoles on PVDF chains, have the lowest amount of β-phase (4.7%). Fig. 10.11 illustrates the possible mechanism for formation of β-phase in composite, as the electron negativity of fluorine atoms is much more strong than carbon and hydrogen atoms, the dipoles in $-CF_2$ interact with π-electrons from the surface of π-electron-rich so that PVDF chains with a zigzag conformation formed easily. Polymer composite with 100% polar phases has been achieved by Xing et al. [113] with the incorporation of only ionic liquid (IL) $[BMIM]^+[PF6]^-$ modified MWCNTs (IL-MWCNT). It is shown that incorporation of IL-modified MWCNTs (1:1) into the PVDF matrix not only accelerates crystallization of PVDF but also induces 100 percent polar crystals simply crystallized from the melt state. Not only the polar content but the crystallization temperature (T_c) and melting temperature (T_m) of the PVDF crystals as a function of the IL to CNTs ratio also changes, as depicted in the Fig. 10.12A. It is considered that both the specific interactions between $>CF_2$ groups in PVDF chains with the planar cationic imidazolium ring wrapped on the MWCNTs surface lead to the full zigzag conformations of PVDF and hence formation of high content polar crystal β and γ forms by subsequent crystal growth from the nuclei, schematically represented in Fig. 10.12B.

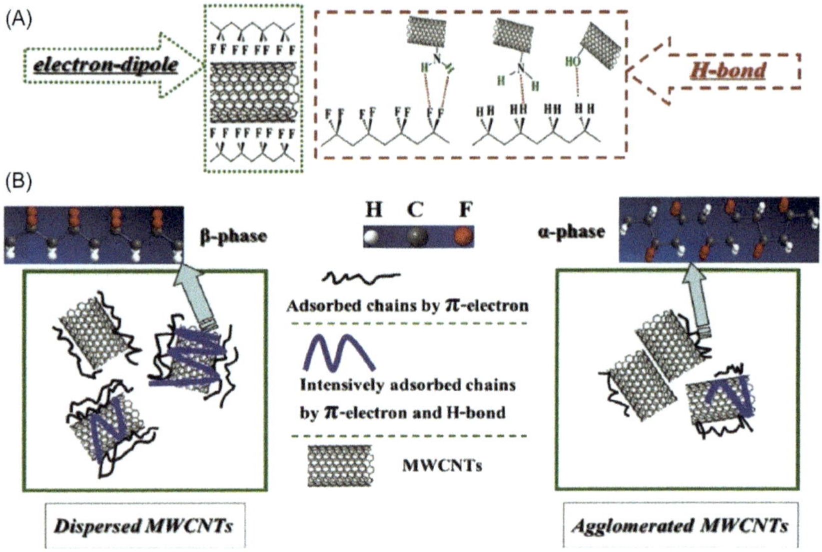

Figure 10.11 Schema reflecting the role of CNTs on the formation of the b-phase in PVDF: (A) the chemical boning between functionalized CNTs and PVDF chains; (B) the adsorbed chains of PVDF on the surface of CNTs influenced by the dispersion of CNTs.
After K. Ke, P. Pötschke, D. Jehnichen, D. Fischer, V. Brigitte, Polymer achieving β-phase poly(vinylidene fluoride) from melt cooling: effect of surface functionalized carbon nanotubes 55 (2014) 611–619.

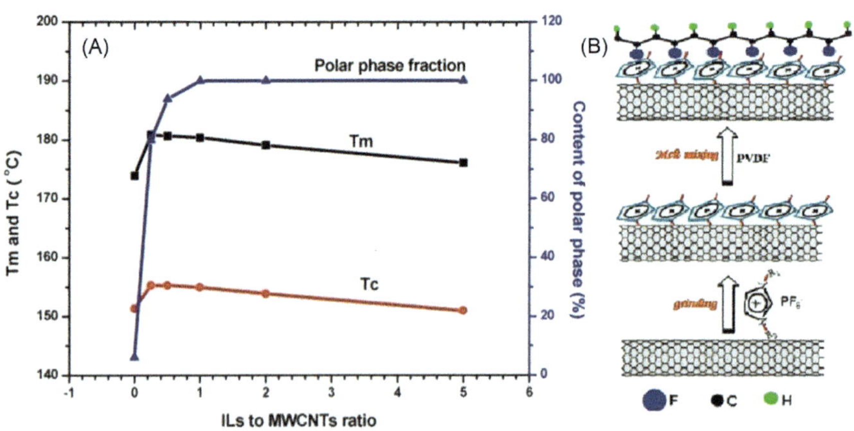

Figure 10.12 (A) Variation of polar fraction, T_c and T_m (calculated from DSC) of PVDF in the nanocomposites as a function of the IL to MWCNTs ratio. (B) Schematic diagram of the linker effect of IL for PVDF and MWCNTs as well as the mechanism for formation of PVDF TT conformations.
After C. Xing, L. Zhao, J. You, W. Dong, X. Cao, Y. Li, Impact of ionic liquid-modified multiwalled carbon nanotubes on the crystallization behavior of poly(vinylidene fluoride). J. Phys. Chem. B 116 (2012) 8312–8320.

10.4 Composites of ferroelectric polymer to enhance dielectric permittivity with low loss

Electrical energy storage devices with high performance, compact as well as low cost, are on the clear demand in portable electronic devices, stationary power systems, and hybrid electric vehicles. Among various energy storage technologies including batteries, fuel cells, capacitors, and supercapacitors, capacitor devices possess the advantage of high power density due to the fast electrical energy storage and discharge capability. Dielectric materials with high dielectric constants are required for such applications. The energy density of a linear dielectric material is given by

$$U_e = \frac{1}{2}\varepsilon E^2 \tag{10.2}$$

where ε is the dielectric constant of the material and E is the magnitude of the applied electric field.

Generally, the energy density U_e of a di-phasic composite is represented by the equation

$$U_e = \alpha_1 U_e^{(1)} + \alpha_2 U_e^{(2)} + \lambda U^{(3)} \tag{10.3}$$

where α_1 and α_2 are volume fractions of the constituent dielectric materials in a composite and $U_e^{(1)}$ and $U_e^{(2)}$ are their corresponding energy densities, $U^{(3)}$ is the energy density associated with interface effects, and λ is proportional to the interfacial area either a positive or negative contribution to energy density. Therefore, it is important to have high energy densities from both phases in order for the nanocomposite to exhibit considerable energy density. However, it has been shown that the filler must have a much greater permittivity than the surrounding polymer matrix to achieve high effective composite permittivity and energy density, as for high dielectric contrast, a "threshold volume fraction" of filler exists that is a function of particle asperity, above which the effective dielectric constant and energy density increase rapidly [114]. But, a high contrast in the dielectric permittivities of two phases leads to a highly inhomogeneous electric fields with local hot spots of increased electric field concentration and reduced dielectric strength, thus reducing the effective breakdown strength of the composite that is a serious drawback of nanocomposite approach [115–118]. Hence, the balancing between the seemingly contradictory criteria of enhancing dielectric constant while maintaining high dielectric strength is the main challenge. Hence, the constituents of the composites as well as the synthetic approaches that enable a homogeneous distribution of each constituent in nanocomposites with a well-defined interface between them are of great importance.

Poly(vinylidene fluoride) (PVDF) and its copolymers, among the other nonlinear dielectric, have been considered for applications as high-density energy storage devices because of their various advantages, such as large dielectric breakdown strength, low cost, light weight and ease of processing into large areas. PVDF-based polymers have high dielectric breakdown but always suffer from low dielectric constant ($\varepsilon \sim 10$) and thus low energy densities. Ceramics, on the other hand, have a high dielectric constant but always suffer from relatively low dielectric breakdown strength. The composite approach takes advantages on the idea that the combination of inorganic materials of large permittivity with polymers of high breakdown strength may lead to a large energy storage capacity. In addition, large interfacial areas in the composites containing nanometer scale fillers promote the exchange coupling effect through a dipolar interface layer and result in high polarization levels and dielectric responses [119]. In order to achieve high dielectric constant ferroelectric high-*k* materials such as $BaTiO_3$, $Pb(Zr,Ti)O_3$ (PZT), $Pb(Mg_{1/3}Nb_{2/3})O_3$-$PbTiO_3$ (PMN-PT), $CaCu_3Ti_4O_{12}$(CCTO) or other ferroelectrics or relaxor ferroelectrics, possess a very large dielectric permittivity are utilized as the ceramic filler because of their high dielectric constant [120–125]. In this chapter the $BaTiO_3$ has been taken as example in the whole discussion of ferroelectric polymer (PVDF) composites. $BaTiO_3$ nanoparticle is one of the most commonly investigated nanofillers due to its ferroelectric property and high dielectric constant. However, the value of dielectric constant varies between 1500–6000 depending on the size of the particles. $BaTiO_3$ exists in various crystallographic forms, with a tetragonal ferroelectric phase at temperature between 0 and the Curie temperature Tc, above which the unit cell of $BaTiO_3$ converts to the paraelectric cubic structure. The tetragonal form as shown in the figure of $BaTiO_3$ exhibits ferroelectric distortions involving the displacement of the cations (Ti^{4+} and Ba^{2+}) relative to the anion (O^{2-}), leading to a net dipole moment [126,127]. However, the ferroelectric ceramics exhibit electromechanical effect, such as the piezoelectric effect, that results in a mechanical resonance in the device during charging and discharging and thus would limit the reliability of the device. Moreover, the dielectric constant of these materials is strongly dependent on temperature [128]. One of the easiest ways to achieve the ferroelectric polymer P(VDF-TrFE) ceramic composite is by dissolving the polymer in methyl ethyl ketone (MEK) or N,N-dimethylacet amide (DMA) [129,130] by heating and stirring for few hours. MEK has a molecular weight of 72.11, a density of 0.81 g/cm^3, a viscosity of 0.40 cps, and a dielectric constant of 18.5 (at 25°C), a boiling point of 79.6°C. Homogeneous solution of $BaTiO_3$ nano-powder with different wt% can be obtained by $BaTiO_3$ with MEK solvent and ultrasonicate the solution for hours. Now, these two solutions of $BaTiO_3$ and P(VDF-TrFE) with common solvent MEK are mixed together to obtain P(VDF-TrFE)/$BaTiO_3$ ferroelectric polymer nanocomposite solutions. The film of P (VDF-TrFE)/$BaTiO_3$ can be prepared on different substrates by spin coating method. The thickness of the composite film can be controlled by rpm coating [131]. The scanning electron microscope (SEM) image shown in Fig. 10.13 illustrates that there is a minimal amount of polymer that plays the role of binder. Moreover, the composites prepared by this two-step mixing method and hot-press processing are dense with few air voids or defects. Fig. 10.14 illustrates comparative morphologies studied by field emission scanning electron microscopic (FESEM) imaging of pure copolymer and optimized concentration of P(VDF-TrFE)-$BaTiO_3$

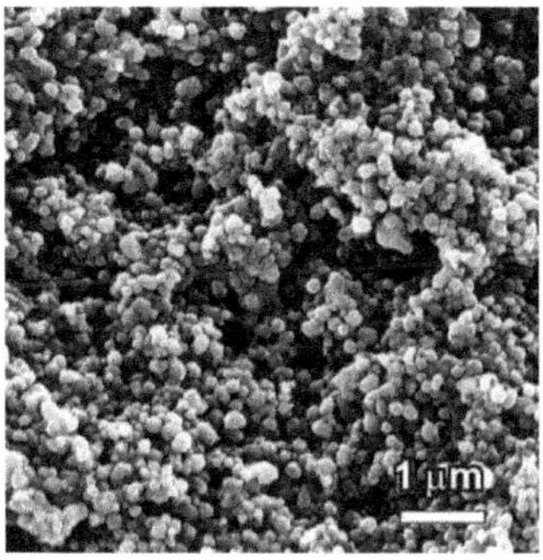

Figure 10.13 Scanning electron micrograph of the cross-section of a typical sample of $BaTiO_3$/PVDF composite with $BaTiO_3$ (particles of 50 nm size).
After Y.P. Mao, S.Y. Mao, Z.-G. Ye, Z.X. Xie, L.S. Zheng, Size-dependences of the dielectric and ferroelectric properties of $BaTiO_3$/polyvinylidene fluoride nanocomposites. J. Appl. Phys. 108 (2010) 014102–014106.

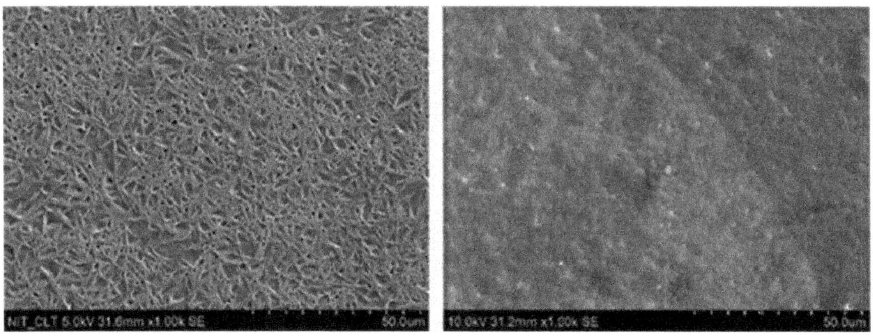

Figure 10.14 FESEM image of P(VDF-TrFE) polymer (left), P(VDF-TrFE)/0.8% $BaTiO_3$ nanocomposite (right) and EDS elemental mapping of P(VDF-TrFE)/0.8% $BaTiO_3$ nanocomposite. Elemental mapping study clearly shows the exact distribution and dispersion of $BaTiO_3$ nanoparticle in the polymer matrix.
After Valiyaneerilakkal et al. [A.K. Singh, C.K. Subash, K. Singh, S.M. Abbas, S. Varghese Polymer Composites-2015].

nanocomposite. Elemental mapping study clearly shows the exact distribution and dispersion of $BaTiO_3$ nanoparticle in the polymer matrix. From AFM topologies, it is seen that the film forming capability of the solution become more difficult after 0.8% composition of $BaTiO_3$ and it also leads to films with high porosity. Such direct solution processing of $BaTiO_3$ and other nanofillers in a polymer host generally results in poor film quality and inhomogeneities, which are mainly caused by agglomeration of the nanoparticles. High surface energy and large surface area nanoparticles cause to form large aggregates that lead to a highly inhomogeneous film when simply blended in a polymer matrix. Moreover, in such systems, residual free surfactant can lead to high leakage current and dielectric loss [132]. It is evidenced that the increase of solvent quantity did not only give positive effect of the increase of capacitance, but also give negative effects of the increase of leakage current. Notably, $BaTiO_3$ possesses a size-dependent dielectric and ferroelectric properties, which becomes one of the biggest issues in the utilization of $BaTiO_3$ nanocomposites. Mao et al. [130] reported that the dielectric constant of the composites with particle sizes over 300 nm become nearly constant ($\varepsilon \sim 65$), while that of the composites with particle sizes below 200 nm increases to 93 with particle size decreasing down to 100 nm. Moreover, the dielectric constant with particle sizes below 100 nm rapidly decreases to 48 with decreasing particle sizes, although the remnant polarization always increases with the particle size as shown in the Fig. 10.15. The nanosized $BaTiO_3$ particles tend to have smaller dielectric constant and depressed ferroelectricity than those of micron-sized and above, and its ferroelectricity disappears when particles are smaller than a certain critical size (17–30) nm depending sensitively on the synthetic techniques [133,134]. Although, in contrast, higher values of dielectric constant have also been noticed. In conclusion, direct solution processing of $BaTiO_3$ particles in a polymer host generally results in poor film quality and inhomogeneities, which are mainly caused by agglomeration of the nanoparticles.

Hence, there are many challenges in realizing high performance nanocomposites using these high-k particles and ferroelectric polymers, including realization of homogeneous nanoparticle dispersions and the tailoring of polymer/nanoparticle interfaces. As an effective route, surface modifications in the ceramic nanofillers modify the interface areas between ceramic nanofillers and the polymer matrix, and hence improves the homogeneity of the nanocomposites. To realize the full potential of nanoparticles to enhance the properties of ferroelectric polymer nanocomposites, significant efforts have recently been devoted to the design and synthesis of core-shell nanoparticles viz., the use of hybrid fillers [135], nanoparticle surface modification by organic molecules such as silanes, [136] phosphonic acid [137], and

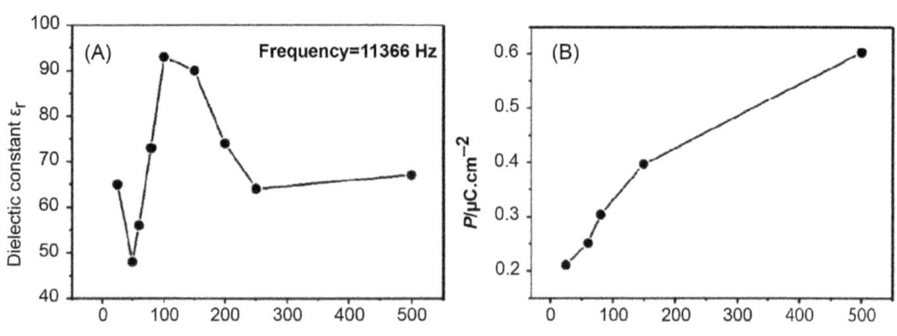

Figure 10.15 Dielectric constant and remnant polarization $BaTiO_3$/PVDF composites with different particle size.
After Y.P. Mao, S.Y. Mao, Z.-G. Ye, Z.X. Xie, L.S. Zheng, Size-dependences of the dielectric and ferroelectric properties of $BaTiO_3$/polyvinylidene fluoride nanocomposites. J. Appl. Phys. 108 (2010) 014102–01410.

ethylene diamine [138], and nanoparticle surface initiated in situ polymerization [139–143], have been developed to achieve nanocomposites with core-shell nanoparticle dispersion in a polymer matrix. Surface modification strategies to prepare surface functionalized fillers have been categorized as, (1) grafting-to, chemically binding the preformed polymers to the nanoparticle surface [139,144–146] and (2) grafting-from, initiating the controlled radical polymerization from the nanoparticle surface functionalized with an initiator [143,147]. And ferroelectric polymer nanocomposites with core-shell nanoparticles are generally processed through different routes viz., blending with polymer, in situ polymerization, direct preparation, etc.

In the grafting-to method, a preformed and end functionalized polymer is attached to the surface of the nanoparticle. The advantage of this strategy is controlling (1) the ability of the molecular composition and (2) the molecular weight of the polymer chains according to the desired performance of the final nanocomposites. Taking $BaTiO_3$ as instance for high-k particles, Fig. 10.16 shows the schematic illustration of the preparation process homogeneous ceramic–polymer nanocomposites treated by polyvinylprrolidone (PVP) and PVDF polymer matrix [148]. Polyvinylprrolidone (PVP) [145], titanate [47] phosphonic acid [137], etc., have also been used to core-shell $BaTiO_3$ nanoparticles following grafting-to method. However, the major disadvantage of grafting-to approach is the incomplete surface coverage of the nanoparticles, and also strongly affected by the stability of surface ligands [143,149]. Grafting-from method relies on the formation of nanocomposites by the in situ polymerization of monomers on initiator-functionalized nanoparticles surfaces. Controlled radical polymerization, viz., reversible addition-fragmentation chain transfer (RAFT) polymerization, and atom transfer radical polymerization

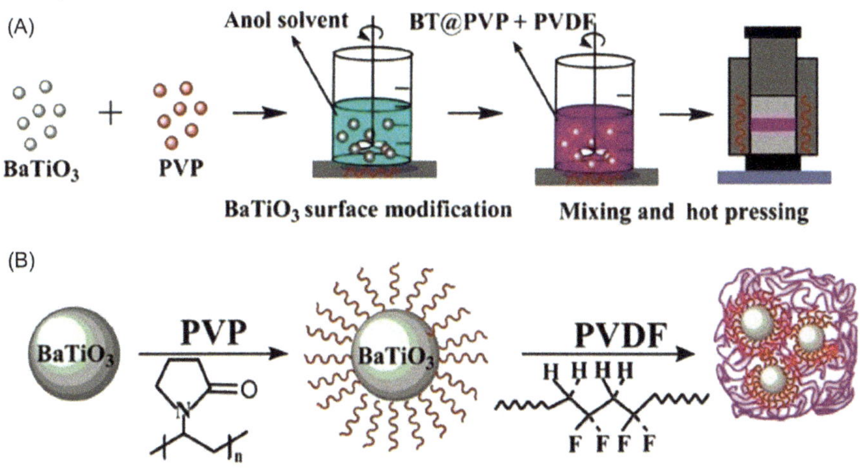

Figure 10.16 Schematic illustration for (A) fabrication process, and (B) modified mechanism of core – shell structured PVP/$BaTiO_3$-PVDF nanocomposites.
After J. Fu, Y. Hou, M. Zheng, Q. Wei, M. Zhu, H. Yan, Interfaces improving dielectric properties of PVDF composites by employing surface modified strong polarized $BaTiO_3$ particles derived by molten salt method. ACS Appl. Mater. 7 (2015) 24480–24491.

(ATRP) are powerful grating-from techniques [142,143,150,151]. For an example, Fig. 10.17 illustrates the preparation of surface modification of $BaTiO_3$ nanoparticles using fluoro-polymer and P(VDF-HFP) nanocomposites [143]. The advantages of this technique are (1) the complete shell layer coated on the nanoparticle surfaces prevents nanoparticles aggregation; (2) voids and pores free nanocomposites can be formed directly from core-shell nanoparticles using the shell layer as a matrix; (3) any polymer chains are robustly bonded on the nanoparticles surfaces, resulting in a strong nanoparticle/matrix interface; and (4) there is a broad range of monomers that can be polymerized. But there are drawbacks of this method: the grafting-from method may be plagued by the low initiation efficiency, and thus low grafting density and yield [151]. SEM images, shown in Fig. 10.18 illustrates that nanocomposites with surface modified nanoparticles are much more homogeneous. These dielectric nanocomposites consist of ferroelectric polymer and surface treated nanofillers can be considered as a three phase material, consisting of a polymer matrix (phase 1), an interfacial phase of fixed thickness l (phase 2), and nanoparticle fillers (phase 3), schematically shown in Fig. 10.19A. For these three phase materials, the effective dielectric permittivity can be expressed as [150,151]

$$\varepsilon_{\text{eff}} = \varepsilon_1 + f_2(\varepsilon_2 - \varepsilon_1)a_2 + f_3(\varepsilon_3 - \varepsilon_1)a_3 \tag{10.4}$$

where a_r is the electric field concentration factor for corresponding phase r, which relates the average electric field in phase r to that applied at boundary, E_0,

$$\langle E_r \rangle = a_r E_0 \ and\ f_2 = \frac{(r+l)^3 - r^3}{r^3} f_3$$

f_3 and r are the volume fraction and radius of nanoparticles, respectively.

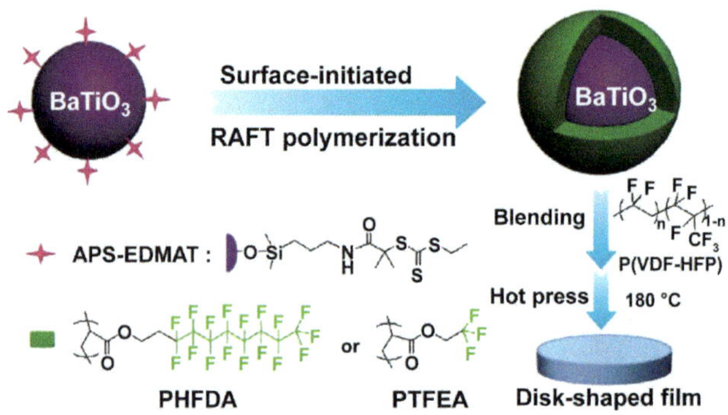

Figure 10.17 Schematic illustration for the preparation of fluoro-polymer-$BaTiO_3$ nanoparticles and P(VDF-HFP) nanocomposite films.
After K. Yang, X. Huang, Y. Huang, L. Xie, P.K. Jiang, Fluoro-polymer $BaTiO_3$ hybrid nanoparticles prepared via RAFT polymerization: toward ferroelectric polymer nanocomposites with high dielectric constant and low dielectric loss for energy storage application. Chem. Mater. 25 (2013) 2327–2338.

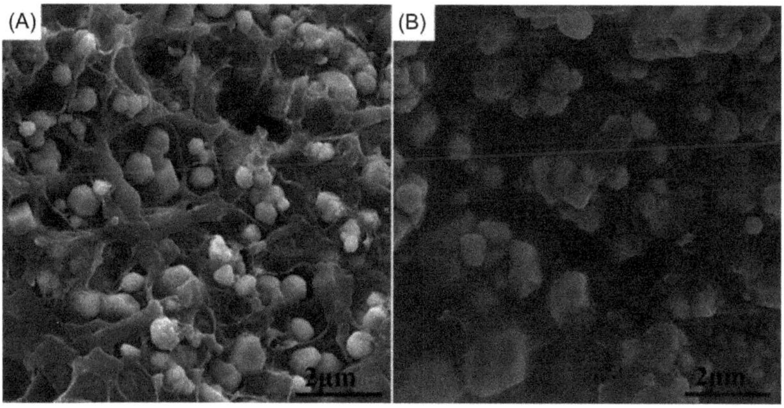

Figure 10.18 SEM images of the freeze-fractured cross-section of the composite filled with (A) the untreated 600 nm sized $BaTiO_3$ and (B) PVP modified 600 nm sized BT at a concentration of 40 vol %.
After J. Fu, Y. Hou, M. Zheng, Q. Wei, M. Zhu, H. Yan, Improving dielectric properties of PVDF composites by employing surface modified strong polarized $BaTiO_3$ particles derived by molten salt method. ACS Appl. Mater. Interfaces 7 (2015) 24480–24491.

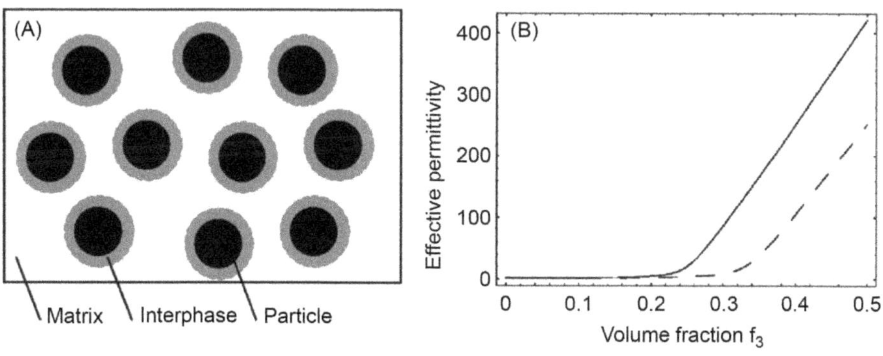

Figure 10.19 (A) Schematic diagram of a dielectric nanocomposite consisting of polymer matrix, nanoparticles, and interfacial phase. (B) Normalized effective permittivity of nanocomposite as function of volume fraction of nanoparticles; the dashed line ignores the interfacial effect, while the solid line takes into account the interphase.
After J.Y. Li, L. Zhang, S. Ducharme, Electric energy density of dielectric nanocomposites. IEEE Electr. Insul. Mag. Lett. 90 (2007) 132901–132902.

The effective permittivity of the nanocomposites, calculated using the Eq. (10.4) with $\frac{\varepsilon_3}{\varepsilon_1} = 1000$, corresponding to typical ratio of permittivity for ceramic and polymer, $\varepsilon_2 = \frac{(\varepsilon_3 + \varepsilon_1)}{2}$, and $l/r = 0.1$, corresponding to particle size around 100 nm for typical exchange length of a few nanometers is shown in Fig. 10.19B. Fig. 10.19B shows an increment in effective dielectric permittivity with concentration of nanofillers in the composite. Moreover, different theoretical models have been proposed for the effective dielectric constant of these composites viz.,

Furukawa model [152], Maxwell–Wagner equation [153], Rayleigh model [154], modified Lichtenecker equation [155] modified Kerner model [156], Yamada model [157], etc. The dielectric permittivity as well as the breakdown voltage of these nanocomposites with surface treated $BaTiO_3$. The dielectric permittivity increases with the volume fraction of $BaTiO_3$ followed by a maximum value around 50–60%, after which dielectric permittivity decreases rapidly with increase in volume fraction of the nanoparticle, as shown in the Fig. 10.20. For higher fractions of $BaTiO_3$ experimental results deviate significantly from the theoretical results. Such deviation is attributed mainly to the porosity in the nanocomposite film. With increasing the volume fraction of $BaTiO_3$, the breakdown field decreases. The breakdown behavior for the volume fraction greater that 50% due to the increase in the volume fraction of air voids, which will significantly lower the breakdown strength of the nanocomposites due to the low breakdown strength of air (3 V/μm) [158]. The breakdown voltage increases for the composites with core-shell nanoparticles as shown in Fig. 10.21.

Fig. 10.22A shows that there are plenty of traps inside the PVDF due to the crystal lattice defect. Current carriers may be trapped and turn into the space charge and the breakdown strength depends on the amount of traps in PVDF as well as on the amount of trapped carriers. Such decrease in the breakdown strength of $BaTiO_3$/PVDF composite with increasing $BaTiO_3$ has been attributed to the gradient in concentration that arises across the interface between $BaTiO_3$ particles and PVDF. This is, however, not the case for coated $BaTiO_3$/PVDF nanocomposite. As shown in Fig. 10.22B, a significant enhancement in dielectric breakdown strength has been observed at BT volume content of about 7%. The dielectric breakdown strength increases when the volume content of coated BT is less than 7%. In

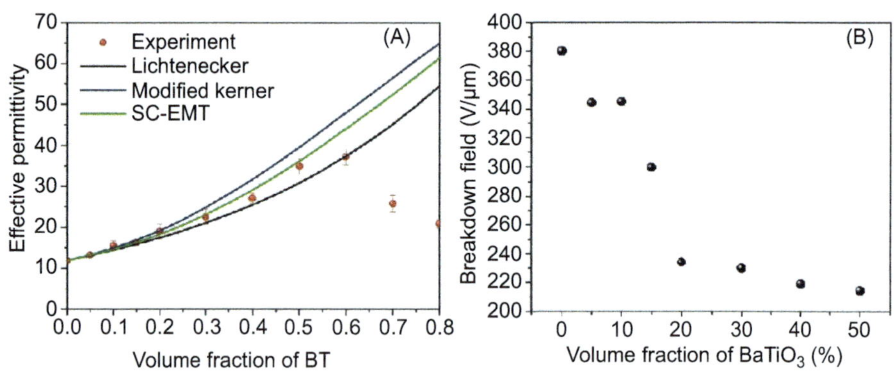

Figure 10.20 (A) Comparison of measured effective relative permittivity (at 1 kHz) of nanocomposites as a function of PFBPA-$BaTiO_3$ nanoparticle volume fraction with predicted values from different theoretical models. (B) breakdown strengths (failure probabilities: 63.2%) at each volume fraction as determined from the Weibull analysis.
After P. Kim, N.M. Doss, J.P. Tillotson, P.J. Hotchkiss, M.J. Pan, S.R. Marder, et al., High energy density nanocomposites based on surface-modified $BaTiO_3$ and a ferroelectric polymer. ACS Nano 3 (2009) 2581–2592.

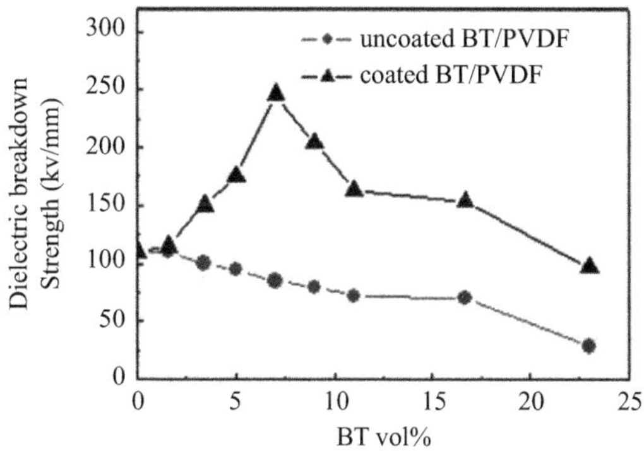

Figure 10.21 Dielectric strength of nanocomposites as a function of $BaTiO_3$ volume fraction.
After X. Dou, X. Liu, Y. Zhang, H. Feng, J.–F. Chen, S. Du, Improved dielectric strength of barium titanate-polyvinylidene fluoride nanocomposite. Appl. Phys. Lett. 95 (2009) 132904-1-3.

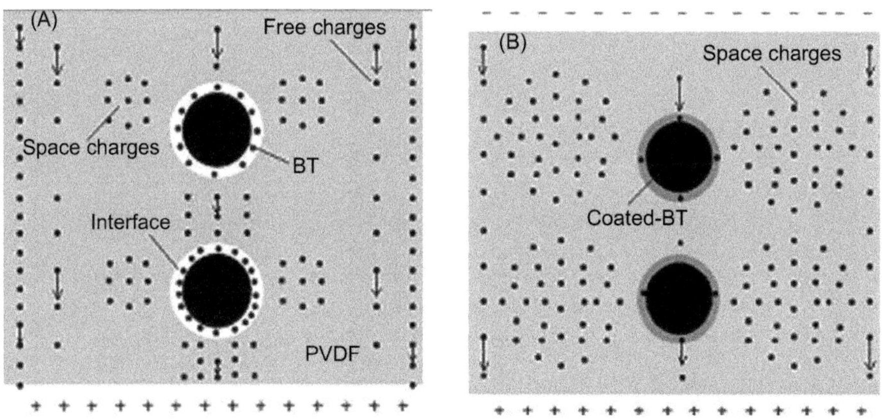

Figure 10.22 (A) and (B) Schematics of the nanocomposites.
After X. Dou, X. Liu, Y. Zhang, H. Feng, J.–F. Chen, S. Du, Improved dielectric strength of barium titanate-polyvinylidene fluoride nanocomposite. Appl. Phys. Lett. 95 (2009) 132904-1-3.

contrast, the breakdown strength decreases when the volume fraction of BT is more than 7%. Therefore, there is a maximum of the dielectric strength in this system. As shown in Fig. 10.22B, a significant enhancement in dielectric breakdown strength has been observed at BT volume content of about 7%. The dielectric breakdown strength increases when the volume content of coated BT is less than 7%. In contrast, the breakdown strength decreases when the volume fraction of BT is more than 7%. Therefore, there is a maximum of the dielectric strength in this system. The ferroelectric and dielectric properties of the composite systems enhanced up to technological favorable specifies. However, in order to achieve such functions, high volume fraction of inorganic ceramics and carbon nanotubes have always to be added into the composites, which suffer from several drawbacks, particularly in terms of high weight, low flexibility, and poor mechanical performance. Moreover, inorganic contents of the composite lead to retreat of constraints of inorganic ferroelectric materials. Very recently, a novel molecular ferroelectric material diisopropylammonium bromide (DIPAB) with spontaneous polarization 23 $\mu C/cm^2$ has been reported [159,160]. The PVDF-DIPAB composite also exhibit polar β-form with high dielectric constant and low loss [161]. PVDF-DIPAB composite has been prepared by dissolving PVDF powder in DMF at 60°C by magnetic stirring following the addition of DIPAB in presence of silicone coupling agent KH-570. The mixture solution was magnetic agitated in a 60°C water bath to ensure the complete dissolution and dispersion of DIPAB. Composites of polymer with organic ferroelectric materials would be able to meet the desired functions. However, more extensive research work is needed.

References

[1] J. Valasek, Piezo-electric and allied phenomena in Rochelle salt, Phys. Rev. 17 (1971) 475–481.

[2] J. Valasek, Piezo-electric activity of Rochelle salt under various conditions, Phys. Rev. 19 (1922) 478–491.

[3] J. Valasek, Properties of Rochelle slat related to the Piezo-electric effect, Phys. Rev. 20 (1922) 639–664.

[4] J. Valasek, Dielectric anomalies in Rochelles salt crystals, Phys. Rev. 24 (1924) 560–568.

[5] M.E. Lines, A.M. Glass, Principles and Applications of Ferroelectrics and Related Materials, Oxford University Press, 1977.

[6] A.K. Batra, M.D. Aggarwal, M. Edwards, A.S. Bhalla, Present status of polymer: ceramic composites for pyroelectric infrared detectors, Ferroelectric Ferroelectrics 366 (2008) 84–121.

[7] R.E. Newnham, Composite electroceramics, Ferroelctrics 68 (1986) 1–32.

[8] K. Mazur, J. Handerek, T. Piech, A Surface effect in ferroelkectric Yttrium manganite, Acta Phys. Polon. 37 (1970) 31–40.

[9] J. Curi, P. Curie, Développement, par pression, de l'électricité polaire dans les cristaux hémièdres à faces inclinées, C. R. Acad. Sci. Paris 91 (1880) 294–295, Comptes Rendus de l'.

[10] E.T. Jaynes, Ferroelectricity, E. T. Jaynes, Princeton University Press, 1953.

[11] J.A. Anderson, Rep. Nat. Res. Council (April 1918).

[12] W.G. Cady, Rep. Nat. Res. Council (May 1918).

[13] G. Busch, P. Scherrer, A New Seignettoelectric Substance, Naturwissenschaften 23 (1935) 737–738.

[14] A. von Hippel, R.G. Breckenridge, F.G. Chesley, L. Tisza, High dielectric constant ceramics, Ind. Eng. Chem. 38 (1946) 1097–1109.

[15] G. Shirane, A. Takeda, Phase transitions in solid solutions of $PbZrO_3$ and $PbTio_3$ (i) small concentrations of $PbTio_3$, J. Phys. Soc. Jpn. 7 (1952) 5–11.

[16] E. Sawaguchi, Ferroelectricity versus antiferroelectricity in the solid solutions of $PbZrO_3$ and $PbTiO_3$, J. Phys. Soc. Jpn. 8 (1953) 615–629.
[17] E. Wainer, A.N. Salomon, Titanium Alloy Mfg. Co. Elec. Report 8 (September 1942).
[18] E. Wainer, A.N. Salomon, Titanium Alloy Mfg. Co. Elec. Report 9, (January 1943).
[19] B.M. Wul, I.M. Goldman, Comtes Rendus dielectric constants of titanates of metals of the seond group 46 (1945) 139-142.
[20] V. Ginsburg, USSR, J. Phys. Chem. 10 (1946) 107.
[21] B.M. Wul, L.F. Vereschagen, Comtes Rendus 48 (1945) 634.
[22] B.M. Wul, USSR, J. Phys. 10 (1946) 95.
[23] B.M. Wul, Dielectric constants of some titanates, Nature 156 (1945) 480–481.
[24] B.M. Wul, Barium titanate: a new ferro-electric, Nature 157 (1946) 808–809.
[25] J.F. Nye, Physical Properties of Crystals, Clarendon Press, Oxford, 1957.
[26] J. Li, Y. Liu, Y. Zhang, H.-L. Caic, R.-G. Xiong, Molecular ferroelectrics: where electronics meet biology, Phys. Chem. Chem. Phys. 15 (2013) 20786–20796.
[27] H. Kawai, The piezoelectricity of poly (vinylidene fluoride), Jpn. J. Appl. Phys. 8 (1969) 975–976.
[28] E. Fukuda, Mechanical deformation and electrical polarization in biological substances, Biorheology 5 (1968) 199–208.
[29] E. Fukuda, Piezoelectricity in polymers and biological materials, Ultrasonics 6 (1968) 229–234.
[30] V.A. Bazhenov, Piezoelectric Properties of Woods, Consultant Bureau, 1961.
[31] Y. Wada, R. Hira, T. Furukawa, Y. Takubo, Rept. Progr. Polym. Phys. Jpn 9 (1966) 435.
[32] E. Fukada, M. Date, N. Hirai, Piezoelectric effect in poly(y-Methyl L-Glutamate), Nature 211 (1966) 1079–1080.
[33] T. Furukawa, Y. Uematsu, K. Asakawa, Y. Wada, Piezoelectricity, pyroelectricity, and thermoelectricity of polymer films, J. Appl. Polyrn. Sci. 12 (1968) 2675–2680.
[34] K. Kasai, Rept. Progr. Polyrn. Phys. Jpn 12 (1969) 393.
[35] J.G. Bergman, J.H. McFee, G.R. Crane, Pyroelectricity and optical second harmonic generation in polyvinylidene fluoride films, Appl. Phys. Lett. 18 (1971) 203–204.
[36] H.S. Nalwa (Ed.), Ferroelectric Polymers: Chemistry, Physics, and Applications, Marcel Dekker, Inc, New York, 1995.
[37] A.J. Lovinger, Unit cell of the γ phase of poly(vinylidene fluoride), Macromolecules 14 (1981) 322–325.
[38] J.B. Lando, H.G. Olf, A. Peterlin, Nuclear magnetic resonance and x-ray determination of the structure of poly(vinylidene fluoride), J. Polym. Sci. Part A 4 (1966) 941–951.
[39] A.J. Lovinger, in: D.C. Basset (Ed.), Developments in crystalline polymers, Applied Science Publishers Ltd, N. J. Englewood, 1982.
[40] S. Chen, K. Yao, F.E.H. Tay, C.L. Liow, Ferroelectric poly(vinylidene fluoride) thin films on Si substrate with the β phase promoted by hydrated magnesium nitrate, J. Appl. Phys. Phys. 102 (2007), 104108-1-9.
[41] S. Fujisaki, H. Ishiwara, Y. Fujisaki, Low-voltage operation of ferroelectric poly(vinylidene fluoride-trifluoroethylene) copolymer capacitors and metal-ferroelectric-insulator-semiconductor diodes, Appl. Phy. Lett. 90 (2007) 162192–162193.
[42] A. Salimi, A.A. Yousefi, Conformational changes and phase transformation mechanisms in PVDF solution-cast films, J. Polym. Sci.: Part B: Polym. Phys. 42 (2004) 3487–3495.
[43] M. Tamura, K. Ogasawara, N. Ono, S. Hagiwara, Piezoelectricity in uniaxially stretched poly(vinylidene fluoride), J. Appl. Phys. 45 (1974) 3768–3771.

[44] R.G. Kepler, R.A. Anderson, Ferroelectricity in polyvinylidene fluoride, J. Appl. Phys. 49 (1978) 1232–1235.
[45] T. Furukawa, Ferroelectric properties of vinylidene fluoride copolymers, Phase Trans. 18 (1989) 143–211.
[46] M.G. Buonomenna, P. Macchi, M. Davoli, E. Drioli, Poly(vinylidene fluoride) membranes by phase inversion: the role the casting and coagulation conditions play in their morphology, crystalline structure and properties, Euro. Polym. J. 43 (2007) 1557–1572.
[47] X. He, K. Yao, Crystaliization mechanism and piezoelectric properties of solution-derived ferroelctric poly(vinylidene fluoride) thin films, Appl. Phys. Lett. 89 (2006), 112909-112909.
[48] P. Sajkiewicz, A. Wasiak, Z. Goclowski, Phase transitions during stretching of poly (vinylidene fluoride), Euro. Polym. J. 35 (1999) 423–429.
[49] T. Furukawa, M. Date, E. Fukada, Y. Tajitsu, A. Chiba, Ferroelectric behavior in the copolymer of vinylidenefluoride and trifluoroethylene, Jpn. J. Appl. Phys. 19 (1980) L109–L112.
[50] T. Yamada, T. Ueda, T. Kitayama, Ferroelectric-to-paraelectric phase transition of vinylidene fluoride-trifluoroethylene copolymer, J. Appl. Phys. 52 (1981) 948–952.
[51] Y. Higashihata, J. Sako, T. Yagi, Piezoelectricity of vinylidene fluoride-trifluoroethylene copolymers, Ferroelectrics 32 (1981) 85–92.
[52] A.J. Lovinger, Ferroelectric transition in a copolymer of vinylidene fluoride and tetrafluoroethylene, Macromolecules 16 (1983) 1529–1534.
[53] Y. Murata, N. Koizumi, Curie transition in copolymers of vinylidene fluoride and tetrafluoroethylene, Polymer J. 17 (1985) 1071–1074.
[54] T. Tanigami, K. Yamaura, S. Matsuzawa, K. Mizoguchi, K. Miyasaka, Structural studies on ethylene-tetrafluoroethylene copolymer 1. Crystal structure, Polymer 27 (1986) 999–1006.
[55] B.L. Farmer, J.B. Lando, J. Macromol, Conformational and packing analysis of the alternating copolymer of ethylene and tetrafluoroethylene, Sci.-Phys B1 1 (1975) 89–119.
[56] S. Tasaka, N. Inagaki, T. Okutani, S. Miyata, Structure and properties of amorphous pizzoelectric vinylidene cyanide copolymers, Polymer 30 (1989) 1639–1642.
[57] M. Kisimoto, K. Nakajima, I. Seo, Structure and physical properties of vinylidene cyanidestraight chain fatty acid vinyl ester copolymers, Polymer Prepare. Jpn. 37 (1988) 2249–2251.
[58] Y.J. Park, S.J. Kang, K.J. Kim, M.S. Lee, W.I. Jeong, I.J. Park, et al., Irrreversible extinction of ferroelectric polarization in P(VDF-TrFE) thin films upon melting and recrystallization, Appl. Phys. Lett 88 (2006) 242908.
[59] P. Ueberschlag, PVDF piezoelectric polymer, Sens. Rev. 21 (2001) 118–126.
[60] R.L. Miller, J. Raisoni, Single crystals of poly(vinylidene fluoride), J. Polym. Sci. Polymer Science Polymer Physics Ed, 1, 4, 1976, pp. 2325–2326.
[61] A.J. Lovinger, Crystallization of the β phase of poly(vinylidene fluoride) from the melt, Polymer 22 (1981) 412–413.
[62] D. Yang, Y. Chen, β-phase formation of poly(vinylidene fluoride) from the melt induced by quenching, J. Mater. Sci. Lett. 6 (1987) 599–603.
[63] A. Salimi, A.A. Yousefi, Analysis method: FTIR studies of β-phase crystal formation in stretched PVDF films, Polym. Test. 22 (2003) 699–704.
[64] R. Hasegawa, M. Kobayashi, H. Tadokoro, Molecular conformation and packing of poly(vinylidene fluoride). Stability of three crystalline forms and the effect of high pressure, Polym. J. 3 (1972) 591–599.

[65] J. Scheinbeim, C. Nakafuku, B.A. Newman, K.D. Pae, High-pressure crystallization of poly(vinylidene fluoride), J. Appl. Phys. 50 (1979) 4399–4405.
[66] X. He, K. Yao, Crystallization mechanism and piezoelectric properties of solution-derived ferroelectric poly(vinylidene fluoride) thin films, Appl. Phys. Lett. 89 (2006) (2006) 112909–112910.
[67] D. Mandal, K. Henkel, D. Schmeißser, Comment on Preparation and characterization of silver poly(vinylidene fluoride) nanocomposites: formation of piezoelectric polymorph of poly(vinylidene fluoride), J. Phys. Chem. B 111 (2011) 10567–10569.
[68] D. Mandal, K. Henkel, D. Schmeißer, The electroactive β-phase formation in poly (vinylidene fluoride) by gold nanoparticles doping, Mater. Lett. 73 (2012) 123–125.
[69] J.I. Scheinbeim, B.A. Newman, A. Sen, Field-induced crystallization in highly plasticized poly(vinylidene fluoride) films, Macromolecules 19 (1986) 1454–1458.
[70] X. Ma, J. Liu, C. Ni, D.C. Martin, D.B. Chase, F. Rabolt, et al., Orientation in electrospun poly(vinylidene fluoride) fibers, Molecular 1 (2012) 428–431.
[71] S.J. Kang, Y.J. Park, I. Bae, K.J. Kim, H.–C. Kim, S. Bauer, et al., Ferroelectric PVDF/PMMA blend films with ultralow roughness for low voltage non-volatile polymer memory, Adv. Funct. Mater. 19 (2009) 2812–2818.
[72] R.E. Bernstein, C.A. Cruz, D.R. Paul, J.W. Barlow, LCST behavior in polymer blends, Macromolecules 10 (1977) 681–686.
[73] R.E. Bernstein, D.R. Paul, J.W. Barlow, Polymer blends containing poly(vinylidene fluoride). Part II: Poly(vinyl esters), Poly. Eng. Sci. 18 (1978) 683–686.
[74] R.E. Balke, I. Cabasso, Poly(vinylidene fluoride)/poly(vinyl acetate) miscible blends: 1. Thermal analysis and spectroscopic (FTi.r.) characterization, Polymer 29 (1988) 1831–1842.
[75] C. Leonard, J.L. Halary, L. Monnerie, Hydrogen bonding in PMMA-fluorinated blends: FTi.r. investigations using ester model molecules, Polymer 26 (1985) 1507–1513.
[76] T. Nishi, T.T. Wang, Melting point depression and kinetic effects of cooling on crystallization in poly(vinylidene fluoride)-poly(methyl methacrylate) mixtures, Macromolecules 8 (1975) 909–915.
[77] T.T. Wang, T. Nishi, Spherulitic crystallization in compatible blends of poly(vinylidene fluoride) and poly(methyl methacrylate), Macromolecules 10 (1977) 421–425.
[78] S.J. Kang, Y.J. Park, J. Sung, P.S. Jo, C. Park, K.J. Kim, et al., Spin cast ferroelectric beta poly(vinylidene fluoride) thin films via rapid thermal annealing, Appl. Phys. Lett. 92 (2008) 012921–012922.
[79] D. Yang, E.L. Thomas, Effect of PMMA on the morphology and $\alpha \rightarrow \beta$ phase transition of oriented PVF_2/PMMA blends, J. Mater. Sci. Lett. 6 (1987) 593–598.
[80] C. Leonard, J.L. Halary, L. Monnerie, Crystallization of poly(vinylidene fluoride)-poly (methyl methacrylate) blends: analysis of the molecular parameters controlling the nature of poly(vinylidene fluoride) crystalline phase, Macromolecules 21 (1988) 2988–2994.
[81] M. Li, N. Stingelin, J.J. Michels, M.-J. Spijkman, K. Asadi, K. Feldman, et al., Ferroelectric phase diagram of PVDF:PMMA, Macromolecues 45 (2012) 7477–7485.
[82] F. Xia, Z. Cheng, H. Xu, H. Li, Q. Zhang, G.J. Kavarnos, et al., High electromechanical responses in a poly(vinylidene fluoride–trifluoroethylene–chlorofluoroethylene) terpolymer, Adv. Mater. 14 (2002) 1574–1577.
[83] J.R. Gregorio, M. Cestari, Effect of crystallization temperature on the crystalline phase content and morphology of poly(vinylidene fluoride), J. Polym. Sci. Part B: Polym. Phys. 32 (1994) 859–870.

[84] R. Gregorio Jr., Determination of the α, β, and γ crystalline phases of poly(vinylidene fluoride) films prepared at different conditions, J. Appl. Polym. Sci. 4 (2006) 3272–3279.
[85] W. Li, Z. Zhang, Q. Meng, W. Xia, Y. Zheng, Z. Xu, Electric energy storage properties of poly(vinylidene fluoride), Appl. Phys. Letts. 96 (2010) 192905–192910.
[86] L. Priya, J.P. Jog, Intercalated poly(vinylidene fluoride)/clay nanocomposites: structure and properties, J. Polym. Sci. Part B: Polym. Phys. 41 (2003) 31–38.
[87] Y.M. Song, Z.D. Zhao, W.X. Yu, B. Li, X.F. Chen, Morphological structures of poly (vinylidene fluoride)/montmorillonite nanocomposites, Sci. China Ser. B: Chem. 50 (2007) 790–796.
[88] D. Mandal, K.J. Kim, J.S. Lee, Simple synthesis of palladium nanoparticles β phase formation and the control of chain and dipole poly (vinylidene fluoride) thin films, Langmuir 28 (2012) 10310–10317.
[89] N. Levi, R. Czerw, S. Xing, P. Lyer, D.L. Carroll, Properties of polyvinylidene difluoride – carbon nanotube blends, Nano Lett. 4 (2004) 1267–1271.
[90] S. Yu, W. Zheng, W. Yu, Y. Zhang, Q. Jiang, Z. Zhao, Formation mechanism of β-Phase in PVDF/CNT composite prepared by the sonication method, Macromolucules 42 (2009) 8870–8874.
[91] S. Manna, A.K. Nandi, Piezoelectric β polymorph in poly(vinylidene fluoride)-functionalized multiwalled carbon nanotube nanocomposite films, J. Phys. Chem. C 111 (2007) 14670–14680.
[92] D. Li, Y. Xia, Electrospinning of nanofibers: reinventing the wheel? Adv. Mater. 16 (2004) 1151–1170.
[93] A. Greiner, J.H. Wendorff, Functional self-assembled nanofibers by electrospinning, Adv. Polym. Sci. 219 (2008) 107–171.
[94] D.H. Reneker, A.L. Yarin, Electrospinning jets and polymer nanofibers, Polymer 49 (2008) 2387–2425.
[95] S. Agarwal, J.H. Wendorff, A. Greiner, Use of electrospinning technique for biomedical applications, Polymer 49 (2008) 5603–5621.
[96] J.D. Schiffman, C.L. Schauer, A review: electrospinning of biopolymer nanofibers and their applications, Polym. Rev. 48 (2008) 317–352.
[97] Z.M. Huang, Y.Z. Zhang, M. Kotaki, S. Ramakrishna, A review on polymer nanofibers by electrospinning and their applications in nanocomposites, Compos. Sci. Technol. 63 (2003) 2223–2253.
[98] B. Sun, Y.Z. Long, H.D. Zhang, M.M. Li, J.L. Duvail, X.Y. Jiang, et al., Advances in three-dimensional nanofibrous macrostructures via electrospinning, Prog. Polym. Sci. 39 (2014) 862–890.
[99] S. Iijima, Helical microtubules of graphitic carbon, Nature 354 (1991) 56–58.
[100] S.J. Tans, M.H. Devoret, H. Dai, A. Thess, R.E. Smalley, L.J. Geerligs, et al., Individual single-wall carbon nanotubes as quantum wires, Nature 386 (1997) 474–477.
[101] E.W. Wong, P.E. Sheehan, C.M. Lieber, Nanobeam mechanics: elasticity, strength, and toughness of nanorods and nanotubes, Science 277 (1997) 1971–1975.
[102] G.H. Kim, S.M. Hong, Y. Seo, Piezoelectric properties of poly(vinylidene fluoride) and carbon nanotube blends: β-phase development, Phys. Chem. Chem. Phys. 11 (2009) 10506–10512.
[103] S.-H. Wang, Y. Wan, B. Sun, L.-Z. Liu, W. Xu, Mechanical and electrical properties of electrospun PVDF/MWCNT ultrafine fibers using rotating collector, Nanoscale Res. Letts. 9 (2014) 522–526.

[104] J.Y. Lim, S. Kim, Y. Seo, AIP Conference Proceedings Enhancement of β-phase in PVDF by electrospinning 1664 (2015) 070006-070007.
[105] M. Sharma, V. Srinivas, G. Madras, S. Bose, Outstanding dielectric constant and piezoelectric coefficient in electrospun nanofiber mats of PVDF containing silver decorated multiwall carbon nanotubes: assessing through piezoresponse force microscopy, RSC Adv. 6 (2016) 6251–6258.
[106] S. Huang, W.A. Yee, W.C. Tjiu, Y. Liu, M. Kotaki, Y.C.F. Boey, et al., Electrospinning of polyvinylidene difluoride with carbon nanotubes: synergistic effects of extensional force and interfacial interaction on crystalline structures, Langmuir 24 (2008) 13621–13626.
[107] B. Mohammadi, A.A. Yousefi, S.M. Bellah, Effect of tensile strain rate and elongation on crystalline structure and piezoelectric properties of PVDF thin films, Polym. Test. 26 (2007) 42–50.
[108] X. He, K. You, Crystallization mechanism and piezoelectric properties of solution-derived ferroelectric poly(vinylidene fluoride) thin films, Appl. Phys. Lett. 89 (2006) 112909–112910.
[109] K. Koga, H. Ohigashi, Piezoelectricity and related properties of vinylidene fluoride and trifluoroethylene copolymers, J. Appl. Phys. 59 (1986) 2142–2151.
[110] S.R. Anton, H.A. Sodano, A review of power harvesting using piezoelectric materials (2003–2006), Smart Mater. Struct. 16 (2007) R1.
[111] R.E. Belke, I. Chabasso, Poly(vinylidene fluoride)/poly(vinyl acetate) miscible blends: 1. Thermal analysis and spectroscopic (FTi.r.) characterization, Polymer 29 (1988) 1831–1842.
[112] K. Ke, P. Pötschke, D. Jehnichen, D. Fischer, V. Brigitte, Achieving β-phase poly (vinylidene fluoride) from melt cooling: effect of surface functionalized carbon nanotubes, Polymer 55 (2014) 611–619.
[113] C. Xing, L. Zhao, J. You, W. Dong, X. Cao, Y. Li, Impact of ionic liquid-modified multiwalled carbon nanotubes on the crystallization behavior of poly(vinylidene fluoride), J. Phys. Chem. B 116 (2012) 8312–8320.
[114] L. An, S.A. Boggs, J.P. Calame, Energy storage in polymer films with high dielectric constant fillers [Feature Article], IEEE Electr. Insul. Mag. 24 (2008) 5–10.
[115] J.Y. Li, L. Zhang, S. Ducharme, Electric energy density of dielectric nanocomposites, IEEE Electr. Insul. Mag. Lett. 90 (2007) 132901–132902.
[116] F. Peruani, G. Solovey, I.M. Irurzun, E.E. Mola, A. Marzocca, J.L. Vicente, Dielectric breakdown model for composite materials, Phys. Rev. E 67 (2003), 066121-1-8.
[117] J.P. Calame, Evolution of Davidson–Cole relaxation behavior in random conductor–insulator composites, J. Appl. Phys. 94 (2003) 5945–5958.
[118] J.Y. Li, C. Huang, Q.M. Zhang, Enhanced electromechanical properties in all-polymer percolative composites, Appl. Phys. Lett. 84 (2004) 3124–3127.
[119] T.J. Lewis, Interfaces: nanometric dielectrics, J. Phys. D: Appl. Phys. 38 (2005) 202–212.
[120] R.E. Newnham, Composite electroceramics, Annu. Rev. Mater. Sci. 16 (1986) 47–68.
[121] D.K. Das-Gupta (Ed.), Ferroelectric Polymer and Ceramic-Polymer Composites, Trans. Tech, Aedermannsdorf, Switzerland, 1994.
[122] R. Gregorio Jr., M. Cestari, F.E. Bernardino, Dielectric behaviour of thin films of β-PVDF/PZT and β-PVDF/$BaTiO_3$ composites, J. Mater. Sci. 31 (1996) 2925–2930.
[123] H.L.W. Chan, M.C. Cheung, C.L. Choy, Study on $BaTiO_3$ /P(VDF)-TrFE 0-3 composites, Ferroelectrics 224 (1999) 113–120.

[124] Y. Bai, Z.-Y. Cheng, V. Bharti, H.S. Xu, Q.M. Zhang, High-dielectric-constant ceramic-powder polymer composites, Appl. Phys. Lett. 76 (2000) 3804–3807.

[125] Z.M. Zhang, Y. Shen, C.W. Nan, Dielectric behavior of three-phase percolative Ni-$BaTiO_3$/polyvinylidene flouride composites, Appl. Phys. Lett. 81 (2002) 4814–4816.

[126] P.K. Dutta, R. Asiaie, S.A. Akbar, W. Zhu, Hydrothermal synthesis and dielectric properties of tetragonal $BaTiO_3$, Chem. Mater. 6 (1994) 1542–1548.

[127] H. Tabata, H. Tanaka, T. Kawai, Formation of artificial $BaTiO_3$/$SrTiO_3$ superlattices using pulsed-laser deposition and their dielectric properties, Appl. Phys. Lett. 65 (1994) 1970–1972.

[128] Z.-Y. Cheng, R.S. Katiyar, X. Yao, A. Guo, Dielectric behavior of lead magnesium niobate relaxors, Phys. Rev. B 55 (1997) 8165–8174.

[129] U. Valiyaneerilakkal, S. Verghese, Poly (vinylidene fluoride-trifluoroethylene)/barium titanate nanocomposite for ferroelectric nonvolatile memory devices, AIP Adv. 3 (2013) 042131–042132.

[130] Y.P. Mao, S.Y. Mao, Z.-G. Ye, Z.X. Xie, L.S. Zheng, Size-dependences of the dielectric and ferroelectric properties of $BaTiO_3$/polyvinylidene fluoride nanocomposites, J. Appl. Phys 108 (2010) 014102–014106.

[131] R. Schroeder, L.A. Majewski, M. Grell, High-performance organic transitors using solution-processed nanoparticle-filled high-k polymer gate insulators, Adv. Mater. 17 (2005) 1535–1539.

[132] S.-D. Cho, K.-W. Paik, Proceedings of 51st IEEE Electronic Components and Technology Conference on 2001, p. 1418.

[133] W.L. Zhong, Y.G. Wang, P.L. Zhang, B.D. Qu, Phenomenological study of the size effect on phase transitions in ferroelectric particles, Phys. Rev. B 50 (1994) 698–703.

[134] S. Wada, H. Yasuno, T. Hoshina, S.-M. Nam, H. Kakemoto, T. Tsurumi, Preparation of nm-sized barium titanate fine particles and their powder dielectric properties, Jpn. J. Appl. Phys. 42 (2003) 6188–6195.

[135] E.A. Stefanescu, X.L. Tan, Z.Q. Lin, N. Bowler, M.R. Kessler, Multifunctional PMMA-Ceramic composites as structural dielectrics, Polymer 51 (2010) 5823–5832.

[136] Z.M. Dang, H.Y. Wang, H.P. Xu, Influence of silane coupling agent on morphology and dielectric property in $BaTiO_3$/polyvvynylidene fluoride conmposites, Appl. Phys. Lett. 89 (2006) 112902–112904.

[137] P. Kim, N.M. Doss, J.P. Tillotson, P.J. Hotchkiss, M.J. Pan, S.R. Marder, et al., High energy density nanocomposites based on surface-modified $BaTiO_3$ and a ferroelectric polymer, ACS Nano 3 (2009) 2581–2592.

[138] J. Li, J. Claude, L. Norena-Franco, S. Seok, Q. Wang, Electrical energy storage in ferroelectric polymer nanocomposites containing surface-functionalized $BaTiO_3$ nanoparticles, Chem. Mater. 20 (2008) 6304–6306.

[139] H.M. Jung, J.H. Kang, S.Y. Yang, J.C. Won, Y.S. Kim, Barium titanate nanoparticles with diblock copolymer shielding layers for high-energy density nanocomposites, Chem. Mater. 22 (2009) 450–456.

[140] N. Guo, S. DiBenedetto, D. Kwon, L. Wang, M. Russell, M. Lanagan, et al., Supported metallocene catalysis for in situ synthesis of high energy density metal oxide nanocomposites, J. Am. Chem. Soc. 129 (2007) 766–767.

[141] L.Y. Xie, X.Y. Huang, C. Wu, P.K. Jiang, Core-shell structured poly(methyl methacrylate)/$BaTiO_3$ nanocomposites prepared by in situ atom transfer radical polymerization: A route to high dielectric constant materials with the inherent low loss of the base polymer, J. Mater. Chem. 21 (2011) 5897–5906.

[142] K. Yang, X.Y. Huang, L.Y. Xie, C. Wu, P.K. Jiang, T. Tanaka, Core-shell structured polystyrene/$BaTiO_3$ hybrid nanodielectrics prepared by in situ RAFT polymerization: a route to high dielectric constant and low loss materials with weak frequency dependence, Macromol. Rapid Commun. 33 (2012) 1921–1926.
[143] K. Yang, X. Huang, Y. Huang, L. Xie, P.K. Jiang, Fluoro-polymer $BaTiO_3$ hybrid nanoparticles prepared via RAFT polymerization: toward ferroelectric polymer nanocomposites with high dielectric constant and low dielectric loss for energy storage application, Chem. Mater. 25 (2013) 2327–2338.
[144] P. Kim, S.C. Jones, P.J. Hotchkiss, J.N. Haddock, B. Kippelen, S.R. Marder, et al., Phosphonic acid modified barium titanate polymer nanocomposites with high permittivity and dielectric strength, Adv. Mater. 19 (2007) 1001–1005.
[145] K. Yu, Y. Niu, Y. Zhou, Y. Bai, H. Wang, C.J. Randall, et al., Nanocomposites of surface-modified $BaTiO_3$ nanoparticles filled ferroelectric polymer with enhanced energy density, J. Am. Ceram. Soc. 96 (2013) 2519–2524.
[146] X. Dou, X. Liu, Y. Zhang, H. Feng, J.–F. Chen, S. Du, Improved dielectric strength of barium titanate-polyvinylidene fluoride nanocomposite, Appl. Phys. Lett. 95 (2009), 132904-1-3.
[147] Z. Li, L.A. Fredin, P. Tewari, S.A. DiBenedetto, M.T. Lanagan, M.A. Ratner, et al., In situ catalytic encapsulation of core-shell nanoparticles having variable shell thickness: dielectric and energy storage properties of high-permittivity metal oxide nanocomposites, Chem. Mater. 22 (2010) 5154–5164.
[148] J. Fu, Y. Hou, M. Zheng, Q. Wei, M. Zhu, H. Yan, Interfaces Improving dielectric properties of PVDF composites by employing surface modified strong polarized $BaTiO_3$ particles derived by molten salt method, ACS Appl. Mater. 7 (2015) 24480–24491.
[149] S.K. Kumar, N. Jouault, B. Benicewicz, T. Neely, Nanocomposites with polymer grafted nanoparticles, Macromolecules 46 (2013) 3199–3214.
[150] S.A. Paniagua, Y.S. Kim, K. Henry, R. Kumar, J.W. Perry, S.R. Marder, Surface-initiated polymerization from barium titanate nanoparticles for hybrid dielectric capacitors, ACS Appl. Mater. Interfaces 6 (2014) 3477–3482.
[151] N. Durand, B. Boutevin, G. Silly, B. Ameduri, Grafting from polymerization of vinylidene fluoride (VDF) from silica to achieve original silica PVDF core shells, Macromolecules 44 (2011) 8487–8493.
[152] T. Furukawa, K. Ishida, E. Fukada, Piezoelectric properties in the composite systems of polymers and PZT ceramics, J. Appl. Phys. 50 (1979) 4904–4913.
[153] Y. Sun, Z. Zhang, C.P. Wong, Influence of interphase and moisture on the dielectric spectroscopy of epoxy/silica composites, Polymer 46 (2005) 2297–2305.
[154] T. Bhimasankaram, S.V. Suryanarama, G. Prasad, Piezoelectric polymer composite materials, Curr. Sci. 74 (1998) 967–976.
[155] Y. Rao, C.P. Wong, J. Qu, T. Marinis, A precise numerical prediction of effective dielectric constant for polymer-ceramic composite based on effective-medium theory, IEEE Trans. Comp. Packaging Technol. 23 (2000) 680–683.
[156] N. Jayasundere, B.V. Smith, Dielectric constant for binary piezoelectric 03 composites, J. Appl. Phys. 73 (1993) 2462–2467.
[157] T. Yamada, T. Ueda, T. Kitayama, Piezoelectricity of a high content lead zirconate titanate/polymer composite, J. Appl. Phys. 53 (1982) 4328–4333.
[158] W.E. Davies, J. Dutton, F.M. Harris, F.L. Jones, Electrical breakdown of air at high voltages, Nature 205 (1965) 1092–1093.

[159] D.W. Fu, H.L. Cai, Y. Liu, Q. Ye, W. Zhang, Y. Zhang, et al., Diisopropylammonium bromide is a high-temperature molecular ferroelectric crystal, Science 339 (2013) 425–428.
[160] A. Piecha, A. Gagor, R. Jakubasa, P. Szklarza, Room temperature ferroelectricity in diisopropylammonium bromide, Cryst. Eng. Commun.. 15 (2013) 940–944.
[161] Y. Sui, W.-T. Chen, J.-J. Ma, R.-H. Hua, D.-S. Liu, Enhanced dielectric and ferroelectric properties in PVDF composite flexible films through doping with disopropylammonium bromide, RSC Adv. 6 (2016) 7364–7369.

Polymer–carbon nanotubes composites obtained via radical polymerization in water-dispersed media

11

Dan Donescu, Mihai C. Corobea, Catalin I. Spataru and Marius Ghiurea
National Institute for Research & Development in Chemistry and Petrochemistry – ICECHIM Bucharest, Bucharest, Romania

Chapter Outline

11.1 Introduction

In 2007, Winey and Vaia published a concise analysis on polymeric nanocomposites [1]. This analysis began with a comparison of nanofillers type (spheres, rod, plate). The primary element for characterization and classification is thought to be the aspect ratio (the ratio between the long and short axis of the particle geometry). The aspect ratio strongly correlates to the ratio between the particle volume and the interfacial surface available for the polymer particle interaction. By reducing the short axis of the particles, both characteristics are drastically increased. Nanowires and nanotubes possess the highest aspect ratios for their geometrical features [2].

Analyzing the dynamics of publications on different types of nanofillers indicates the most interest is shown for carbon nanotubes (CNT) [1,2,3]. This particular subject has recently generated a large number of review papers that attempt to systemize the most important elements of polymer-CNT nanocomposites. The main direction of recent CNT research [4] is to develop more concrete mechanical and electrical properties, which makes them better polymer nanocomposites at low filler loadings.

Hybrid Polymer Composite Materials: Processing. DOI: http://dx.doi.org/10.1016/B978-0-08-100789-1.00011-3

To obtain a theoretically better volume ratio $V_{interface}/V_{particle}$ (to increase the available surface area for polymer interactions), the filler dispersion should be near 100% elementary particles in the polymer matrix [1,2]. This task is difficult because of the high aspect ratio and hydrophobic character of CNTs. Polymer matrixes are known for their hydrophobicity (depending on their structure), which is less than that of CNTs. For these reasons, there is a weak adhesion of the interfacial polymer to the CNT surface, which produces van der Waals forces that induce certain CNT states (ropes and bundles) [5]. Therefore, these forces should be equilibrated by those between the CNTs and the polymer matrix to obtain a good dispersion.

Like any physical and chemical phenomena, CNT dispersions in polymer matrixes are affected by the partner's nature and the processing conditions. The main features affecting the CNTs nature are the aspect ratio and surface modification via different functionalizations.

Ideally, CNTs with higher aspect ratio should produce associated composites with better mechanical and electrical properties.

There are authors who, after studying commercially available CNTs, have concluded that longer and cheaper CNTs are required. However, we should allow for even a more difficult dispersibility and higher probability of fracturing when processing longer CNTs.

CNTs decorated via the covalent or noncovalent attachment of several functional groups can increase the dispersibility and processability profile for obtaining nanocomposites. Covalent functionalization via opening $\pi-\pi$ bonds in a CNT promotes defects in the CNT's surface. When increasing the polymer matrix properties, a compromise must be made to achieve good compatibility with the polymer without decreasing the initially desired properties. An example of this problem was published by Ford et al. using grafted copolymers based on polystyrene and CNTs [6]. Even if the percolation concentration of the physical mixtures is very similar (grafted and ungrafted), the absolute conductivity of the final nanocomposites decreases. This phenomenon was attributed to an insulating barrier in the grafted CNTs even when in the percolated state.

With regard to the nature of the polymer matrix, there are some reviews that follow a comparative approach. These studies conclude that the chemical nature of the matrix directly influences the electrical and mechanical properties without accounting for the different manners (synthesis and processing) through which the materials were obtained. Just like CNTs, polymers can be modified to obtain favorable interactions. For example, the addition of pyrenyl groups can increase the interactions with neat CNTs [7]. All of the analyzed reviews mention a key question for improving the overall composite material properties: finding an appropriate experimental procedure to ideally disperse the CNTs (100% dispersed particles). A complex and systemic analysis of almost all experimental procedures was conducted by Grady [8]. He was able to identify five main methods: dispersion–reaction, dissolution–dispersion–precipitation, dispersion–dispersion–precipitation, melt mixing, and no fluid mixing. Intense dispersion via ultrasonication or mechanical means can damage the CNTs by forming fractures on their surface and throughout their entire body. This diminished CNT length, closely related to its strong self-aggregation and body flexibility, drastically decreases the number of available CNTs with the desired aspect ratio. Observing the dispersibility in organic fluids should be very useful. For in situ polymerizations, the initial monomer can spread across the CNT's surface [9]. However, a large number of papers dealing with CNT states in nanocomposites using sophisticated analysis methods can offer an answer to the famous question: "How nano are nanocomposites?" The answer is "not very." [10–14]. In the spirit of the above mentioned findings, a new question arises: "Why is a good dispersion of CNTs in nanocomposites so important?" The answer is "because we need an efficient concentration of CNTs to achieve the desired performance." There are quite a lot of papers indicating the negative effects of CNT agglomeration on the mechanical properties [12,14–19]. Therefore, methods that offer good dispersibility promote better mechanical properties. The mechanical properties are also controlled by the interfacial interactions between the phases. For this reason, a reinforcing effect was evidenced for hard polymers [20], but also for elastomeric polymers at temperatures above the Tg [21,22].

The nature of CNTs allows their associated polymer composites to offer good electrical properties compared to other filler composites [23]. The most important property for this effect is the percolation concentration. The conductivity abruptly increases (order of magnitude) at certain values of this concentration. Comparing the literature data for CNTs with a similar nature and polymer matrixes indicates a certain lack of reproducibility. These differences in the experimental results are attributed to the involvement of the CNT's aspect ratio in the development process and initial treatment. It is estimated that the percolation threshold is inversely proportional to the aspect ratio.

An important element for the electrical properties of nanocomposites based on CNTs and polymers is the contact of the filler between elementary particles. In this context, the ideal dispersion is no longer an absolute target for obtaining a low resistance to current.

Even if the actual results did not match those expected (from 20 years ago), several opportunities have been created to improve the properties for applications in different areas. These improvements are why these types of nanocomposites are already functioning in the automotive industry, the aerospace sector, for coating materials, antistatic applications, sensors, actuators, membrane production, and in the fibers industry, and others [24]. The problem of extending the areas of application is the price of CNTs, which must attain a similar price to other nanofillers already in use by the nanocomposite industry for a sustainable development.

11.2 CNT/polymer nanocomposites obtained from water dispersions

The ability of CNTs to disperse in low viscosity liquids has been previously reported. If these liquids are monomers of interests, it is believed that a good dispersion would also be achieved in the polymer phase after polymerization. In this context the evolution of ISI publications (Figs. 11.1 and 11.2) highlights the intensity of research in this direction on the last decade.

The highest interest involved by the in situ direction (Hirsch index 96 and more than 53,000 citations) was quite expected from technological reasons, but very few examples can be found concerning the unmodified CNTs by chemical methods. Moreover, the number of publications also includes the indirect process for the suspension and emulsion process like latex or dispersion blending route (Fig. 11.1). Therefore we believe that a technological breakthrough toward even a larger application spectrum will be possible when CNTs polymer composites will be produced even more easily (e.g., CNTs as catalyst substrate for the polymer synthesis), without additional functionalizing steps. This breakthrough will first influence the masterbatch market with several benefits also for the environmental and health issues concerns. The "safe by design" concept starts to become a more and more adopted approach, and this can be seen in the evolution of the wet process involved by the polymerization process in suspension, dispersion (Hirsch index 81 with over 35,000 citations), or emulsion (Hirsch index 34 and over 4100 citations). An interesting aspect of the state of the art is the fastest increase of the emulsion polymerization process with CNTs when compared with suspension polymerization process (Hirsch index 40 and over 7000 citations).

Miniemusion and microemulsion polymerization processes can also be considered (in a reductionist manner) such as particular cases of the emulsion polymerization process. But strictly from the polymer synthesis approach, there are particular cases with completely different mechanisms and phase organization controls. The

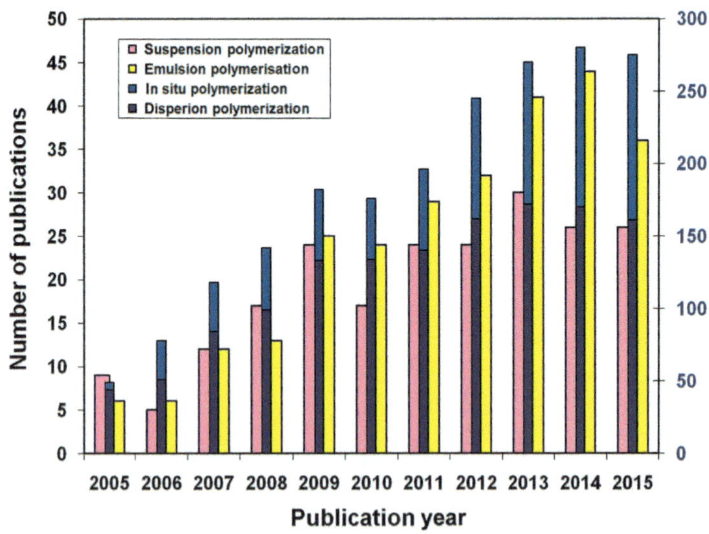

Figure 11.1 Publication evolution of ISI references, during the last decade, on the topic of CNTs and different polymerization routs for the matrix.

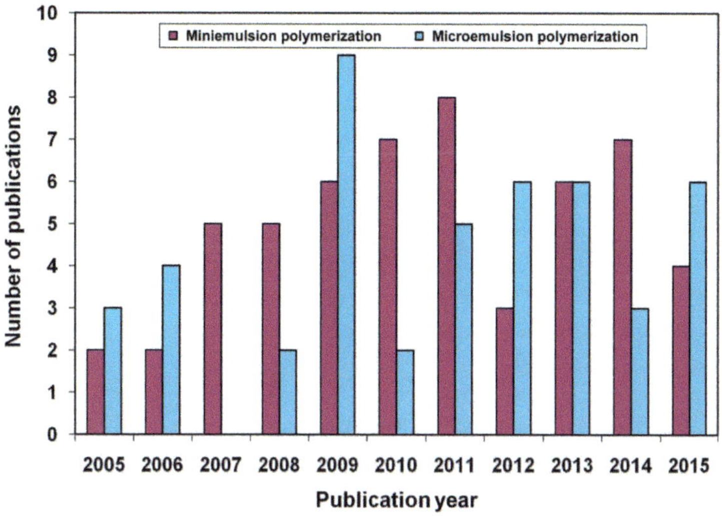

Figure 11.2 Evolution of the ISI publications (between 2005 and 2015) for CNTs polymer composites using miniemulsion and microemulsion polymerization.

huge reduction of the interfacial tension (on particles/media interface) involves an ideal approach for dispersing (nano)fillers in a monomer/polymer matrix and can assist the polymer growing on the filler surface like in an encapsulation process. The state of the art (Fig. 11.2) involving CNTs used in microemuslion polymerization highlights a Hirsch index of 21 with over 1700 citations, meanwhile for the CNTs and miniemulsion polymerization there's a Hirsch index of 18 with over 1300 citations. Both processes are still in the emerging area of the in situ polymerization but with a high potential for new research. The miniemulsion polymerization process involves less surfactant concentrations when compared with microemulsion by offering in the same time a low interfacial tension and organizing possibilities for free radical polymerization, able to be used in the CNTs decoration with polymer networks. The statistics of papers between 2005 and 2015 (Fig. 11.2) highlighted a number of publications 20% larger for the miniemulsion polymerization when compared with microemusion polymerization for obtaining CNTs polymer composites.

For the final material products, the nanocomposites should be in a state that is easy to isolate from the polymerization media. An ideal media for such a purpose is water. Water has poor interactions with both CNTs and monomers because of their polarity differences. At the same time, water is a nontoxic and environmentally friendly polymerization media with a low viscosity and adequate thermal conductivity for the exothermal reactions involved in polymerization. Varying the monomer/water ratio as well as the concentration and nature of the stabilizers allows for a large range of options for radical polymerization: suspension polymerization, emulsion, microemulsion or miniemulsion polymerization. Based on initial analyses, the most favorable polymerization conditions should be the first and last ones.

Suspension polymerization seems efficient because, the large final particles are several microns in diameter, which is close to the CNT length. Miniemulsion is also promising, because the hydrophobic agent could assist the stabilizer in increasing the polymer interactions with the CNTs [25]. Based on the morphology of the micron-sized particles obtained via suspension polymerization, two important elements can be observed: first, the mechanical stirring and sonication processes reported for the CNTs are much shorter than for the initial state (used in microemulsion polymerization); second, CNTs can be found both on the particle surface and inside the particles with some exceptions [17,26].

Grafting polymer on CNTs often can result in centers with different reactivity on CNTs body with involvement in both structure and morphology. For example polymethylmethacrylate (PMMA)–multiwalled carbon nanotube (MWCNT) composites can adopt shish-kebab morphology with the polymer phase attached to the nanofiller body. FTIR analysis shows a new peak (around 1630 cm^{-1}) specific to the grafted methyl methacrylate (MMA) on the CNTs [19,27,9].

A very important observation is the possibility of identifying these grafted CNTs via selective solubilization. Grafted CNTs can form stable solutions (with the polymer solvent) at low concentrations. At the same time, the grafted concentration can influence the molecular weight of the polymer (polymerization with benzoyl peroxide as the initiator). The bimodal distributions found in the GPC curves were attributed to oligomer formation during the transfer processes [28–30].

Emulsion polymerization was also used for polymer-CNT hybrids. This polymerization process is the simplest route for obtaining submicron polymer particles and consists of a monomer, surfactant, water, and initiator (usually potassium or ammonium persulfate). To obtain the CNT hybrids, the nanofiller can be added either directly to the water phase-in water and monomer phase or after the formation of the monomer-water dispersion-as a monomer, solvent, and water phase mixture [31–35].

In several cases, CNTs are pretreated with strong oxidizing agents such as HNO_3 or H_2O_2. The polar groups formed during this process provide a certain stability to the monomer-water dispersions even without a surfactant. Some interesting results were previously published that showed initial emulsions are more stable and have finer particles when untreated CNTs are dispersed in the monomer, while modified CNTs are preferred for dispersion in the water phase. The data mentioned above indicating treating CNTs to increase their hydrophilic character can be accomplished using oxidizing agents. During the emulsion polymerization process, the persulfate initiators are also oxidized. These parallel reactions prompted the authors to investigate whether the ammonium persulfate (APS) was able to modify the hydrophilic character of the CNTs [36].

Our data involving miniemulsion polymerization of different monomers without preliminary chemical modification of CNTs have proven that treating MWCNT with APS at 65°C under ultrasonication (2 h) produces a stable dispersion in water. In contrast to neat CNT dispersions, the water dispersions after APS treatment are stable for over 5 months. Further data will determine the chemical structure of the grafted groups on the MWCNT (1–2 wt%) that are providing this stability [28,29].

Sonication was used to better disperse the initial mixtures. Because of the radicals generated during ultrasonication, emulsion polymerization can be performed without supplementary radical initiators [37,38]. During the polymerization, grafted copolymers form on the CNT and were identified by IR specific peaks [39–43].

Electron microscopy images demonstrated that the CNTs are wrapped by the polymer nanoparticles [44–47].

The thermal stability of the nanocomposites generally increased upon inclusion of the CNTs. However, in some cases, the thermal stability decreased, which was explained by the CNT content increasing the thermal conductivity and thereby inducing a faster heating rate in the polymer chain [48–51].

The electrical and mechanical properties of the CNT-polymer composites obtained via the discussed processes increased because of the presence of the CNTs. In particular cases (large number of CNTs), the mechanical properties can be decreased as a consequence of aggregate formation (stress concentrators) [52–55].

The monomers studied up to now for polymer-CNT hybrids formed via radical emulsion polymerization were styrene (St), butyl acrylate (BuA), methyl methacrylate, acrylonitrile (AcN), and aniline [28,29,33,37,42].

Particular interest was shown in the so-called "latex technology" involving polymer-CNT nanocomposites obtained from water dispersion media. In this method, polymer latexes are mixed with small concentrations of CNTs in a water-surfactant media to restrict agglomerate formation. This route uniformly deposited the CNTs filler onto the latex particle surfaces [41,45,50,51,56,57,58].

Film formation of polymer materials is a complex phenomenon. A compact film can be obtained using the above mentioned technique after water evaporation. This elimination involves three main events: concentration, particle deformation, and the coalescence of the polymer particles [25,59]. The presence of CNTs at the polymer–water interface can influence all three stages. Film-forming materials possess an important characteristic: a minimum temperature of film formation (TMFF). This temperature is strongly dependent on the glass transition temperature (Tg) of the polymer matrix. Film formation occurs under the Tg through the three above mentioned stages [25]. Several polymer-CNT latexes were studied using latex technology processes based on different polymers: polystyrene (PSt), PMMA, poly(styrene-co-butylacrylate), poly(methylmethacrylate-co-butylacrylate), natural rubber, and carboxilated butadiene styrene copolymers [41,45,50,51,58].

A supplementary stage at elevated temperatures was used to create compact films (and promote coalescence) of thermoplastics with high Tg, the so-called annealing process. Reducing the TMFF was accomplished for some systems by using surfactants with a plasticizer effect capable of modifying the polymer matrix [60]. These are also called penetrating emulsifiers.

In general, the interactions between the polymer latex particles and CNTs from the diluted surfactant solutions are driven by van der Waals forces. By modeling the electrical charges of the latex particle surfaces and CNT-stabilizer complexes, hybrids can be obtained where the interactions between the matrix and nanofillers are electrostatic in nature [61–63]. The overall interactions are strongly dependent on the matrix polarity [28,29]. It was proven through a comparative study on PMMA and PSt that PSt assures a superior debundling of the CNT agglomerates due to its more pronounced hydrophobic character. Another key element observed in the same study was that the surfactant can act as mediator on the interface between the polymer particles and CNTs. Jin et al. used four types of surfactants: cetyl trimethyl ammonium bromide (CTAB), sodium dodecyl benzene sulfonate (DBS), sodium dodecyl sulfate (SDS), and Triton X100 [40]. CNT stabilization in water depended on the degree of coverage of the polymer particles for both PSt and PMMA and occurred in the following hierarchy: CTAB > DBS > SDS > Triton X100. The mentioned studies used a large variety of stabilizers for the polymer-CNT hybrids: anionic, SDS, DBS, poly(sodium-4styrene sulfonate) (PSS); cationic, CTAB, poly(diallyldimethyl ammonium chloride) (PDDA); nonionic, Triton X100, poly(vinylpyrrolidone) (PVP).

Another important stage of latex technology is the means of dispersing the CNT and surfactant solution into the water. The general approach consists of use of ultrasound. As shown above, this manner has a severe risk, the shortening of the CNTs. Only an optimal amount of energy should be delivered during ultrasonication to avoid these side effects and preserve the CNTs' aspect ratio. Following modification in the UV–VIS spectra, it was proven that the required energy decreased upon increasing the surfactant concentration (SDS) [45]. However, a drastic increase in the surfactant concentration should be avoided to prevent decreasing the electrical conductivity.

Islam et al. proved that DBS (which posses a high capacity for stabilizing CNTs) promotes a reduced shortening of CNTs relative to SDS and Triton X100 for identical ultrasonication processes [64]. The same study showed a greater damaging effect for high power tip sonicators relative to bath sonicators.

Grossiord et al. [65,66] determined the critical energy required for destroying aggregates and debundling single wall carbon nanotubes (SWCNT) for use in nanofiller technology. SWCNTs possess weaker van der Waals attractions between the CNT and require less energy to achieve the maximum debundling [67]. It was shown that this energy increases rapidly with the concentration of the CNTs.

These studies on water dispersions of CNTs are noteworthy for the latex technology approach for understanding both the shortening side effects and surfactant interactions at the interface as well as their involvement in managing the initial aspect ratio of the CNTs.

The ultrasonication process can even affect CNTs confined on the polymer particle surfaces. For example, MWCNTs purified with HNO_3 and PSt particles modified sequentially with different PDDA/PSS/PDDA charges were affected by ultrasonication [68]. The sonication treatment can detach the CNTs from the latex particle surface.

Morphological studies revealed the influence of the processing conditions and partner nature. The following convergent elements should be mentioned: oxidatively modified CNTs are easier to disperse; CNT nanofillers are flexible structures capable of bending into curved shapes; inorganics (catalysts, inorganic oxides) induce CNT associations. Latex particles are capable of assembling CNTs on the surface and, the associations made using specific techniques can possess superhydrophobic properties. A relevant study on the morphology of such systems based on colloidal crystal films was published by Jurewicz et al. [55]. Using ionic surfactants, such as SDS and CTAB, produced the more disordered and less perfect CNT networks. If a penetrating surfactant such as Triton X100 is used, the final morphology is honeycomb like. Dalmas et al. studied how the preparation methods related to their final morphology [69]. The mixtures obtained from a latex St-BuA (Tg = 266°K) and MWCNTs were prepared via two methods: drying at 35°C (E) or freeze drying and then pressing at 100°C (FB). The first method (E) assumes Brownian motion, which favors the reaggregation of CNTs after latex coalescence and results in the entanglement of the nanofillers particles. The second method (FB) assumes water sublimation occurs and the CNTs are kept in the initial dispersed state. During hot pressing, the nanofiller rearrangements are limbed because of the polymer viscosity. These diverse routes imply different morphology and affect both the mechanical and electrical properties. The effects produced by CNTs in the associated hybrids were mainly observable above Tg.

The presence of CNTs throughout the physical mixture increased the conductivity relative to the neat matrix much less than theoretically expected. The conductivity of the CNT-polymer nanocomposites can be maintained or increased [40] if repulsive electrostatic forces are suppressed and contact between the dispersed CNTs is available. Dalmas et al. emphasized the critical importance of charge transport through the CNT network [69]. This is the reason why nanocomposites obtained by method E have higher conductivities and lower percolation thresholds than those obtained by method FB. At the same time, the use of DBS as the surfactant (strong interactions with CNTs) reduces the conductivity of MWCNTs, even in binary mixtures.

The conductive properties of the mentioned hybrids can promote the electrorheological properties under particular conditions (mixed with silicon oil) [70].

Another method for obtaining polymer-CNT nanocomposites is radical polymerization in miniemulsions [71]. In contrast to conventional emulsion polymerization, the initial polymerization system contains a costabilizer, which is an ultrahydrophobic material soluble in the monomer phase and insoluble in water phase. Homogenization of the entire initial mixture is achieved through mechanical stirring and sonication. Before polymerization, the monomer-water dispersions are formed as nanometric droplets. The superhydrophobic component assures the dispersion stability by suppressing the Ostwald ripening, which prevents monomer diffusion through the water phase. Because of the droplets' hydrophobic character (monomer and stabilizer), the interactions beteen the dispersed CNTs is controlled by the competition between the van der Waals forces and surfactant affinity for both entities. This heterogeneous system or miniemulsion has the following primary characteristic: the existence of monomer nanodroplets that are efficiently stabilized and serve directly for the radical polymerization. That is why they are called nanoreactors. For obtaining polymer-CNT hybrids, the variety and versatility of the partners is quite large. Several monomers have already been used: styrene, copolymers of styrene with isoprene, with divinyl benzene, acrylonitrile, butyl acrylate, methyl methacrylate, and butyl methacrylate [28,29,72–74].

The surfactants used up to now have belonged to the three main classes: aninonics, cationics and nonionics. The ultrahydrophobic agents used for stabilization were: hexadecane, pentanol, PSt-$AlCl_3$, complex polyurethane, and polystyrene [28,29,75–77].

A morphologic study of the nanocomposites obtained via miniemulsion free radical polymerization, highlighted important differences compared with the classical emulsion polymerization. Nanocomposites obtained via miniemulsion polymerization possessed both a polymer phase generated on the CNT surface and polymer particles attached to the CNTs. Formation mechanism schemes were developed to explain these unique morphologies, which were driven by the interactions of the CNTs with the nanodroplets containing the monomer and hydrophobe [28,29,77,78]. After contact between the water phase (CNTs and surfactant) and the organic phase, a preferential adsorption of the nanodroplets to the nanofiller occurs. As a result, the surfactant is transferred, and the CNT surface can interact with the monomer molecules through either the aromatic double bonds in styrene ($\pi-\pi$ interaction) or the hydrophobic interactions between nonpolar monomers. If these interactions are not strong enough, the CNTs can aggregate again at the end of the polymerization process and the matrix repels the nanofillers.

These general phenomena were observed during the miniemulsion polymerization of butyl methacrylate (BuMA) and styrene in presence of carbon black (CB) nanoparticles [78]. CB was used because of its reduced aspect ratio and contained a polymer phase on its surface. If an excess of the monomer was used, the CB particles should be encapsulated by the polymer phase. In contrast to CNTs, integrated particles (like the Kebab structure) are not possible with CB. BuMA has a higher encapsulation capacity for CB due to degree of the initial monomer adsorption relative to styrene.

An example of selective interactions in polymer-CNT hybrids was published by Lee et al. [79] using a PMMA melt blending (200–230°C) with a styrene-acrylonitrile copolymer (SAN) and MWCNT. The final mixtures indicated an affinity of the CNTs for the SAN phase due to the preferential interactions between the phenyl groups of the styrene units and the graphenic structure of the MWCNT.

Miniemulsion polymerization creates an opportunity for both chemical bonding and physical attraction between partners (simultaneously), which is quite a feature for a nanoscale process. These chemical bonds promote the formation of grafting copolymers on the CNTs through radical reaction obtained upon initiator decomposition and macroradical attacks on the CNT's wall defects. These grafted copolymers assure the stability of the nanocomposite in polymer solvents. The reaction of the CNTs with the primary radicals formed during either the initiator decomposition (AIBN) or the growing macroradicals is specific to mass polymerization or solution polymerization. The existence of such transfer reactions for nanofillers effects the growth of the polydispersity index (PDI = Mw/Mn) in both miniemulsion polymerization and mass polymerization.

The effect of CNTs on polymerization kinetics of BuA was investigated in a complex study. Capek and Kocsisova [77] studied the process using all three types of surfactant (anionic, cationic, and nonionic). With one exception (cetylpyridiniumbromide), the maximal polymerization rate increased in the presence of the CNTs. This kinetic phenomenon is explained by the strong interactions between the CNTs and the π electrons of the pyridinium group. This monomer/polymer particle complex reacts with the formed radicals and decreases the radical concentration available to the nanodroplets.

A comparative analysis of the superficial tension and electrical conductivity between the hybrid latexes and those without CNTs reveals that the surfactant interactions compete between the nanofillers and monomer droplets, which is in good agreement with the schemes provided by other authors [72,73,76,77,79]. The nanocomposites obtained via miniemulsion polymerization were analyzed in terms of their mechanical and electrical properties, which agreed well with the previously analyzed data. The mechanical properties showed some improvements related to the degree of filler dispersion. If the CNT concentration was increased without stabilization, aggregation quickly occurs and the property response is negative. CNT agglomerates act as stress centers that initiate the material failure. The reinforcing effect of CNTs was only observed by Dynamic Mechanical Analysis (DMA) in the high elasticity state (above Tg). The electrical properties increased proportionally with the CNT concentration. However, this increase was below that expected (theoretical) due to the reduced aspect ratio and polymer grafting, both of which reduce the electrical contact probability.

11.3 New results involving CNT nanocomposites obtained by miniemulsion polymerization

The latest results allow an extended evaluation of the monomer polarity using the same miniemulsion polymerization formulation over CNTs concentrations (0–2 wt%). The data indicate that specific monomer behaviors offer selective interactions between the CNTs (St, AcN, AcN-9-vinylcarbazole, St-AcN) in contrast to less competitive ones, such as MMA [28]. All of these factors can be reflected by the grafting capability and general kinetic polymerization profile, which was proven to be the direct consequence of the monomer polarity.

When using a classical approach from AcN industrial processing (using both a monomer concentration above the water solubility level and redox initiating system), a drastic expulsion of the CNTs from the polymer phase occurs even during the first stages of the polymerization process (Fig. 11.1). This phenomenon should be considered as a homogenous dispersion of the CNT within the initial monomer, which can be obtained without sonication with stability profiles exceeding 10 months at −4°C (Fig. 11.3).

Moreover, the so-called compatibility via favorable chain molecule interactions is described in literature. In fact, this process was thought to promote filler dispersions similar to suspension or dispersion polymerization, and thus encapsulate the CNTs in large polyacrylonitrile (PAN) particles, based on the affinity between the monomer and CNTs. Another element intended to promote the anchoring of the polymer chains to the CNT wall was the use of a free radical initiator to attack

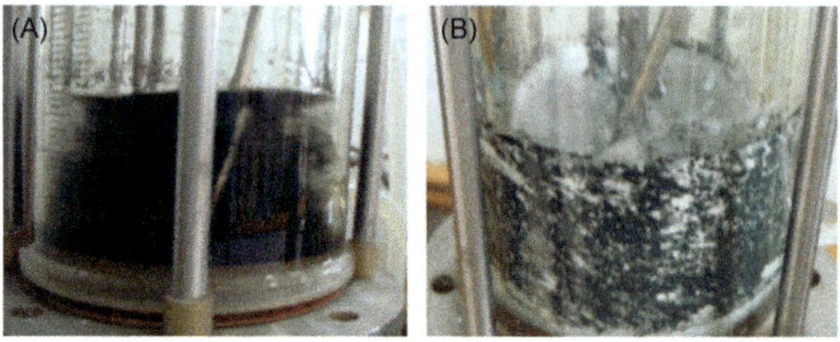

Figure 11.3 AcN polymerization in water at the start of the polymerization (A) and after a 3 h reaction time (B) (wt% = 84.65 DW, 14.94 AcN, 0.03 MWCNT, 0.15 sodium methabisulfite, 0.15 APS, 0.08 H_2SO_4).

Figure 11.4 Initial dispersion (before polymerization) obtained via the miniemulsion method (left), and the final hybrid polymer latex with the MWCNT (right) (from AcN, MWCNT, DBS, hexadecane, AIBN and water).

CNT wall defects in the filler phase. The final results indicated an increased tendency for monomers desorption from the nanotubes into the water followed by rapid precipitation and then crystallization of the polymer formed. The expulsion of the filler begins 1–2 min after adding the initiator (even when a staged methodology is used); moreover, the high exothermic AcN polymerizations begins at 46°C, which favors the expulsion of the filler from the polymer phase. These processes occur in the same manner for St, SAN, and MMA with the exception of the isotherm recovery.

Using a different approach, namely miniemulsion polymerization, allows the system compartments to be better controlled by the surfactant and costabilizer and represents a viable solution to CNT functionalization via the direct covalent attachment of the polymer phase to the side walls. This method produced stable latexes with up to 2% CNTs relative to the initial monomer (Fig. 11.4) [28].

CNT-polymer hybrids obtained by miniemulsion polymerization are still being pioneered but have the potential to be homogenous materials obtained via in situ methods (for generating the polymer) other than solution polymerization. These methods rely on different polymer phase confinements with the benefits of CNT debundling, polymer grafting without functionalization, and decorating latex particles on the CNT structures. Additionally, unmodified CNTs are very hydrophobic particles that tend to aggregate spontaneously, which is a phenomena increased by the high aspect ratio. In essence, a highly hydrophobic partner (like CNTs) could be a promising candidate for the costabilizer (hydrophobe) in a miniemulsion polymerization with a good potential for reducing the Ostwald ripening process. However, their solid nature, dimensions, and bendability are serious drawbacks despite their impressively high aspect ratio during both monomer droplet formation and nucleation. The use of a surfactant can be more or less subtle based on the following: the ionic surfactants (i.e., CTAB) can stabilize both the hydrophobic interactions and ionic charge compensation, while the surfactants offering $\pi-\pi$ interactions promote the stacking of the hydrophobic tail on the CNT surface. These surfactant-CNT interactions are widely discussed in the present literature; however, one thing is certain: the longer the tail and higher the unsaturated carbon-carbon bond density. In this context the increase in stabilization involves the agglomerate size decreases. The subtlety comes when the surface of the CNT is modified by the surfactant interaction, which results in a modulated behavior towards the other partners in the miniemulsion polymerization (monomer, radicals, hydrophobe, and so on). The CNTs' interaction with the polymers can primarily act via hydrophobic interactions and from the perspective of free radical polymerization (the mostly commonly used): The most active CNT sites can be found in their end caps rather than within the wall segments, which have the lowest reactivity. For MWCNTs, the wall defects generated during both the synthetic process and subsequent processing events (ultrasonication) can drastically increase the potential for available chemical radicals that are more prone to attack. For the general case of CNTs with less active sites (wall segments), the macroradicals are easily deactivated in this region and there is an inhibitive character for the nanotubes on the free radical polymerization.

A miniemulsion is an emulsion with particle sizes ranging between 50 and 500 nm [80]. Miniemulsions are more stable than conventional emulsions that have lower interfacial tension. Polymerization processes conducted in miniemulsions offer the possibility of obtaining the number of latex particles as there are initial monomer droplets. This phenomenon was demonstrated through several kinetic profiles and characterizations [80].

The main feature of miniemulsions versus macroemulsions is the drastic reduction of their Ostwald transfer (transition of mass from small particles to larger ones); however, Ostwald migration is not completely extinguished. The Ostwald phenomenon is reduced because of system compartmentation and by the monodispersity of the initial monomer droplets. CNT miniemulsion polymerizations are a very particular case known in the literature to differ from classical miniemulsions. The presence of the CNTs perturb the monodispersity of the initial particles before polymerization, and their ability to remain confined within or on the monomer droplets is almost entirely restricted by their geometric characteristics (tens of nanometers in width and tens of microns in length) [81,28,29]. However, this aspect does not exclude the coexistence of micro-scale and miniemulsion forms (Fig. 11.5).

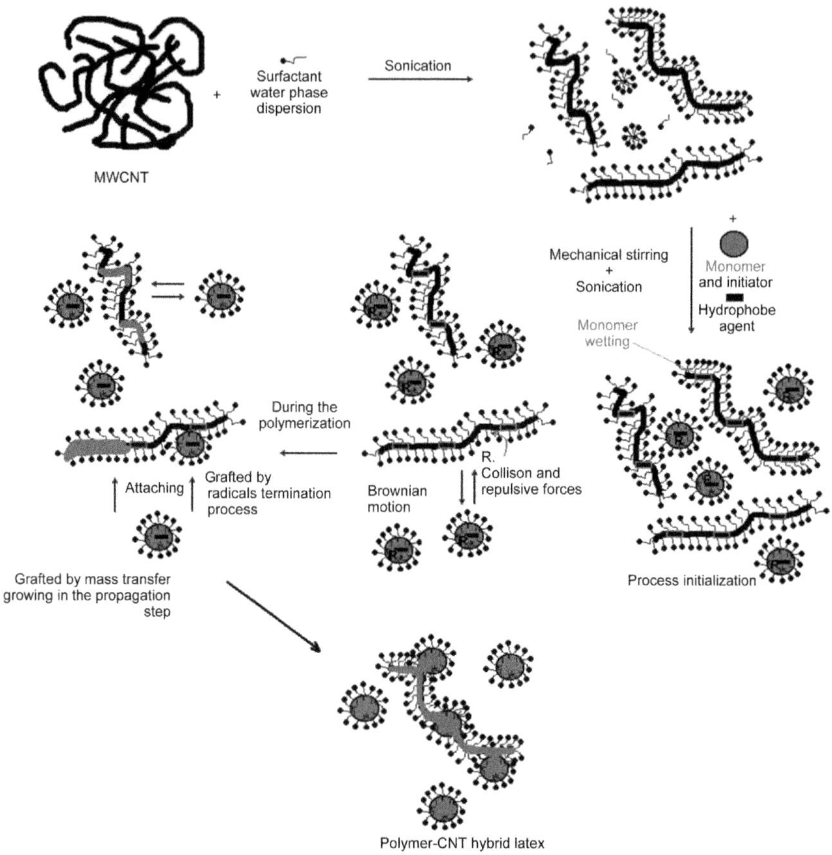

Figure 11.5 Idealized formation mechanism for Polymer-CNT latex particles by miniemulsion polymerization approach by likely events described in literature.

The coexistence of CNTs next to miniemulsion dispersions implies several things regarding their hydrophobic character and specific surface, which can act in several ways (interactions with the monomer, surfactant, hydrophobe, initiator, oligo, and macroradicals). In conventional miniemulsions, the stabilization is managed by the synergetic effect of the surfactant and costabilizer (the hydrophobe). The fine dispersion of the initial particles (monomer droplets), low interfacial tension and large interface allows for radical transfer.

To depict this formation mechanism, we should consider a particular case of miniemulsion polymerization during which the monomer can be adsorbed onto the CNT based on the strong hydrophobic interactions and surfactant adsorption (Fig. 11.5). In an extensive study, Tiarks et al. [78] proved the capability existed for even polar monomers (butyl acrylate) to wet and be adsorbed onto a carbon black surface during a miniemulsion polymerization processes [78]. Moreover, the hydrophobe favors the monomer carbon particle equilibrium and prevents the formation of rigid aggregates. The local desorption of the surfactant from the CNTs provides anchoring sites for both the nanodroplets and final latex particles. The CNT debundling effect in the final structures is an indirect consequence of this equilibrium and the favorable formation of polymer on the nanotubes. The proposed mechanistic scheme is based on the below data.

All thermoplastics possess the same kinetic profile when obtained via miniemulsion radical polymerization in the presence of MWCNTs. The CNTs appear to be debundled with polymer particles decorated upon them with grafted shells on the tube surface. Recent studies have also shown a discontinuity in the surfactant layer on the CNT surface (when using SDS) [72].

CNTs consume a part of both the surfactant and the hydrophobe agent, and grafting occurs even if the oil phase is prepared separately (Fig. 11.5).

The highest amount of grafting (per unit) occurs at low MWCNTs concentrations when the nanotubes are well dispersed and the grafting access is more favorable. In this context, a certain scavenger effect for the CNTs should be considered. When the CNT concentration increases (Fig. 11.6), the ability for grafting decreases on a weight unit basis because of the formation of aggregates, which reduces the interfacial surface. Large amounts of the grafting phase quickly decrease the final particle sizes; however, a slight increasing tendency occurs as the grafting process becomes restricted by aggregate formation. Another peculiarity induced by CNTs in miniemulsion polymerizations, which "consume" a part of the monomer (by grafting), is the aggregate formation which reduces the access to the CNT and restricts transfer. Another important clue for the overall mechanism can be found in the kinetic data provided by Capek et al. [77] for the first time for such a process, which indicates (when using CTAB as the surfactant) an identical decrease in particle sizes in the presence of CNTs (also with increasing particle numbers, which is another deviation from conventional miniemulsion polymerization). Because CNTs introduce a "supplementary" interface to classical miniemulsion polymerizations, assisting this

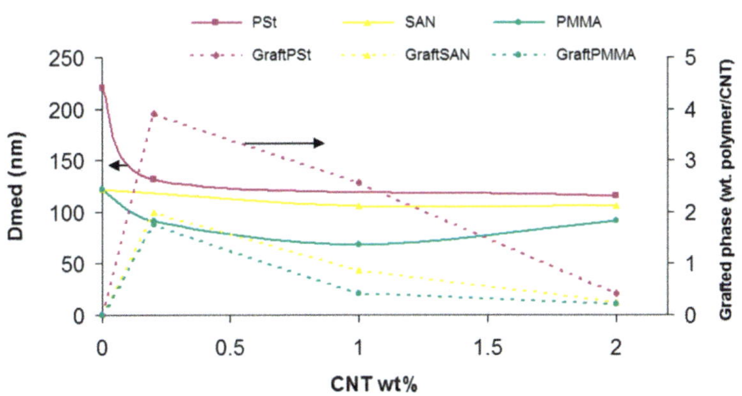

Figure 11.6 Influence of MWCNT's concentration on final particles sizes and grafted polymer (using a constant concentration of monomer, initiator BPO and stabilizers CTAB, hexadecane).

interface using the surfactant relationship can drastically influence the polymerization kinetics. The maximal polymerization rate increases in the presence of CNTs in same order as for interaction with the surfactant: Triton X-405 (Tr), Tween 60 (Tw), dioctyl sodium sulfosuccinate (AOT), Dowfax 2A1 (DW), cetyl trimethylammonium bromide (CTAB). All data indicated in the literature suggested cationic emulsifiers as the first choice in order to create the most active minimeulsion polymerization in presence of CNTs.

In the classical polymer nanoparticle (PNP) approach, the (monomer) droplet nucleation assumed the entry of the radicals within the particles followed by a sequential growing process and final polymer particle formation. In fact, the polymerization has a pseudo-bulk kinetic for the formation of the radicals in the monomer phase (monomer/polymer nanoparticles—MPNP approach). CNTs offer several directions for influencing the overall formation mechanism of the miniemulsion radical polymerization. One direction is through the interactions of the initiator, such as AIBN, with the π-conjugated system, which can increase the decomposition and radical concentration of the polymerization system. In contrast, the cage effect for particles (monomer/polymer with initiator) without CNTs can decrease the primary radical's concentration. The absorption of radicals by CNTs disfavors biradical termination inside free particles. The second direction is represented by the capability of the active sites of the CNTs to influence the total radical concentration. In both cases, the available data indicate an increase in the particle number and average number of radicals per particle (which is abnormal according to the classical miniemulsion polymerization theory) [28,29,80].

Another element in the formation mechanism was discussed by Capek et al. [77] who observed the monomer preferential partitioned between the interior of the emulsifier/CNT associates and the emulsifier molecules from the droplet surface during the homogenization of the reaction mixture (and also from the kinetic profiles).

The monomer polarity is another key factor for both mechanism and CNT involvement in the overall polymerization process [28,29,80]. Less polar monomers like styrene, offer good hydrophobic interaction besides the $\pi-\pi$ staking on nanotubes. The reduced access to the active sites in CNTs and the accessibility of the CNT to the monomer is restricted by aggregation effects and influences the final conversion in a similar manner to the final particles size and grafted amount per wt. unit (Figs. 11.6 and 11.7). These phenomena are unique for PAN and poly(acrylonitrile-co-9-vinylcarbazole) copolymer (PAN-VK) with, PMMA, SAN, and PSt. Moreover, the high polydispersity index in the presence of CNTs is not greatly affected by the monomer polarity [28,29].

Polar monomers are affected by the transfer to the MWCNTs, but as the MWCNT aggregates are formed, the access to the MWCNTs is much lower for the polar monomers than for the nonpolar ones (i.e., styrene), and the conversions are increasing. Meanwhile, more hydrophobic monomers still have access to the MWCNTs, even at high concentrations of the nanotubes [28,29].

Through thermal treatment in the oven (180°C) for 3 h, CNTs can be annealed (Figs. 11.8 and 11.9). After the annealing process, the CNTs appear even more

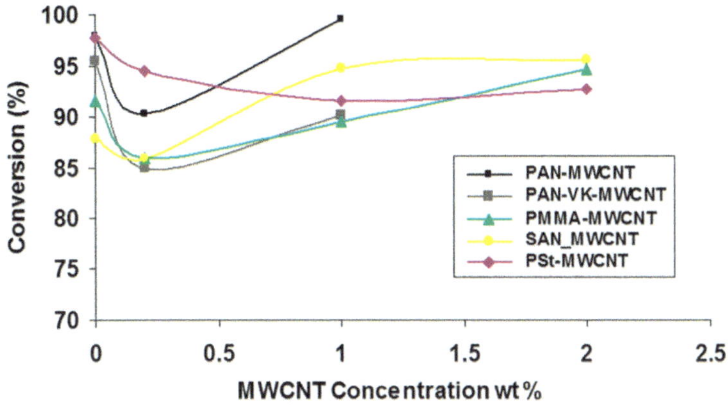

Figure 11.7 Concurrence between the monomer/CNTs and the stabilizer (reflected by monomer conversion).

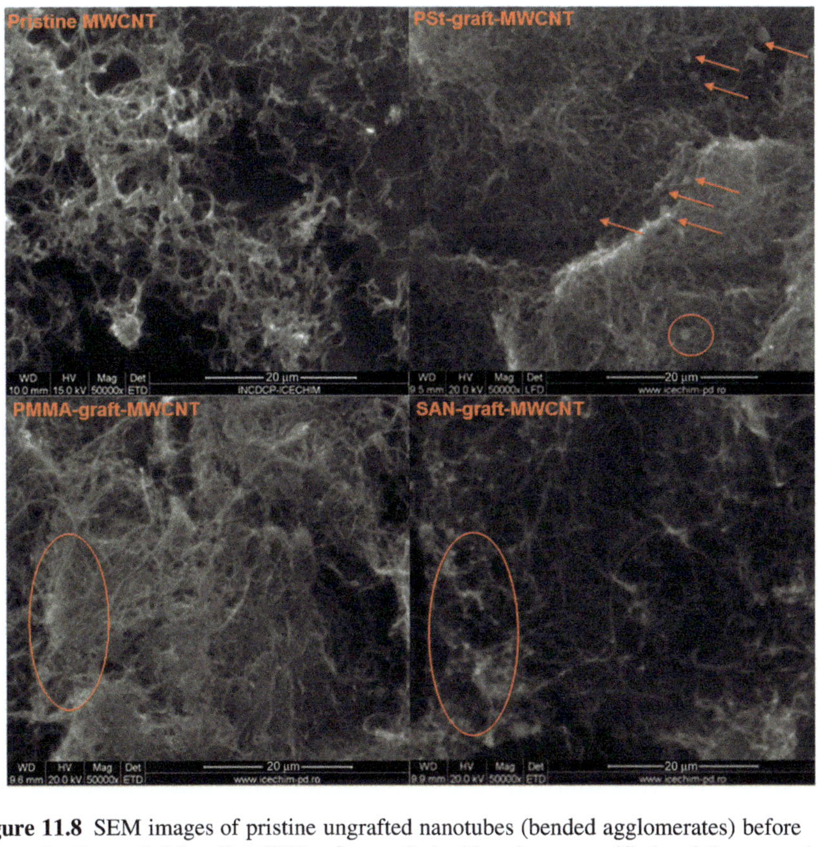

Figure 11.8 SEM images of pristine ungrafted nanotubes (bended agglomerates) before polymerization and debundled CNTs after grafted with polymer, purified and then annealed at 180°C. Grafting by miniemulsion polymerization (2 wt% MWCNT).

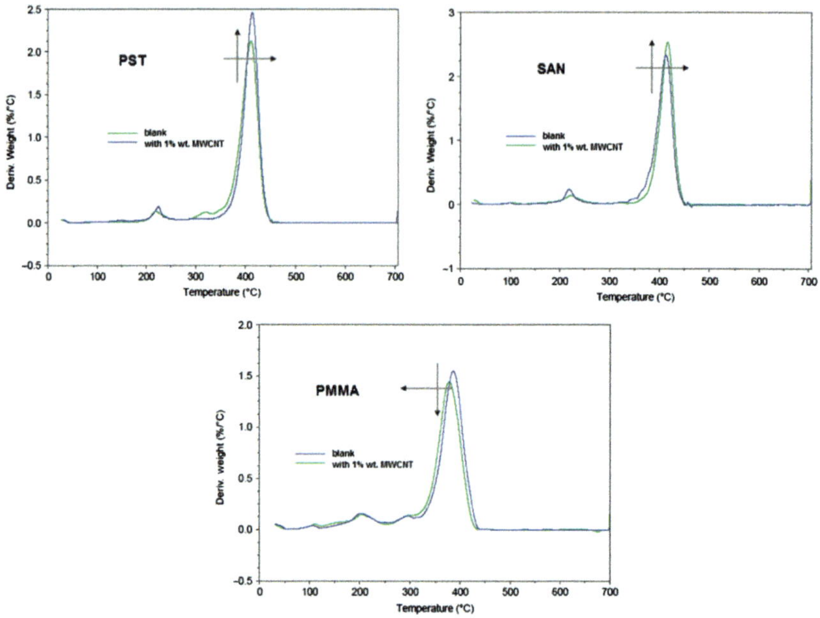

Figure 11.9 Polymer-CNTs nanocomposites obtained thermal stability compared with neat polymers when obtained by miniemulsion polymerization.

debundled, indicating (in thermoplastics above the Tg value) a similar state observed in rubbery state (for elastomers). The CNTs are stable in N-methyl-2-pyrrolidone solutions after polymerization, indicating the grafted polymer on CNTs particles [28,29].

Recent results showed the competitive contribution in the thermal stability (Fig. 11.9) of two phenomena: (1) the interaction of the CNTs with the polymer phase (PSt $\pi-\pi$ stacking, ability to be grafted during the polymerization process) and (2) the increase in thermal conductivity brought by the presence of CNTs in the final material (the heat reaches the polymer chains faster and promotes the decomposition of the material) [28,29]. When PMMA is grafted onto the CNTs, the amounts of the grafted phase as well as the available interactions are minimal, which is why, in the presence of CNTs, the thermal stability decreases. Meanwhile, when PSt and SAN are grafted onto the CNTs, the interaction with CNTs and the amount of grafted phase can increase the thermal conductivity of the CNTs, indicating a major contribution in the overall material of the confined phase on surface of the CNTs [29].

In terms of recent development with influence on applications side, miniemulsion allows compared with conventional emulsion polymerization method a drastically decrease of the polymer composite sheet resistance at the same CNT loads. Moreover, starting from a PSt matrix the sheet resistance can be decreased by more than five folds compared with the neat matrix; experimental results highlighted materials with over 200% increase in the dynamic storage modulus [73]. Other applications with impact on the industrial sector indicated MWCNT as filler and polypyrrole (PPy), polyaniline (PANI) or polythiophene as matrix obtained by miniemulsion, inverse microemulsion, and microemulsion approach for supercapacitor obtaining [82]. PMMA or PSt as matrix with MWCNT were obtained by RAFT mediated emulsion and W/O emulsion respectively for separation and purification applications with high selectivity for organic solvents (superhydrophobic membranes for oil/water separation) with over 99.9% efficiency and high flux capacity (5000 L m^{-2} h^{-1} bar^{-1}) [82,83]. The use of hybrid fillers CNT/titania or Si allowed important development in Li ion battery, photocatalysis, or sensor applications using PPy or PANI as matrix and emulsion and microemulsion route combined in particular cases with vapor deposition [82,84–86].

The versatility of the polymer-CNTs obtained by minemulsion polymerization was used for MWCNT modification with PANI and then used for composites with poly(ethylene terephthalate) (PET). Ultrathin films of PANI were achieved on MWCNT with no loss of morphological profile and the electric conductivity [87]. The further MWCNT modified with PANI were used for in situ polymerization technique for PET/MWCNTs the antistatic properties were achieved at only 1 wt% PANI modified MWCNT. The potential of the method for PET matrix can be extended also to other polymers like polyamide (PA) and PMMA. The applications can be driven in different areas like: antistatic, capacitor, conveyer belts, or electromagnetic shielding. In the case of PA66, PANI presence on MWCNTs obtained by miniemulsion polymerization improves both conductivity (over eight folds) and the dispersability in the final PA66 matrix [88]. Miniemulsion polymerization allows the obtaining of hybrid conductive membranes for purification and separation process based on PAN and MWCNT or PSt, SAN, PMMA with MWCNT hybrid foams for electrostatic dissipation or electromagnetic shielding [28,29]. At the same time, the molecular weight of the in situ synthesized polymers, and the mechanical and thermal properties are higher in the presence of CNTs.

Even if the CNT-polymer composites and nanocomposites are still in an emerging spot on some application the potential is very high given the fact some industrial applications already exist [24]. For automotive and aerospace industry but also for separation, microelectronic devices, several solutions are available for inks, conductive coatings, optical coatings, coating deposition for antistatic, displays, electromagnetic shielding, thermal sensing and imaging, antiicing solutions, membranes, lightning strike protection, anticorrosion coatings, chemical sensing and diagnostics. Companies from almost all continents are involved in end products or are selling technologies in this sector: Adidas, ANS, Axson, Baltic, BASF, Borsig, Canatu, Eagle Windpower, Eikos, Emerson & Cuming Microwave Products Inc., FutureCarbon, General Electric, General Nano, GMT Membrantechnik, Hexcel, Hyperion Catalysis, Iljin Nanotech, Meijo-nano carbon, Nanocyl S.A., NanoIntegris, Nantero, NEC Corp., Nokia, Panasonic Boston Labs, Plasan Ltd., Porifera, Q-flo, Renegade, Samsung, Seldon, Takiron Co., Teco Nanotech Co. Ltd., Tesla NanoCoatings Inc., Top Nanosys, Toray, Unidym Inc, etc. [24].

11.4 Future perspectives

The miniemulsion polymerization approach to obtain CNT-composites has a great potential due to the use of a wet media (CNT particles can create health and environmental issues) by: the absence of solvents, the possibility of grafting without supplementary technological steps (acid treatment, energy consumptions), the ability to design new materials with nanoscale structure, the creation of new water-born materials, the ability to manipulate new structures (films and layer-by-layer configurations) directly from the latex form, the ability to embed the polymer, creating possibilities for masterbatches, the ability to obtain protective coatings for electromagnetic shielding, new foam type materials, membranes, ultrafiltration applications, and many others. Comparing this method with microemulsion and macroemulsion polymerization, two major highlights can be drawn: the amounts of emulsifier needed are considerably lower in the miniemulsion method and almost all of disadvantages of utilizing macroemulsions are drastically reduced.

Miniemulsion especially, but also the rest of the polymerization in dispersed media still have a high potential for original research papers. The research risks are lower compared with other polymerization techniques considering the interface control (between CNTs and the in situ growing polymer). These types of polymerization process allow high molecular weights for the final polymer and their publication impact could be elevated given the interest for the MWCNT used in the in situ polymerization (number of publication and citations). Polymerization in dispersed media for obtaining nanocomposites opens a large spectrum of research for nanomaterials based on the concept "safe by design." The particles covered with polymer are easier to handle not only because of a large "know how" from the polymer science but also for the drastic reduction of health issues and for unknown potential risks. The recycling of polymers topic can also help the future of these materials, which for the moment are less discussed but in the future will represent a hot topic for both researchers and society policy. Given the fact that industry is already producing a lot of materials and new technologies based on polymer-CNT systems, a certain need will exist for an environmentally safer process. One suggestive argument comes from the history of the inks, paints, and coating industry, which once developed large quantities of solvents but then had to reshape their water-born products in order to yield a low impact on the environment.

The CNTs are fillers with high aspect ratio compared with 2-D (i.e., grapheme) and 1-D (i.e., fullerene). The polymerization in dispersed media for obtaining polymer nanocomposites is more difficult to control as the aspect ratio increases as we showed above. Therefore, the knowledge transfer from other nanofillers is constrained. But in the same context the existing results (for CNT) can be translated easily to 1-D and 3-D nanofillers. To conclude in a reflexive manner, life appeared in water and nanotechnology can be reshaped or reborn in water.

References

[1] K.I. Winey, R.A. Vaia (Eds.), Polymer nanocomposites, MRS Bull. 32 (4) (2007) 314–322.
[2] D.W. Schaefer, R.S. Justice, How nano are nanocomposites? Macromolecules. 40 (24) (2007) 8501–8517.
[3] T. Wang, J.L. Keddie, Design and fabrication of colloidal polymer nanocomposites, Adv. Colloid. Interface. Sci. Vols. 147–148 (2009) 319–332.
[4] S. Iijima, Helical microtubules of graphitic carbon, Nature. 354 (1991) 56–58.
[5] J.-H. Du, J. Bai, H.-M. Cheng, The present status and key problems of carbon nanotube based polymer composites, eXPRESS Polym. Lett. 1 (5) (2007) 253–273.
[6] W.T. Ford, Polymers grafted to single-walled carbon nanotubes by radical polymerization, Macromol. Symp. 297 (1) (2010) 18–24.
[7] E.Y. Choi, L.W. Choi, C.K. Kim, Noncovalent functionalization of multi-walled carbon nanotubes with hydroxyl group-containing pyrene derivatives for their composites with polycarbonate, Carbon N. Y. 95 (2015) 91–99.
[8] B.P. Grady, Recent developments concerning the dispersion of carbon nanotubes in polymers, Macromol. Rapid Commun. 31 (3) (2010) 247–257.
[9] M. Lahelin, et al., In situ polymerized nanocomposites: polystyrene/CNT and Poly (methyl methacrylate)/CNT composites, Compos. Sci. Technol. 71 (6) (2011) 900–907.
[10] L. Bokobza, Multiwall carbon nanotube elastomeric composites: a review, Polymer. (Guildf). 48 (17) (2007) 4907–4920.
[11] W. Bauhofer, J.Z. Kovacs, A review and analysis of electrical percolation in carbon nanotube polymer composites, Compos. Sci. Technol. 69 (2009) 1486–1498.
[12] J. Hu, M. Chen, L. Wu, Organic–inorganic nanocomposites synthesized via miniemulsion polymerization, Polym. Chem. 2 (2011) 760–772.
[13] A.M.K. Esawi, M.M. Farag, Polymer nanotube composites: latest challenges and applications, in: second ed., Vikas Mittal (Ed.), Polymer Nanotubes Nanocomposites: Synthesis, Properties and Applications, 12, Schriverner Publishing - Wiley, 2014, p. 429.
[14] Z. Spitalsky, et al., Carbon nanotube–polymer composites: chemistry, processing, mechanical and electrical properties, Progr. Polym. Sci. 35 (3) (2010) 357–401.
[15] P.-C. Ma, et al., Dispersion and functionalization of carbon nanotubes for polymer-based nanocomposites: a review, Compos. Part A Appl. Sci. Manuf. 41 (10) (2010) 1345–1367.
[16] A.L. Martinez-Hernandez, C. Velasco-Santos, V.M. Castano, Carbon nanotubes composites: processing, grafting and mechanical and thermal properties, Curr. Nanosci. 6 (1) (2010) 12–39.
[17] E.A. Zaragoza-Contreras, et al., Evidence of multi-walled carbon nanotube fragmentation induced by sonication during nanotube encapsulation via bulk-suspension polymerization, Micron (Oxford, England: 1993) 40 (5–6) (2009) 621–627.
[18] S.-M. Kwon, et al., Polystyrene composites containing crosslinked polystyrene-multiwalled carbon nanotube balls, J. Appl. Polym. Sci. 110 (6) (2008) 3737–3744.
[19] S.J. Park, et al., Electrorheology of multiwalled carbon nanotube/poly(methyl methacrylate) nanocomposites, Macromol. Rapid Commun. 26 (19) (2005) 1563–1566.
[20] S. Rana, R. Alagirusamy, M. Joshi, A review on carbon epoxy nanocomposites, J. Reinf. Plast. Compos. 28 (4) (2009) 461–487.

[21] G.-B. Lee, et al., Wrinkled elastomers for the highly stretchable electrodes with excellent fatigue resistances, Polym. Test. 53 (2016) 329–337.
[22] S. Liu, et al., Aligned carbon nanotubes stabilized liquid phase exfoliated graphene hybrid and their polyurethane dielectric elastomers, Compos. Sci. Technol. 125 (2016) 30–37.
[23] S. Ata, et al., Scalable, solvent-less de-bundling of single-wall carbon nanotube into elastomers for high conductive functionality, Polymer. (Guildf). 55 (20) (2014) 5276–5283.
[24] M.F.L. De Volder, et al., Carbon nanotubes: present and future commercial applications, Science 339 (2013) 535–539.
[25] J.-C. Daniel, C. Pichot, Les latex synthétiques. Élaboration, Propriétés, Applications, Lavoisier, Tec & Doc, Paris, 2006.
[26] C.N. Lopes, et al., Effect of initiator on the incorporation of graphite into polymer matrix during suspension polymerization, Macromol. Symp. 229 (1) (2005) 72–80.
[27] S.T. Kim, et al., Dispersion-polymerized carbon nanotube/poly(methyl methacrylate) composite particles and their electrorheological characteristics, Macromol. Chem. Phys. 208 (5) (2007) 514–519.
[28] D. Donescu, et al., Nanocomposites based on MWCNT; polystyrene, styrene-acrylonitrile copolymer, or polymethylmethacrylate, obtained by miniemulsion polymerization, J. Appl. Polym. Sci. 131 (2014) 23. Available from: http://dx.doi.org/10.1002/app.41148.
[29] D. Donescu, et al., Polyacrylonitrile–MWCNT hybrids obtained by free radical polymerization in miniemulsions, J. Polym. Res. 20 (9) (2013) 1–10.
[30] M.S.P. Shaffer, K. Koziol, Polystyrene grafted multi-walled carbon nanotubes, Chem. Commun. 18 (2002) 2074–2075.
[31] H. Xia, Q. Wang, G. Qiu, Polymer-encapsulated carbon nanotubes prepared through ultrasonically initiated in situ emulsion polymerization, Chem. Mater. 15 (20) (2003) 3879–3886.
[32] Z.-Y. Wang, et al., Structure and properties of polystyrene-encapsulated multi-walled carbon nanotubes composites prepared through ultrasonically initiated in situ emulsion polymerization, Chem. J. Chin. Univer. 28 (2007) 571–574.
[33] H. Xia, G. Qiu, Qi Wang, Polymer/carbon nanotube composite emulsion prepared through ultrasonically assisted in situ emulsion polymerization, J. Appl. Polym. Sci. 100 (4) (2006) 3123–3130.
[34] H.-X. Wu, et al., Polymer-wrapped multiwalled carbon nanotubes synthesized via microwave-assisted in situ emulsion polymerization and their optical limiting properties, Carbon N. Y. 45 (15) (2007) 2866–2872.
[35] Z. Cheng, Q. Pan, G.L. Rempel, Modification of multiwall carbon nanotubes via soap-free emulsion polymerization of acrylonitrile, J. Polym. Sci. Part A Polym. Chem. 48 (10) (2010) 2057–2062.
[36] J.Y. Xie, et al., Concentration and temperature controlled oxidation and cutting of single-walled carbon nanotubes by ammonium persulfate, Sci. China Chem. 53 (9) (2010) 2026–2032.
[37] J. Deng, et al., Carbon nanotube–polyaniline hybrid materials, Eur. Polym J. 38 (12) (2002) 2497–2501.
[38] T. Jeevananda, et al., Polyaniline-multiwalled carbon nanotube composites: characterization by WAXS and TGA, J. Appl. Polym. Sci. 109 (1) (2008) 200–210.

[39] H. Zhang, et al., The synthesis of polyacrylonitrile/carbon nanotube microspheres by aqueous deposition polymerization under ultrasonication, Carbon N. Y. 48 (3) (2010) 688–695.
[40] H.J. Jin, et al., Carbon nanotube-adsorbed polystyrene and poly(methyl methacrylate) microspheres, Chem. Mater. 17 (16) (2005) 4034–4037.
[41] O. Regev, et al., Preparation of conductive nanotube–polymer composites using latex technology, Adv. Mater. 16 (3) (2004) 248–251.
[42] K. Zhang, et al., Core–shell structured carbon nanotube/poly (methyl methacrylate) composite and its electrorheological characteristics, Diamond Relat. Mater. 17 (7-10) (2008) 1604–1607.
[43] R. Verdejo, M. Shaffer, A. Bismarck, Particle-stabilized surfactant-free medium internal phase emulsions as templates for porous nanocomposite materials: poly-Pickering-Foams, Langmuir. 23 (5) (2007) 2398–2403.
[44] Joachim Loos, et al., Visualization of single-wall carbon nanotube (SWNT) networks in conductive polystyrene nanocomposites by charge contrast imaging, Ultramicroscopy. 104 (2) (2005) 160–167.
[45] J. Yu, et al., Characterization of conductive multiwall carbon nanotube/polystyrene composites prepared by latex technology, Carbon N. Y. 45 (15) (2007) 2897–2903.
[46] J. Loos, et al., On the fate of carbon nanotubes: morphological characterisations, Compos. Sci. Technol. 67 (5) (2007) 783–788.
[47] Y. Li, et al., Superhydrophobic bionic surfaces with hierarchical microsphere/SWCNT composite arrays, Langmuir. 23 (4) (2007) 2169–2174.
[48] M.A. Correa-Duarte, et al., Layer-by-layer assembly of multiwall carbon nanotubes on spherical colloids, Chem. Mater. 17 (12) (2005) 3268–3272.
[49] J.E. Peters, D.V. Papavassiliou, B.P. Grady, Unique thermal conductivity behavior of single-walled carbon nanotube – polystyrene composites, Macromolecules. 41 (20) (2008) 7274–7277.
[50] D.K. Woo, S.J. Lee, Effect of treatment method on rheological properties of latex-blended polystyrene/multi-walled carbon nanotube nanocomposites, Korea-Aust. Rheol. J. 22 (3) (2010) 219–227.
[51] S. Kara, et al., Electrical and optical percolations of polystyrene latex–multiwalled carbon nanotube composites, J. Colloid. Interface. Sci. 344 (2) (2010) 395–401.
[52] F. Dalmas, et al., Viscoelastic behavior and electrical properties of flexible nanofiber filled polymer nanocomposites. Influence of processing conditions, Compos. Sci. Technol. 67 (5) (2007) 829–839.
[53] M. Li, et al., Preparation, solubility, and electrorheological properties of carbon nanotubes/poly(methyl methacrylate) nanocomposites by in situ functionalization, Compos. Part A Appl. Sci. Manuf. 40 (4) (2009) 413–417.
[54] J. Lu, et al., Chemo-sensitivity of latex-based films containing segregated networks of carbon nanotubes, Sens. Actuators B Chem. 155 (1) (2011) 28–36.
[55] I. Jurewicz, et al., Colloid-assisted self-assembly of robust, three-dimensional networks of carbon nanotubes over large areas, Macromol. Rapid Commun. 31 (7) (2010) 609–615.
[56] C. Pichot, Introduction à la polymérisation en milieu hétérogène et à la polymérisation radicalaire., in: J.-C. Daniel, C. Pichot (Eds.), Les latex synthétiques. Élaboration, Propriétés, Applications, Lavoisier, 2006, pp. 183–206.
[57] P. Kruss, Sonochemical initiation of polymerization, in: T. Mason (Ed.), Advances in Sonochemistry, 2nd edition, JAI Press, London, 1991, pp. 2–21.

[58] Z. Peng, et al., Self-assembled natural rubber/multi-walled carbon nanotube composites using latex compounding techniques, Carbon N. Y. 48 (15) (2010) 4497–4503.
[59] Y. Holl, Mécanismes de formation de films minces polymères à partir de latex. [ed.] Jean-Claude Daniel and Christian Pichot, s.l. Les latex synthétiques. Élaboration, Propriétés, Applications, Lavoisier, 2006, pp. 105–148.
[60] B.R. Vijayendran, T. Bone, C. Gajria, Some studies of vinyl acrylic latex – surfactant interaction, in: Mohamed S. El-Aasser, John W. Vanderhoff (Eds.), Emulsion Polymerization of Vinyl Acetate, Applied Science Publishers, London, New Jersey, 1981.
[61] S.M. Fatemi, M. Foroutan, Recent developments concerning the dispersion of carbon nanotubes in surfactant/polymer systems by MD simulation, J. Nanostruct. Chem. 6 (1) (2016) 29–40.
[62] I. Szleifer, R. Yerushalmi-Rozen, Polymers and carbon nanotubes—dimensionality, interactions and nanotechnology, Polymer. (Guildf). 46 (19) (2005) 7803–7818.
[63] L. Vaisman, H.D. Wagner, Gad Marom, The role of surfactants in dispersion of carbon nanotubes, Adv. Colloid. Interface. Sci. Vols. 128–130 (2006) 37–46.
[64] M.F. Islam, et al., High weight fraction surfactant solubilization of single-wall carbon nanotubes in water, Nano. Lett. 3 (2) (2003) 269–273.
[65] N. Grossiord, et al., Conductive carbon-nanotube/polymer composites: spectroscopic monitoring of the exfoliation process in water, Compos. Sci. Technol. 67 (5) (2007) 778–782.
[66] N. Grossiord, et al., Time-dependent study of the exfoliation process of carbon nanotubes in aqueous dispersions by using UV – visible spectroscopy, Anal. Chem. 77 (16) (2005) 5135–5139.
[67] J. Yu, et al., Controlling the dispersion of multi-wall carbon nanotubes in aqueous surfactant solution, Carbon N. Y. 45 (3) (2007) 618–623.
[68] F.P. Araujo, D.F.S. Petri, Ana M. Carmona-Ribeiro, Colloid stability of sodium dihexadecyl phosphate/poly(diallyldimethylammonium chloride) decorated latex, Langmuir 21 (21) (2005) 9495–9501.
[69] F. Dalmas, et al., Multiwalled carbon nanotube/polymer nanocomposites: processing and properties, J. Polym. Sci. Part B Polym. Phys. 43 (10) (2005) 1186–1197.
[70] K. Zhang, J.H. Choi, Smart polymer/carbon nanotube nanocomposites and their electrorheological response, Materials 1 (5) (2014) 3399–3414.
[71] K. Landfester, M. Barrère, Polymérisation en miniémulsion, in: Jean-Claude Daniel, Christian Pichot (Eds.), Les latex synthétiques. Élaboration, Propriétés, Applications, Lavoisier, Paris, 2006, pp. 833–869.
[72] Barraza, H.J., et al., Utilization of SWNT-filled polystyrene obtained by miniemulsion polymerization as a modifier for high-performance epoxy resins NSTI, 2003. Technical proceedings of the 2003 nanotechnology conference and trade show, 3, pp. 538–541. ISBN: 0-9728422-2-5.
[73] H.T. Ham, et al., Singlewall carbon nanotubes covered with polystyrene nanoparticles by in-situ miniemulsion polymerization, J. Polym. Sci. Part A Polym. Chem. 44 (1) (2006) 573–584.
[74] H.F. Lu, et al., Synthesis and lubricating performance of a carbon nanotube seeded miniemulsion, Carbon N. Y. 45 (5) (2007) 936–942.
[75] A.S. Patole, et al., Effective in situ synthesis and characteristics of polystyrene nanoparticle-covered multiwall carbon nanotube composite, J. Polym. Sci. Part B Polym. Phys. 47 (15) (2009) 1523–1529.

[76] M.L.P. Ha, et al., Composites of single-walled carbon nanotubes and styrene-isoprene copolymer latices, Macromol. Chem. Phys. 2008 (5) (2007) 446–456.
[77] I. Capek, T. Kocsisova, On the preparation of composite poly(butyl acrylate)/carbon nanotube nanoparticles by miniemulsion polymerization of butyl acrylate, Polym. J. 43 (2011) 700–707.
[78] F. Tiarks, K. Landfester, M. Antonietti, Encapsulation of carbon black by miniemulsion polymerization, Macromol. Chem. Phys. 202 (1) (2001) 51–60.
[79] M. Lee, et al., Morphology and electrical properties of polymethylmethacrylate/poly (styrene-co-acrylonitrile)/multi-walled carbon nanotube nanocomposites, J. Appl. Polym. Sci. 121 (2) (2011) 743–749.
[80] F.J. Schork, et al., Miniemulsion polymerization, in: Masayoshi Okubo (Ed.), Polymer Particles, 175, Springer, Berlin Heidelberg, 2005, pp. 129–255.
[81] Y. Reyes, M. Paulis, J.R. Leiza, Modeling the equilibrium morphology of nanodroplets in the presence of nanofillers, J. Colloid. Interface. Sci. 352 (2) (2010) 359–365.
[82] S.-H. Park, J. Bae, Tailoring environment friendly carbon nanostructures by surfactant mediated interfacial engineering, J. Industrial Eng. Chem. 30 (2015) 1–9.
[83] J. Gu, et al., Robust preparation of superhydrophobic polymer/carbon nanotube hybrid membranes for highly effective removal of oils and separation of water-in-oil emulsions, J. Mater. Chem. A 2 (37) (2014) 15268–15272.
[84] Y. Chen, et al., High capacity, stable silicon/carbon anodes for lithium-ion batteries prepared using emulsion-templated directed assembly, ACS Appl. Mater. Interfaces 6 (7) (2014) 4678–4683.
[85] S.J. Gao, et al., Photoinduced superwetting single-walled carbon nanotube/TiO_2 ultrathin network films for ultrafast separation of oil-in-water emulsions, ACS Nano 8 (6) (2014) 6344–6352.
[86] Y. Wang, et al., Microemulsion-assisted synthesis of ultrafine $Li_4Ti_5O_{12}$/C nanocomposite with oleic acid as carbon precursor and particle size controller, J. Power Sources 246 (2014) 213–218.
[87] R.Xi Wang, et al., Antistatic poly(ethylene terephthalate)/polyaniline-coating multiwalled carbon nanotubes nanocomposites, Adv. Mater. Res. 549 (2012) 553–557.
[88] H. Wang, et al., Design and synthesis of the polyaniline interface for polyamide 66/multi-walled carbon nanotube electrically conductive composites, Colloid Polym. Sci. 291 (4) (2013) 1001–1007.

Temperature effect in polyurethane/graphene/PMMA nanocomposites using quantum mechanics and Monte Carlo for design of new materials

12

Norma-Aurea Rangel-Vázquez[1], *Juan-Ramón Campos-Cruz*[1], *José-Enrique Jaime-Leal*[1] *and Ricardo Rangel-Vázquez*[2]
[1]Instituto Tecnológico de Aguascalientes, Aguascalientes, Mexico, [2]PCC. Real de Haciendas S/N, Aguascalientes, Mexico

Chapter Outline

Hybrid Polymer Composite Materials: Processing. DOI: http://dx.doi.org/10.1016/B978-0-08-100789-1.00012-5

12.1 Introduction

12.1.1 Molecular modelation

In principle, molecular modeling was the name given to the application of molecular mechanics for structural determination. Today it is included in the methods of quantum chemistry and is characterized by using graphical techniques to represent the structure and the behavior of molecules.

Computational chemistry allows researchers to study, characterize and predict the structure and stability of chemical systems: studying energy differences between different states to explain spectroscopic properties and reaction mechanisms at the atomic level. The ultimate purpose in terms of practical application is the reduction of costly experiments in economic terms and time. Not all molecular models are quantitative, only those that allow you to assign numerical quantities associated with certain system properties values, and as a result an image model of their behavior is obtained are quantitative. By handling procedures molecular geometry of different conformations of a system are studied and assembled with calculation methods determining the study of molecular properties. Graphical methods present an attractive visual analysis of molecular conformations and properties immediately.

Most molecular modeling studies involves the choice of model to describe intra- and intermolecular interactions in the system. The two most common are molecular mechanics and quantum mechanics. These models are used in order to calculate the energy of any arrangement of atoms and molecules in the system and to determine how energy varies by changing the position of the particles. Another step is the calculation of the energy minimization or conformational search. Molecular mechanics equations using a force field of classical mechanics to describe the potential energy surfaces and physical properties of molecules without explicitly considering the electrons. A molecule is described as a set of atoms that interact and this description is called a force field. In Eq. (12.1) the force field that includes the following terms are observed:

$$U = \sum U_s + \sum U_b + \sum U_t + \sum U_{vdw} \tag{12.1}$$

where, U is the total potential energy of the molecule and is the sum of the tension energy (U_s) of the individual bonds, the strain energy of the angle (U_b), torsion energy (U_t) and interactions van der Waals energy (U_{vdw}). The geometry optimization is to find the coordinates of a molecular structure representing a minimum potential energy, that is, to obtain a value of zero.

An algorithm of downstream stages is generally used for geometry optimization. In this algorithm the first derivative of the potential energy with respect to the Cartesian coordinates is used. The advantage of molecular mechanics is that it allows the modeling of large molecules such as proteins and DNA segments; the disadvantage is that there are many chemical properties that are not defined yet in this method, for example electronic excitation states. When working with extremely long and complicated systems, software packages for molecular mechanics are often the most powerful and easy to use. However, sometimes it is easy to use, but not necessarily a good way to describe the system. Quantum mechanics is based on Schrödinger's equation to describe a molecule with a direct treatment of the electronic structure.

Most of the applications of quantum mechanics that separates the motion of the nuclei of the movement of electrons is called the Born–Oppenheimer approximation. This is the first of several approaches to simplify Schrödinger's equation solution. Quantum mechanics is divided into methods ab initio ("from the beginning") and semi-empirical methods; ab initio is the name of the calculations that are performed directly from theoretical principles without including any experimental information.

The most common type of ab initio calculation is called Hartree Fock (HF), which does not take into account Coulomb repulsion specifically electron–electron, but takes into account the net effect on the calculation. Various types of calculations begin with the HF calculation and are then corrected for the electron–electron repulsion explicit. Some of these methods are the perturbation theory of Mohlar–Plesset (MPN, where *n* is the order of correction), the method Generalized Valencia Bond (GVB), Multi-Configurations Self-Consistent Field (MCSCF), Configuration Interaction (CI), and Coupled Cluster theory (CC). As a group, these methods are known as correlated calculations. The advantage of ab initio methods is that they eventually converge on the exact solution; the disadvantage is that it is expensive in the computing resources and time required. Semi-empirical calculations are configured with the same general structure of HF calculation. In order to correct the errors introduced, the method is configured so that a curve fit parameter is made to agree with the experimental data.

Among these methods is, the method of Complete Suppression of Differential Overlap (CNDO), in this method used some integrals as treatments ab initio calculation in where measure are ignored. Here matrix density test and is used in the matrix Fock and diagonalization is performed from another matrix density and consequent other matrix Fock, this process successively performed until the matrix Fock not change and generate a matrix consisting density. There is also the Suppression Intermedia Differential Overlap method (INDO) which is an improvement to the CNDO; the differential overlap between atomic orbitals of the same atom does not despises electron repulsion integrals of a center where the orbital functions are centered on the same atom, but is neglected in electron repulsion integrals of two centers.

The method of Intermediate Suppression of Modified Differential Overlap (MINDO) is the result of a series of modifications to the INDO method, the most important difference being the origin of the parameters; for example, instead of using atomic spectra, Urs data is defined as an adjustable bioelectronics integral. The method Zerner of Intermediate Depression Differential Overlap (ZINDO) is an improved INDO method while INDO is restricted only to organic molecules containing atoms of boron to fluorine; ZINDO covers a wide range of the periodic table including elements transition. The advantage is that semi-empirical calculations are much faster than ab initio and may be useful in moderate size molecules both organic and inorganic. The disadvantage is that the results can yield erratic molecules significantly if the study is very different from the molecules in the database [1].

12.1.1.1 Quantum mechanics

Quantum mechanics began in 1905 when Einstein performed the application of the concepts of the photoelectric effect, and Planck the electromagnetic theory, which found new corpuscular properties of radiation. Later, Bohr was able to determine the existence of the various energy levels of atoms. That is, quantum mechanics carries out the study of physical phenomena at the subatomic level, comprising nuclei, atoms, and molecules in order to develop new technologies in the materials of the future based on the calculation of physicochemical properties and molecular vibrations in the spectra of FTIR, NMR, and UV−VIS, respectively [2]. Quantum mechanics deals with the behavior of matter and radiation in the atomic and subatomic scales.

Thus attempts are made to describe and explain the properties of molecules, atoms, and their constituents: electrons, protons, neutrons. The development of the basic ideas of quantum mechanics began early last century as a result of a series of discoveries and observations that presented evidence for serious problems of classical physics concerning the interpretation of the properties of the atom and its constituent parts as well as properties of electromagnetic radiation and its interaction with matter. According to the hypothesis that electrons have, wave properties may be possible to describe the mathematical equations of waves, namely, the standing waves; so-called wave function describing the evolution of electron motion is thus established [3]. Today quantum mechanics is considered as a key simulation tool in the design of new materials due to the nanometric scale in addition to considering the basis of the Schrödinger's equation (see Eq. 12.2), unlike classical mechanics which applies Newton's laws, that is, where the atoms are considered as particles.

$$\hat{H}\psi = E\psi \tag{12.2}$$

where, $\hat{H}$ represents the Hamiltonian operator or operator of total energy, Ψ is the wave function, which determines the value of the dynamic variables of a system under study [4−5]. This equation has only appropriate solutions when E has values from $E1$, $E2$ to $E\infty$, which are referred to as Eigenvalues of the equation 12.2. While the wave function (Ψ) are attributed direct E_k values, which represent total energy of the system of the form:

$$\psi_k(x,t) = \psi_k(x)\varphi_k(t) = \psi(x)e^{-E_{k}t/h} \tag{12.3}$$

This Eq. (12.3) is known to standing waves, i.e., the probability distribution ($\Psi_k{}^*\Psi_k$) is independent of time, therefore the Eq. (12.4) is obtained, in where:

$$\psi_k^*(x,t)\psi_k(x,t) = \psi_k^*(x)^{-iE_{k}t/h}\psi_k(x)^{-iE_{k}t/h} = \psi_k^*(x)\psi_k(x) \tag{12.4}$$

Another solution of the Schrödinger's equation is represented by Eq. (12.5), where there is a combination of independent coefficients of time [6].

$$\psi(x,t) = \sum_{n} c_n \psi_n(x,t) = \sum_{n} c_n \psi_n(x) e^{-iE_n t/h} \tag{12.5}$$

AM1 comes from a parameterization performed from the model equations MNDO. Because of certain drawbacks, the high repulsion between atoms 2–3 Å, producing the high activation energy, was simultaneously had. So in 1985, Dewar et al., developed the Austin model 1 (AM1) adding Gaussian functions, and the parameterization whole model [7]. Also, performing parameterization atoms as Zn, Hg, by adding exponents for the atoms of the molecule in addition to a function of repulsion shape of the Eqs. (12.6) and (12.7) [8] leads to:

$$f_{AB}^{AM1} = f_{AB}^{MNDO} + \frac{C_A C_B}{R_{AB}/A}\left(\sum_{k} \alpha_{KA} \exp[-b_{KA}(R_{AB} - C_{KA}]^2 + \sum_{k} \alpha_{KB} \exp[-b_{KB}(R_{AB} - C_{KB}]^2\right) \tag{12.6}$$

In where:

$$f_{AB}^{MNDO} = C_A C_B (S_A S_A | S_B S_B)(e^{-\alpha_A K_{AB}} + e^{-\alpha_B K_{AB}}) \tag{12.7}$$

f (MNDO, AB) represents the core, C_A and C_B correspond to loads of atoms A and B and finally, the terms α_A, α_A, αK_A, bK_A and cK_A are the parameters. This simulation method is widely used when having molecules of organic compounds, because it calculates energy geometry optimization and the heat formation with greater precision than the MNDO model, thus the results generated provide greater accuracy in these compounds [9].

The Monte Carlo method of molecular simulation, a probabilistic or stochastic model, where the main function is to evaluate the averages of dynamic magnitudes $<A>$ of a system (see Eq. 12.8) and provide information for those systems that have obtained thermodynamic equilibrium. Knowing that A is a function of the coordinates and the kinetic term of (K) Hamiltonian is the square of p moments, an integral dependent of the coordinate's q is obtained:

$$<A> = \frac{\int dq^N A(q) Z(q)}{\int dq^N Z(q,p)} \tag{12.8}$$

where: $Z(q) = \int dq^N p(q,p)$ is a function of interaction potential $V(q)$, N represents the number of particles while, $p(q,p)$ is assigned reduced phase-space values. The Monte Carlo method will get random configurations or points system based on the probability distribution $Z(q)$ to give the Monte Carlo integration shown in Eq. 12.9:

$$\langle \mathcal{A} \rangle = \lim_{n\to\infty} \frac{1}{n-n_0} \sum_{i=n_0+1}^{n} \mathcal{A}(q_{\mathrm{i}}^{\mathrm{N}}) \approx \frac{1}{n-n_0} \sum_{i=n_0+1}^{n} \mathcal{A}(q_{\mathrm{i}}^{\mathrm{N}}) \tag{12.9}$$

where, not for the number of steps of the equilibrium stage and the second term is only valid if you have a large number of steps, n [10].

12.1.2 Nanotechnology and nanoscience

Nanoscience is the science of all objects at the nanoscale, i.e., on a scale from 1 to 100 nanometers. As a consequence, it has a wide range of applications, for example, in chemical, colloids, polymers, copolymers, and semiconductor quantum dots. In physics it is used to understand the movement of electrons and photons in various structures, while in biology and biochemistry the objective is to understand and design new nanostructures to remove various viruses [11].

Nanomedicine, through the application of nanotechnology, is used to improve the quality of life of human beings via the use of computer algorithms to predict the behavior of drugs in the human body [12] through new molecules and mechanisms of absorption and pharmaceutical delivery, as well as to provide treatments for cancer and HIV and tissue engineering due to improvements in the release of the active ingredient and the amount required in the human body. These may be micelles, liposomes, dendrites, nanoparticles, nanotubes, or conjugated polymers, to name a few [13].

In order to attack viruses and cancer cells, organ cleansing is being carried out the development of nanorobots [14].

12.1.2.1 Nanocomposites

Composites that are reinforced with particles of various materials are perhaps the most used in everyday materials. The particles are added to improve the elastic modulus of matrix, fire behavior, heat resistance and yield strength, among others. To scale the particle size to the nanometer scale, it has been shown that new properties can be obtained in materials [15]. Nanocomposites are compounds in which at least the size of one of the phases is in the nanometer range (1 nm = 10^{-9} m). A nanocomposite matrix is divided into load, the load being located in the nanometric size. The nanocomposites materials can be classified according to the matrix material: ceramic matrix nanocomposites (CMNC), metal matrix nanocomposites (MMNC), and polymer matrix nanocomposites (PNC) [16]. Typically loads more spiked to polymeric matrices are particles: spherical (silica, metal, and other organic and inorganic particles), fibers (nanofibers and nanotubes) and lamellar (carbon, graphite, silicates and laminates aluminosilicates and other sheet materials) [17].

The polymers are considered good host matrices for composite materials because they can be easily adapted to produce multiple physical properties. Choosing a polymer matrix is usually and mainly guided by their mechanical, thermal, electrical, optical, and magnetic behavior. Moreover, other properties such as hydrophobicity/hydrophilicity, chemical stability, biocompatibility, and optoelectronic properties and chemical functionalities are considered. The polymers can also allow easier forming and better processing of composite materials [18].

By incorporating conductive fillers in the polymer matrix as a second phase, electrical conductivity is achieved in the resulting composite materials [19]. Despite being a relatively new technology, promising applications of nanocomposites are numerous, including both the generation of new materials such as improving the performance of known devices. Some of these applications are: high performance catalysts, technological data storage, optical fibers, chemical sensors, magnetic devices, photo-electrochemical applications, conversion of light energy, aerospace and aeronautical materials, and many more [20].

12.1.3 Polymers

12.1.3.1 Polyurethane (PU)

Polymers are macromolecules of high molecular weight between 10^3 and 10^4, where the carbon atoms are attached by covalent bonds and are constituted by the repetition of small units called monomers [21–22]. Polyurethane (see Fig. 12.1) is a polymer obtained from the reaction from an isocyanate group and a long or short chain polyol, but can also be used as fillers or other resins agents [23–24]. During this reaction other more complex reactions in order to obtain the urethane group are carried out. PU may be flexible or rigid, which depends on the reagents used in the synthesis [24]. From their synthesis there have been a wide variety of applications such as patches, catheters, dressings, microcapsules, to name a few [23–24]. Being the main components:

- *Glycols*. They are compounds with two hydroxyl groups in their structure, the most important are diethylene glycol and glycerol. They are used as polyester and polyether: polyesters are compounds obtained from the condensation of a carboxylic acid and an alcohol while polyethers are used due to resistance to hydrolysis reactions.

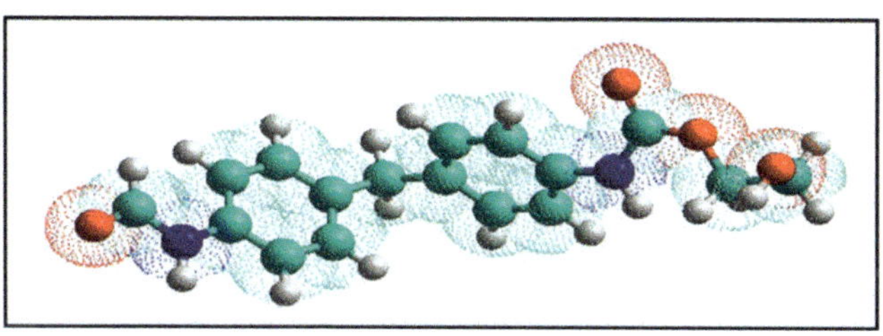

Figure 12.1 PU structure, where: ● carbon, ○ hydrogen, ● oxygen and, ● nitrogen, respectively.

- *Isocyanates.* They are organic compounds classified as aromatic and aliphatic. Aromatic isocyanates generate the yellow color in the PU foam due to the attack of UV radiation. Aliphatic isocyanates are resistant to hydrolysis and thermal degradation; cyclohexyl diisocyanate or HDI are the most frequently used [25].
- *Catalysts* are used because they have more control and balance throughout the reaction; they can be aliphatic or alkanolamines [26].
- *Chain extenders.* These are intended for the separation of soft and hard segments, which are segmented polyurethanes that have a diol low molecular weight [27].

Currently, the PUs are used in a wide variety of biomedical applications. For example in the cardiovascular field they are used because of its excellent physicochemical, mechanical properties and biocompatibility. They are also used for ventricular pacing leads and devices; however, due to the chemical properties, they are used in tissue engineering of vascular grafts and heart valves [28] in addition to the design of blood bags [29]. Another important application are synthetic injectable bone fillers because the material is implemented as a reactive liquid mixture of two components as a gel in situ [30].

12.1.3.2 Polymethyl methacrylate (PMMA)

PMMA (see Fig. 12.2) is a thermoplastic polymer produced by methylmethacrylate monomer through free radical polymerization. The polymer has no atoms of the methyl group in the main chain which causes a displacement between the chains. However, it has excellent mechanical properties and is highly resistant to rupture [31–32]. Among the main applications include:

- *Optics:* magnifiers, sunscreen, watch glasses
- *Vehicles:* headlights, triangles signal
- *Electric engineering:* parts for switches, control panels, dials
- *Medicine:* packaging of drugs, pills, capsules, suppositories [33], respiratory equipment and accessories catheters, heart valves, blood bags [34] micro-sensors, dental and bone prostheses [35].

In the medical area it is used as a bone cement in order to maintain the position of the implant in the human body [36].

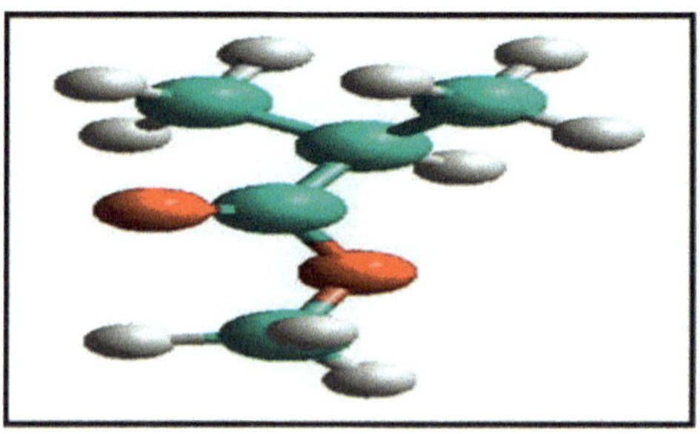

Figure 12.2 PMMA structure, where, ○ hydrogen, ● carbon and, ● oxygen.

The PMMA is resistant to high strain, i.e., the mechanical properties are similar to those of cortical bone and cancellous, for example, the flexural modulus of PMMA is 1–3 GPa and 10–20 both bones GPA and 10–2000 MPa, respectively. In the compressive strength, the PMMA is between 85–110 MPa and the cortical bone has a range of 133–193 MPa. The tensile strength of PMMA is 30–50 MPa and the cortical bone is 50–133 MPa, respectively [37].

12.1.4 Graphene

Graphene is composed of a hexagonal network of carbon atoms [38–39] where the electronic structure of graphene is constituted by six cones with a distance between them of 0.142 nm. Graphene undoped has a density of states of zero generating that the electrical conductivity is relatively low in the order of $\sigma \sim e2/\mathrm{h}$. However, this value changes when graphene is doped (n, p) although it depends on the polarity the field. A graphene doping mechanism may involve the adsorption of water or ammonia; generating electrical conductivity is increased even more than copper. Doping of graphene is carried out using nitrogen or boron atoms in addition to organic molecules, polymers or inorganic compounds. Applications range from sensors, actuators, transistors, solar cells, batteries, superconductors, adsorption of drugs, prostheses, etc. [40].

Excitations graphene are determined by the model of Dirac, where there are shifts at a constant speed, and the attachment points of the cones are referred to as Dirac points where the particles have a maximum energy of 1 eV showing that the Hall's effect has unusual behavior in the structure of graphene [39]. Furthermore, the ratio of the energy with respect to time is given by the equation $E = vP$, where v corresponds to the speed of Fermi–Dirac, in the graphene the speed is 300 times less than the speed of light [41].

Hybridization is the type sp^2 generating the hexagonal structure of graphene. The band indicates the robustness of the network, i.e., by the Paulin exclusion principle; these bands have a deep valence band, where p orbital forms a covalent bond perpendicular to the planar structure because they have an extra electron [42], while sigma bonds are generated due to the formation of three bonds with each nearest neighbor atom to atom, and the latter is used to form electron in the 2pz state in the conductive strip π [43]. Graphene has excellent thermal conductivity of about 5000 W/mK, the electrical conductivity is 108 S/m, strength of 42 N/m, Young's modulus of 1 tPA, and a surface area of $2.63 \times 106\ m^2/kg$ [40].

In order to study the thermal properties of graphene molecular modeling is used for computational methods such as molecular dynamics (MD), Greens functions without equilibrium (NEGF), and the Boltzmann equation. Molecular dynamics simulations have shown that you can change the flow of heat by introducing vacancies, grain boundaries, isotopic impurities, edge roughness, etc. [44]. At present it has generated increased interest for using graphene for adsorption and controlled release of drugs due to interactions generated through graphene and hydrophobic drugs. For example, Sunda et al. synthesized a graphene/carbon nanotube where graphene worked well as a therapeutic agent. These nanoparticles represent a new way for pharmacokinetic analysis, pharmacodynamics, and biorecognition with the aim of increasing the efficiency and effectiveness of various drugs [45].

12.1.4.1 Medical applications

When foreign material comes into contact with the human body, the first reaction is generated protein adsorption, because these have a direct influence on blood compatibility. However, if there is a negative reaction of the material simultaneous deformed platelets are activated. So the main properties that should have a biomaterial is the area and surface charge, the hydrophilic/hydrophobic character, etc. [46].

The new graphene-based materials have presented excellent results in the area of tissue engineering, e.g., Kim et al, synthesized GO/$CaCO_3$ where, they found that this nanomaterial had excellent bone bioactivity and compatibility with osteoblasts. Other experiments reported the use of graphene for the synthesis of nanosheets applied scaffolding with the aim of increasing MG63 cells. Obtaining covalent scaffolds of chitosan/graphene have resulted in an increase in cell attachment and cell reproduction preosteoblast MC3T3-E1 and mouse slow degradation of chitosan, this was due to hydrophilicity of graphene oxide. Finally, we studied the combination of nano chitosan fibers, PVA and graphene where achievement get complete healing of the dermis at the end of the tenth day, this was attributed to the ability of graphene to generate new cells [47].

12.1.5 Prosthesis

A prosthesis is an artificial device which needs various components: a power source, power transmission system, control system, and a gripper. The main types of prostheses are type, mechanical, electrical, myo-electric, hybrid and pneumatics, and mainly hand. However, an implant may have different drawbacks such as poor mechanical strength or poor integration with the human body. If this occurs, the implant is removed. The latest technological advances in bioengineering have allowed the use of computer simulation in medicine for the characterization of biological, mechanical, and chemical behavior of structures by using computational chemistry models.

12.2 Methodology

12.2.1 Geometry optimization

In order to obtain geometry optimization, i.e., solving the Schrödinger's equation and get the minimum energy, the AM1 model was used. Polak Ribiere algorithm with a conjugate gradient 0.0001 kcal/(Å mol) were applied for molecules of PU, G, and PMMA, respectively. Table 12.1 shows that the Gibbs free energies (ΔG) were negative indicating that these materials react spontaneously to absorption of

Table 12.1 **Thermodynamic properties**

Properties	Polyurethane (PU)	Graphene (G)	Polymethylmetacrylate (PMMA)
ΔG (kcal/mol)	− 4423.0563	− 5409.1243	− 1601.5369
Log P	1.80	7.02	1.05

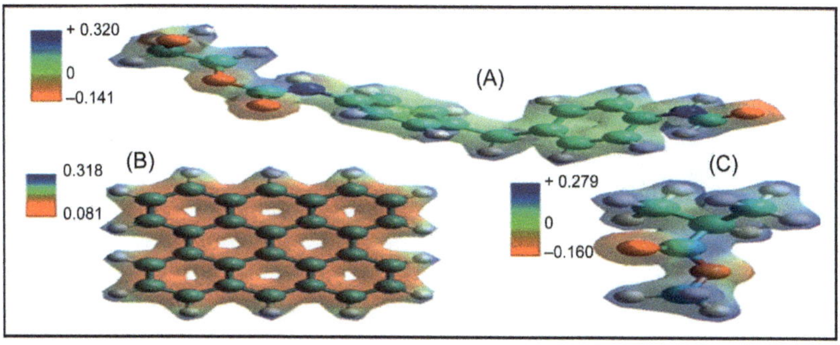

Figure 12.3 MESP of, (A) PU, (B) Graphene, and (C) PMMA, respectively.

PMMA in the PU/G nanocomposite. It is also noted that the partition coefficients (log P) are positive, i.e., have a hydrophobic property, so they are only soluble in nonpolar solvents, i.e., there is no diffusion in the human body therefore medical applications increase. The log P is obtained from the QSAR properties.

12.2.2 *Obtaining electrostatic potential map*

HyperChem shows the electrostatic potential map (MESP) to select the Compute/ Plot molecular graphs of the menu option; later in the dialog window molecular properties option are chosen. Fig. 12.3 shows the MESP about of PU, G, and PMMA respectively, where we see that areas rich in electrons (red) correspond to the bonds CO and C = O, while the nucleophilic or positive zones (blue) are attributed to CH and NH finally, neutral potential, that is, where a load balancing are located in CC bonds. Note that in the molecule graphene CC bonds and C = C have negative charges due to sp^2 hybridized permitting flow of electrons through the chain.

12.2.3 *FTIR analysis*

To determine the wavelengths of each molecule the compute/vibration rotation analysis/vibrational spectrum is selected. Spectra were analyzed in various vibrations selecting a particular frequency and applying the animate vibrations, so that the bond type corresponds to the vibrational mode that displays its command.

12.3 Results and discussions

12.3.1 Crosslinking: polyurethane–graphene (PU/G)

12.3.1.1 Geometry optimization

By model AM1, the optimal ratio for crosslinking PU/G was calculated due to the electronegativity the carbonyl group observed in Fig. 12.4. It was verified that there is the formation of hydrogen bonds between the CH bond of graphene and the C = O of PU as shown in Fig. 12.4. As shown in Table 12.2 the Gibbs free energy shows that there is a spontaneity in the several relationships of crosslinking. We chose to use the 2:2 due to the influence of the carbonyl group in the reaction mechanism in addition to the vulnerability of being attacked by nucleophilic atoms like hydrogen [48]. The partition coefficient (log P) has a positive value of 20.99, i.e., as individual components, also of hydrophobicity. Therefore the PU/G nanocomposite will not have decomposition reaction contact water present in the human body.

12.3.1.2 Electrostatic potential map (MESP)

Fig. 12.5 shows the electronic distribution of PU/G crosslinking in where the electrophilic zones corresponded to the C = O and CO groups and the CH bonds were assigned nucleophilic areas. It is shown that the interaction energy of the molecules is accomplished by forming hydrogen bonds, that is, an overlap of atomic orbitals that generates the molecular orbitals are assigned to both the molecule PU as graphene.

It is noteworthy that the bigger the overlap is there will be less energy bonding the molecular orbital; with this phenomenon the crosslinking process of polymeric nanocomposite (PU/G) is verified. Finally, it is appreciated that doping of graphene generates an increase in electrical conductivity and electron density, which is attributed to the Fermi surface of graphene located at the junction point between the valence and conduction bands, respectively.

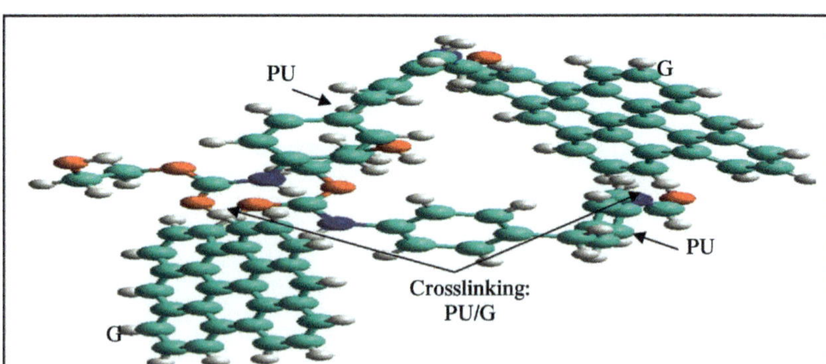

Figure 12.4 Molecular geometry optimized of PU/Graphene nanocomposite (PU/G), where: ○ hydrogen, ● carbon, ● oxygen and, ● nitrogen.

Table 12.2 **Gibbs free energy for established systems**

Relation: PU/G	ΔG (kcal/mol)
1-1	− 9,839.0279
1-2	− 15,246.9720
2-2	− 19,616.7780

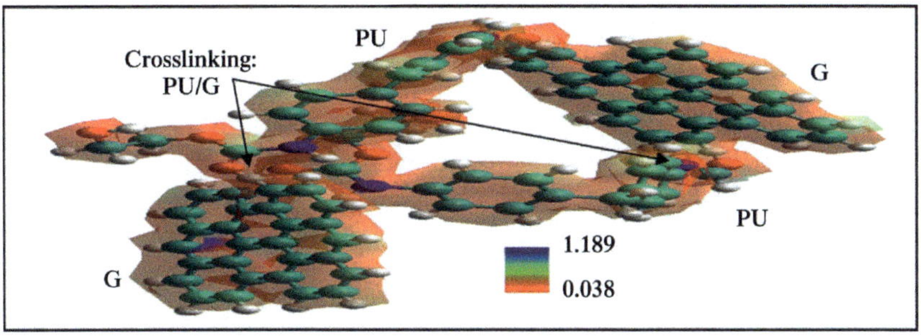

Figure 12.5 MESP of PU/Graphene nanocomposite, where: ○ hydrogen, ● carbon, ● oxygen and, ● nitrogen.

Table 12.3 **Vibrational analysis of PU/G nanocomposite**

Bond	Molecule	Frequency (cm^{-1})
OH stretching	PU	3805, 3233
NH stretching	PU	3335–3238
CH asymmetric and symmetric	PU	3595–3083
N = C = O stretching	PU	2543, 2151
C = N (NCO)	PU	2464, 2021
C = O stretching	PU	1750–1635
C = O, CH (crosslinking)	PU, G	3512, 1732, 1644
CN stretching, NH bending	PU	1585
CH vibration	PU	1557–1372
CC	PU	917–884
CH	G	3756–2901
CC	G	2415–1430
Benzene rings	G	1535–1525
CC	G	1379–503

12.3.1.3 FTIR

Table 12.3 shows the results vibrational crosslinking of PU/G in where the characteristic absorptions peaks of the PU are observed at 3805 and 3233 cm^{-1} was attributed at the OH stretching. The vibrations CH asymmetric and symmetric were localized between 3595–3083 cm^{-1}, at 3335–3238 cm^{-1} was attributed to NH stretching [49]. The isocyanate group was observed to 2543, 2464, 2151, and

2021 cm^{-1}, the carbonyl group corresponded in the range of 1750–1635 cm^{-1}, the CH and NH were appreciated at 1557–1372 cm^{-1} [50–51]. The CN stretching and NH bending were assigned at 1585 cm^{-1} [49]. The peak at 3512 cm^{-1} can be attributed to the OH stretching vibrations of the C–OH groups. The absorption bands at 1535–1525 cm^{-1} can be ascribed to benzene rings [52–53]. Also in the vibrational analysis a region in which a simultaneous vibration of C = O of PU and CH-graphene in the region between 1732 and 1644 cm^{-1} is presented found, whereby crosslinking is checked by hydrogen.

12.3.2 Adsorption of PMMA

12.3.2.1 Geometry optimization PMMA minimum adsorption and partition coefficient

In order to determine the optimal ratio for adsorption of PMMA in the polymer nanocomposite (PU/G), the calculations by Gibbs free energy (ΔG) shown in Table 12.4 was observed that the relationship 2-2-2 had the minimum required energy to perform the spontaneity of the reaction, i.e., adsorption acrylate. The optimized molecular geometry was obtained in where the adsorption of PMMA is shown in Fig. 12.6, where an attraction of the carbonyl group of the PMMA shown to CH bonds (graphene), generating new hydrogen bonds. The partition coefficient (log P) was 24.67, i.e., which has no affinity for polar substances such as water, so it is a system with hydrophobic tendency. It also indicates that the prosthesis constituted by PU/G/PMMA will not be affected by contact with water in the body, thus generating a prolonged life.

12.3.2.2 Electrostatic potential map (MESP)

The MESP shown in Fig. 12.7 shows the areas that have a slightly lower electron density in the CH bonds of PU/G. This is represented by the blue coloration. However, there's a wide electrophilic area (red) that's attributed to the resonance electronics of the chains in the graphene structure and the electronegative character of the carbonyl group of PU and PMMA, respectively.

Table 12.4 Determination of the Gibbs free energy of different amounts of PMMA in PU/G nanocomposite

PU/G/PMMA	ΔG (kcal/mol)
2-2-1	+ 244,876.23
2-2-2	− 22,876.4608
2-2-3	− 24,478.7738
2-2-4	− 26,080.9105
2-2-5	− 27,682.9463
2-2-6	− 29,286.8231

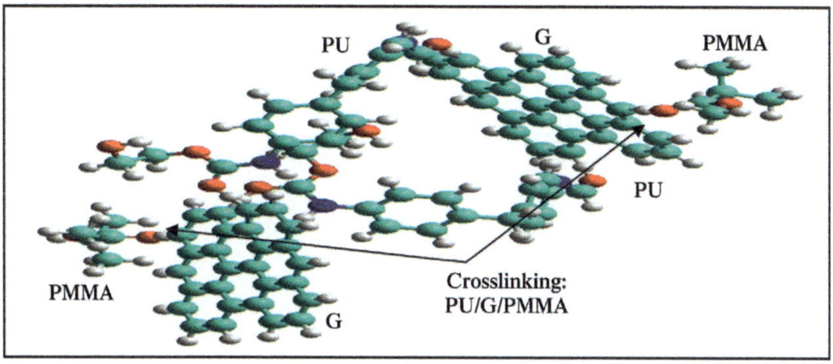

Figure 12.6 Molecular geometry optimized of PMMA adsorption in PU/G nanocomposite, where: ○ hydrogen, ● carbon, ● oxygen and, ● nitrogen, respectively.

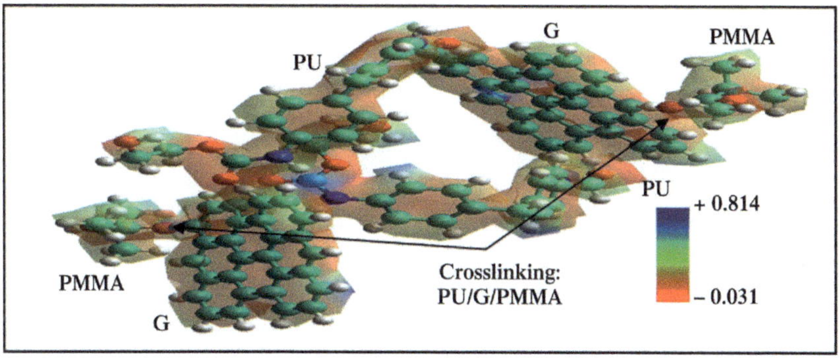

Figure 12.7 MESP about of PMMA adsorption in PU/G nanocomposite, where: ○ hydrogen, ● carbon, ● oxygen and, ● nitrogen, respectively.

12.3.2.3 FTIR

Table 12.5 shows the main allocations corresponding to adsorption of PMMA on PU/G nanocomposite. The main bands of PMMA were appreciated in which the methyl groups were localized in the range of 3157–3020 and 1443–983 cm^{-1} [54]. An intense peak appeared at 2075–2069 and 1724–1603 cm^{-1} were attributed to carbonyl group stretching. The broad peak ranging from 1327–1221 cm^{-1} can be attributed to CO (ester bond) stretching vibration. At 1075 cm^{-1} corresponded to OCH_3 bond [55–56]. In the meanwhile, the nanocomposite presented the bands between 3222 and 3169 cm^{-1} and are attributed at the existence of hydrogen bonds due to the crosslinking G/PMMA. The methyl group corresponds to symmetric and asymmetric stretching vibrations and were appreciated at 3225–2995 cm^{-1}, although the C = C bonds and carbonyl of quinine at 1674–1422 cm^{-1} were localized [57]. The hydroxyl group presents at 3485–3453 cm^{-1} an indication of the presence of

Table 12.5 FTIR results of adsorption of PMMA in PU/G nanocomposite

Bond	Molecule	Frequency (cm^{-1})
O–H stretching	PU/G/PMMA	3485–3453
N–H	PU	3469–3435
C–H symmetric and asymmetric stretching	PU, G, PMMA	3225–2992
CHO stretching	G/PMMA	3222–3169
C–H stretching	PMMA	3157–3020, 1443–983
C=O	PU	2056–1908
C–N	PU	1908, 1715
C–C	PU	1796–1737
C=C stretching	G	1674–1422
C–O–H	PU	1544, 1525
N–H	PU	1493–1422
C–H	PU	1429–873
C–H	G	1403–783
C=O stretching	PMMA	2075–2069, 1724–1603
C–O vibration	PMMA	1327, 1221
O–CH_3 stretching	PMMA	1075
C–C	PU, PMMA, G	785
N–H	PU	469–388

inter and intra molecular hydrogen bonds, verification that the adsorption the PMMA was realized between CH bonds of graphene [58] due to the change in the sp^2 hybridized carbons to sp^3 hybridized carbons, which could be generated for the covalent attachment of the graphene with PMMA.

12.3.3 Temperature effect in PU/G/PMMA nanocomposites

12.3.3.1 Geometry optimization

Monte Carlo modeling determines the changes in Gibbs free energy with respect to temperature seen in Table 12.6, which are caused by the increased binding energy and entropy decrease in order to maintain the hydrogens bonds in thermodynamic equilibrium. In addition, it is appreciated that the temperature slightly modifies the density of materials. It also shows that there is a change of energy between 0.102 and 0.035% at a temperature range between 308.15 and 313.15K, respectively. With respect to body temperature (310.15K), this process is attributed to the excellent resistance intragranular avoiding degradation of PU/G/PMMA nanocomposite. The formation of hydrogen bonds during adsorption of PMMA produces large changes in enthalpy, being more negative as well as entropy, which arises due to the hydrogens in position "to," which are more resistant compared to the "ortho" [59] position.

Table 12.6 Temperature effect on the thermodynamics properties of PU/G/PMMA nanocomposite

Properties/ temperature	298.15 (K)	308.15 (K)	310.15 (K)	313.15 (K)
ΔG (kcal/mol)	− 22,833.149	− 22,537.575	− 22,560.761	− 22,552.692
Log *P*	24.67	24.77	24.77	24.77

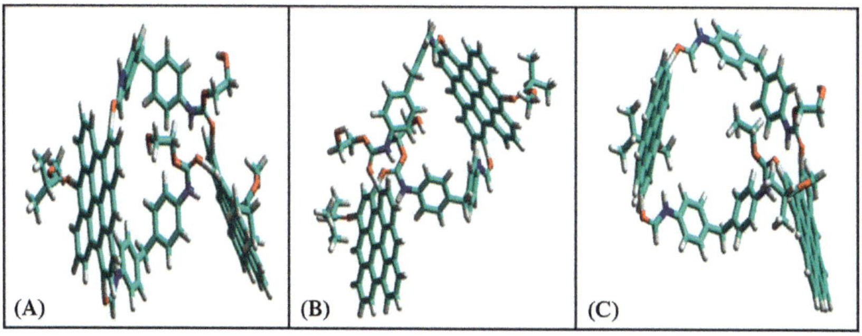

Figure 12.8 Molecular geometry optimized of PU/Graphene/PMMA with several temperatures: (A) 308.15, (B) 310.15, and (C) 313.15K, where: ○ hydrogen, ● carbon, ● oxygen, and ● nitrogen, respectively.

In Fig. 12.8 the geometry optimization is observed after calculating the minimum energy at different temperatures using Monte Carlo modeling, where rearrangements of the structures caused by temperature changes are seen. However, the new hydrogen bonds generated are not affected. It is also appreciated that graphene atoms have an increased mobility at 313.15K favoring generate more reactive sites for the adsorption of PMMA [60].

Dipole moments values calculated were of 2076, 3377, and 3868 Debyes at 308.15, 310.15, and 313.15K, respectively, where the forces of intermolecular attractions in the PU/G/PMMA nanocomposite are strong regardless of temperature changes.

12.3.3.2 Electrostatic potential map (MESP)

Fig. 12.9 shows the electrostatic potential determined by Monte Carlo that were analyzed at several temperatures. The limits of the isodensity surfaces were observed, i.e., the distribution of the electron density includes in negative regions of carbonyls bonds using the red color, which indicates the π interactions because the main electronic transitions in the system are of the type $\pi \rightarrow \pi^*$, $n \rightarrow \pi^*$, π^* (function as electron acceptors) and interaction π (is a donor). While the positives are assigned to the CH bonds, it also shows that the potential provides electronic restructuring caused by different molecular interactions in the effective point load

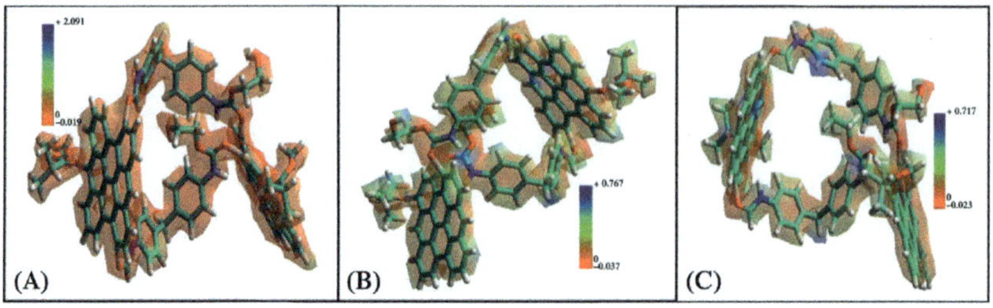

Figure 12.9 MESP of PU/Grafeno/PMMA, to different temperatures: (A) 308.15, (B) 310.15, and (C) 313.15K, where: ○ hydrogen, ● carbon, ● oxygen and, ● nitrogen, respectively.

with respect to temperatures [61–63]. Finally, it is noted that the positive value of the MESP decreases which is due to rearrangement of orbitals in the PU/G/PMMA nanocomposite. However, the material is hydrophobic, i.e., there is no affinity to polar solvents of the human body generating degradation enzyme with which the prosthesis had losses of mechanical properties.

12.3.3.3 FTIR

Table 12.7 shows the FTIR signals obtained at different temperatures using the Monte Carlo simulation, where it is seen that there are shifts in frequencies caused by the effect of temperature. However, they are not of great concern due to the appreciative change between 0102 and 0035% compared to human body temperature. At 3682–2660 cm^{-1} were appreciated; the stretching vibrations of CH groups corresponds to graphene [64]. These bands originated during the crosslinking and adsorption of PU and PMMA on graphene structure and were corroborated at 1437–693 cm^{-1} due to the H-termination of graphene structure [65]. The range between 2244 and 1563 cm^{-1} was attributed to a CN vibration. The intense bands at 2183–1857 cm^{-1} are unequivocally due to the stretching vibrational mode of the carbonyl due to PU and PMMA.

The peak presents a minimum change with the temperature that is not affected by the structure of PU/G/PMMA besides these vibrations of the carbonyl groups within the chains in which nanocomposites are organized [66]. The band at 1395–1215 cm^{-1} were assigned to OH stretching vibrations due to adsorption PU/G and PMMA/G, respectively. The PMMA analysis shows that the CH symmetric and asymmetric stretching were attributed at 3517–2548 cm^{-1}; also, the C = C stretching vibration was assigned to 1638–1578 cm^{-1}; at 1478–864 cm^{-1} were appreciated the CH vibrations of PU and PMMA, respectively [67]. The signal between 3799 and 3210 cm^{-1} corresponds to crosslinking PU/G in where only shifts in wavelength are present due to interactions $\pi\pi$ characteristics of graphene structure and the carbonyl group of the PU that are modified by temperature effects. Finally, the adsorption of PMMA was verified in the range of 3208–389 cm^{-1}.

Table 12.7 FTIR results of the temperature effect on the thermodynamics properties of PU/G/PMMA nanocomposite

Bond	Molecule	308.15 (K)	310.15 (K)	313.15 (K)
N–H	PU	3699, 3292	3317–3220	3591–3225
C–H	PU	3646–2776	3461–2752	3652–2710
C–H	PMMA	3517–2548	3664–2655	3522–2787
O–H	PU/G	3488–3210	3582–3429	3799–3321
C–H	G	3556–2719	3682–2660	3494–2868
C–H (Aromático)	PU	3481–2757	3659–2935	3210–2326
CHO	G/PMMA	3315–3208	3364–3321	3389–3354
C–N	PU	2244–1665	2418–1595	2452–1563
C = O	PU	2183–1857	2091–1977	2055–1984
C = O	PMMA	2029	2160–1988	2165–1933
C–C	PU	1891–1536	1847, 1650, 1513, 1493	1810–1550
C–C	G	1747–1368	1811–1352	1792–1369
C–O	PU	1661–1264	–	1563–1184
O–H	PU	1601–1498	1546	1572–1468
C = O	PMMA	1638–1578	1638–1597	1638–1594
N–H	PU	1676–1443	1655–1482	1692–1449
C–H	PMMA	1462–958	1478–925	1477–864
C–H	PU	1426–1345	1465–889	1483–794
C–H (Aromático)	PU	1415–907	1436–842	1465–871
C–H	G	1417–890	1400–831	1437–693
O–H	G	1395–1215	1259–1293	1278–1325
C–C	PU	1134–649	1045–989	1069–785
C–C	PMMA	1019–947	1025–967	1034–981
C–C	G	1185–467	1017–431	984–415
C–C (Aromático)	PU	1067–531	1005–437	1036–457

12.4 Conclusions

The PU/G crosslinking was carried out by forming hydrogen bonds in CH (graphene) and C = O (PU), which was verified by FTIR vibration at 3512, 1732, and 1644 cm^{-1}. Also, in the MESP the electronic distribution of loads were appeared on these bonds. It was determined that the optimum ratio for the absorption of PMMA in the PU/G by the calculations of Gibbs free energy (ΔG), where the value of the ΔG was $-22{,}876.4608$ kcal/mol because it represents the minimum energy for the spontaneous reaction. The partition coefficient (log P) indicates a positive value, thus it is verified that the material structure PU/G/PMMA nanocomposite will not have an affinity to polar solutions in the human body. PMMA adsorption was analyzed by FTIR where adsorption was appreciation through the C = O and

CH (graphene) in the range of 3222–3169 cm^{-1} in addition to the symmetric and asymmetric stretching vibrations of the methyl groups from 3225 to 2992 cm^{-1}; this is attributed to the sp^2 hybridizations characteristics of grapheme, while the MESP determines the distribution of nucleophilic and electrophilics in the adsorption zones.

Monte Carlo modeling along the analysis of PU/G/PMMA nanocomposite at 308.15, 310.15, and 313.15K, respectively, determines the effect of temperature on the structural properties. ΔG values determine a change between 0.035% and 0.102% with respect to the human body temperature, indicating resistance to degradation by polar solutions as the partition coefficient again presented a positive value, i.e., character hydrophobic. By the FTIR, it was determined that there are small shifts in the wavelengths due to rearrangement of the structure at different temperatures. The MESP showed an increase in the nucleophilic areas, which coincides with changing temperatures since they are directly related. Thus, these results are excellent because the prosthesis will be able to retain their structural properties regardless of the temperature presented by the human body.

References

[1] E. Lewars, Computational Chemistry. Introduction to the Theory and Applications of Molecular and Quantum Mechanics, second ed, Springer, Netherlands, 2011.

[2] D.J. Griffiths, The time-independent Schrodinger equation, in: D.J. Griffiths (Ed.), Introduction to Quantum Mechanics, Prentice Hall, USA, 1995, pp. 20–68.

[3] C.F. Lizalda-Aponte, Modelado molecular de la interacción enzima-ligando (docking) entre la ciclooxigenasa 2 y flavonas y flavonoles de *Swinglea glutinosa*, Bs. Thesis, Universidad Tecnológica de Pereira, España, <http://repositorio.utp.edu.co/dspace/bitstream/11059/3045/1/5727L789.pdf>.

[4] R. Müller, H. Wiesner, Teaching quantum mechanics on an introductory level, Am. J. Phys 70 (2002) 200–209.

[5] R.C. Greenhow, J.A.D. Matthew, Continuum computer solutions on the Schrödinger equation, Am. J. Phys 60 (1992) 655–663.

[6] D. Zorrilla-Cuenca, Cálculo de propiedades atómicas mediante bases no estándar, Ph.D thesis, Universidad de Cadiz, 1999, <http://www2.uca.es/dept/quimica_fisica/documentos/CTPM/Tesina1.pdf>.

[7] F. Jensen, Introduction to Computational Chemistry, second ed, Wiley, England, 2007.

[8] D. Frenkel, B. Smith, Understanding Molecular Simulation: From Algortinms to Applications, second ed, Academic Press, USA, 2002.

[9] D.C. Young, Computational Chemistry, A Practical Guide for Applying Techniques to Real-World Problems, Wiley-Interscience, New York, 2001.

[10] G. Orozco-Hernández, J.J. Olaya-Florez, E. Restrrepo-Parra, Fundamentals of materials simulation using Monte Carlo method, Adv. Inv. Ing 11 (2014) 100–110.

[11] G.M. Whitesides, Nanoscience, nanotechnology and chemistry, Small 1 (2005) 172–179.

[12] C.P. Poole, F.J. Owens, Introduction to Nanotechnology, first ed, Wiley-Interscience, Canada, 2003.

[13] D. Blanco-Bea, A. Pérez-Tejeda, A. Acuña-Pardo, J. Carreño-Cuador, Nanomedicina: aspectos generales de un futuro promisorio Nanomedicine: general aspects of a promissory future, Rev. Hab. Cienc. Med 10 (2011), <http://scielo.sld.cu/scielo.php?script=sci_arttext&pid=S1729-519X2011000300018>.
[14] S. Fernandez-Gil, Rev. Cienc (2014), <http://revistacmc.jgcalleja.es/?p=571>.
[15] E.T. Thostenson, C. Li, T.W. Chou, Nanocompoistes in context, Comp. Sci. Techn 65 (2005) 491–516.
[16] P.H. Cury-Camargo, K.G. Satyanarayana, F. Wypych, Nanocomposites: synthesis, structure, properties and new application opportunities, Mater. Res 12 (2009) 1–39.
[17] E.A. Moncada-Acevedo, R. Quijada, M.Y. Pedram, Use of PP grafted with itaconic acid as new compatibilizer for PP/clay nanocomposites, Macromol. Chem. Phys 207 (2006) 1309–1320.
[18] I.Y. Jeon, J.B. Baek, Nanocomposites derived from polymers and inorganic nanoparticles, Materials 3 (2010) 3654–3674.
[19] M. Castelain, H.J. Salavagione, G. Martinez, Funcionalizacion de nanotubos de carbono y grafeno con polimeros mediante quimica CLICK, Rev. Iberoamer. Polim 12 (2011) 239–254.
[20] L. Delva, K. Ragaert, J. Degrieck, L. Cardon, The effect of multiple extrusions on the properties of montmorillonite filled polypropylene, Polymers 6 (2014) 2912–2927.
[21] R.J. Young, P.A. Lovell, Concepts and nomenclature. In: R.J. Young, P.A. Lovell (Eds.), Introduction to Polymers, CRC Press, United Kingdom, pp. 3–14.
[22] J. Mondragon-Aguilar, Fibras textiles (2002) 1–17. <http://www.artisam.org/descargas/pdf/POLIMEROS.pdf>.
[23] B.S. Gregori, Y. Hoz, L. Alba, M. Guerra, E. Ochoa, Sintesis de poliuretanos, caracterizacion por espectroscopia infrarroja, ICIDCA XLII (2008) 3–7.
[24] J. Vega-Baudrit, M. Sibaja-Ballesteros, M.E. Hernandez-Hernandez, P. Alvarado-Aguilar, Sintesis y caracterizacion de redes elastomericas de poliuretano (PU) utilizado en la elaboracion de calzado, II, Utilizacion de modelos matematicos, Rev. Iberoamer. Polim 7 (2006) 174–198.
[25] Materias primas para la fabricacion de poliuretanos. <http://www.eis.uva.es/~macromol/curso05-06/pu/materiasprimas.htm>.
[26] Catalizadores de amina para poliuretano: pautas de manipulacion segura American Chemistry Council, *ISSUE*, AX173 (2011) 1–19. <https://polyurethane.americanchemistry.com/Resources-and-Document-Library/12017.PDF>.
[27] P.C. Caracciolo, G.A. Abraham, Poliuretanos biomédicos: síntesis, propiedades, procesamiento y aplicaciones, in: H.C. Sousa, A. Brag, A. Sosnik (Eds.), Biomateriais Aplicados ao Desenvolvimento de Sistemas Terapeuticos Avanzados, Universidad de Coimbra, 2015, pp. 147–181.
[28] J.V. Cauich-Rodríguez, L.H. Chan-Chan, F. Hernandez-Sánchez, J.M. Cervantes-Uc, Degradation of polyurethanes for cardiovascular applications, INTECH (2013) 51–82.
[29] V. Kanyanta, A. Ivankovic, Mechanical characterisation of polyurethane elastomer for biomedical applications, J. Mech. Behav. Biomed. Mat 3 (2010) 51–62.
[30] N.A. Rangel, H.E. De-Alva, J. Romero, J.L. Rivera, A. Álvarez, E. García, Síntesis y caracterización de materiales reforzados (composites) de poliuretano poroso/hidroxiapatita, Rev. Iberoamer. Polim 8 (2007) 99–111.
[31] D.W. Van-Krevelen, Their correlation with chemical structure; their numerical estimation and prediction from additive group contributions, in: D.W. Van-Krevelen (Ed.), Properties of Polymers, Elsevier, Netherlands, 1997, pp. 1–898.

[32] Y. Kobayashi, A. Kurosawa, D. Nagao, M. Konno, Fabrication of barium titanate nanoparticles-PMMA composite films and their dielectric properties, Polym. Eng. Sci 49 (2009) 1069–1075.
[33] W.R. Zeng, S.F. Li, W.K. Chow, Preliminary studies on burning behavior of polymethylmethacrylate (PMMA), J. Fire Sci 20 (2002) 297–317.
[34] L.W. McKeen, Plastics used in medical devices, in: Modjarrad, Ebnesajjad (Eds.), Handbook of Polymer Applications in Medicine and Medical Devices, 1st edition, Elsevier, USA, 2013, pp. 21–53.
[35] S.J. Kim, B. Choi, K.S. Kim, W.J. Bae, S.H. Hong, J.Y. Lee, et al., The potential role of polymethyl methacrylate as a new packaging material for the implantable medical device in the bladder, BioMed Res. Int 2015 (2015) 1–8.
[36] S.M. Zebarjad, S.A. Sajjadi, T.E. Sdrabadi, A. Yaghmaei, B. Naderi, A study on mechanical properties of PMMA/hydroxyapatite nanocomposite, Engineering 3 (2011) 795–801.
[37] J. Todd, Polymethylmethacrylate: properties and contemporary uses in orthopaedics, J. Am. Acad. Orthopaedic Surg 18 (2010) 297–305.
[38] A.H. Castro-Neto, F. Guinea, N.M.R. Peres, K.S. Novoselov, A.K. Geim, The electronic properties of grapheme, Rev. Mod. Phys 81 (2009) 109–162.
[39] J.H. Son, T.R. Rybolt, Force field based MM2 molecule-surface binding energies for graphite and grapheme, Graphene 2 (2013) 18–34.
[40] J. Liu, Y. Xue, M. Zhang, L. Dai, Graphene-based materials for energy applications, MRS Bull 37 (2012) 1265–1272.
[41] A.H. Castro-Neto, F. Guinea, N.M.R. Peres, Drawing conclusions from graphene, Phys. World (2006) 1–5.
[42] G. Chen, W. Weng, D. Wu, C. Wu, J. Lu, P. Wang, Preparation and characterization of graphite nanosheets from ultrasonic powdering technique, Carbon 42 (2004) 753–759.
[43] J. Scott, Mechanical and electrical properties of graphene sheets, Ph. D thesis, Cornell University, 2008. <http://www.mceuengroup.lassp.cornell.edu/sites/mceuen/files/publications/Thesis_Bunch.pdf>.
[44] E. Pop, V. Varshney, A.K. Roy, Thermal properties of graphene: Fundamentals and applications, MRS Bull 37 (2012) 1273–1281.
[45] T.K. Das, S. Prusty, Recent advances in applications of graphene, Int. J. Chem. Sci. Appl 4 (2013) 39–55.
[46] W. Paul, C.P. Sharma, Blood compatibility and biomedical applications of graphene, Trends Biomater. Artif. Organs 25 (2011) 91–94.
[47] D. Bitounis, H. Ali-Boucetta, B. Hee-Hong, D.H. Min, K. Kostarelos, Prospects and challenges of graphene in biomedical applications, Adv. Mater 25 (2013) 2258–2268.
[48] B. Fernández-d'Arlas, L. Rueda, K. Kaba, I. Mondragon, A. Eceiza, Microdomain composition and properties differences of biodegradable polyurethanes based on MDI and HDI, Polym. Eng. Sci 48 (2008) 519–529.
[49] S. Ibrahim, A. Ahmad, N.S. Mohamed, Characterization of novel castor oil-based polyurethane polymer electrolytes, Polymers 7 (2015) 747–759.
[50] K.C. Pradhan, P.L. Nayak, Synthesis and characterization of polyurethane nanocomposite from castor oil-hexamethylene diisocyanate (HMDI), Adv. Appl. Sci. Res 3 (2012) 3045–3052.
[51] D. Paul, S. Paul, N. Roohpour, M. Wilks, P. Vadgama, Antimicrobial, mechanical and thermal studies of silver particle-loaded polyurethane, J. Funct. Biomater 4 (2013) 358–375.

[52] S. Bykkam, V. Rao, K. Shilpa-Chakra, T. Thunugunta, Synthesis and characterization of graphene oxide and its antimicrobial activity against klebseilla and *Staphylococus*, Intl. J. Adv. Biotechnol. Res 4 (2013) 142–146.

[53] F.Y. Ban, S.R. Majid, N.M. Huang, H.N. Lim, Graphene oxide and its electrochemical performance, Int. J. Electrochem. Sci 7 (2012) 4345–4351.

[54] E.D. Emmons, R.G. Kraus, S.S. Duvvuri, J.S. Thompson, A.M. Covington, High-pressure infrared absorption spectroscopy of poly(methyl methacrylate), J. Polym. Sci. Pol. Phys 45 (2007) 358–367.

[55] N.D. Singho, N.A. Che-Lah, M.R. Johan, R. Ahmad, FTIR studies on silver-Poly (methylmethacrylate) nanocomposites via in-situ polymerization technique, Int. J. Electrochem. Sci 7 (2012) 5596–5603.

[56] E.M. Abdelrazek, A.M. Hezma, A. El-khodary, A.M. Elzayat, Spectroscopic studies and thermal properties of PCL/PMMA biopolymer blend, Egypt. J. Basic Appl. Sci 3 (2016) 10–15.

[57] S. Mohamadi, Preparation and characterization of PVDF/PMMA/Graphene polymer blend nanocomposites by using ATR-FTIR technique, *INTECH*, pp. 213–232.

[58] L. Gong, I.A. Kinloch, R.J. Young, I. Riaz, R. Jalil, K.S. Novoselov, Interfacial stress transfer in a graphene monolayer nanocomposite, Adv. Mater 22 (2010) 2694–2697.

[59] I.I. Geru, Water structure, quantum nature of hydrogen bonds, in: G. Duca (Ed.), Management of Water Quality in Moldova, Water Science and Technology library, Springer International Publishing, Switzerland, 2014, pp. 1–34.

[60] A. Mohsin, Graphene synthesis and characterization on copper, M. Sci. thesis, University of Iowa, 2012. <http://ir.uiowa.edu/etd/3354>.

[61] W.J. Hehre, A Guide to Molecular Mechanics and Quantum Chemical Calculations, first ed, Wavefunction, USA, 2003.

[62] L. Jun-na, C. Zhi-Rang, Y. Shen-Fang, J. Zhejiang, Study on the prediction of visible absorption maxima of azobenzene compounds, J. Zhejiang Univ. Sci 6B (2005) 584–590.

[63] W.N. Leng, J. Grunden, G.P. Bartholomew, G.C. Bazan, A.M. Kelley, Vibratonal and electronic spectroscopy of a donor-acceptor substituted distyrylbenzene andits covalent diers, J. Phys. Chem. A 108A (2004) 10050–10059.

[64] M. Abha, T. Pawan, M.K. Singh, D.S. Misra, FTIR studies of nitrogen doped carbon nanotubes, Diamond Relat. Mater 15 (2006) 385–388.

[65] Y. Xue, D. Yu, L. Dai, R. Wang, D. Li, A. Roy, et al., Three-dimensional B,N-doped graphene foam as a metal-free catalyst for oxygen reduction reaction, Phys. Chem. Chem. Phys 15 (2013) 12220–12226.

[66] P.N. Simoes, I. Reva, L.M. Pedroso, R. Fausto, A.A. Portugal, Low-temperature FTIR spectroscopic and theoretical study on an energetic nitroimine: dinitroammeline (DNAM), J. Phys. Chem. A 112 (2008) 3432–3443.

[67] T. Otsuka, Y. Chujo, Poly(methyl methacrylate) (PMMA)-based hybrid materials with reactive zirconium oxide nanocrystals, Polym. J 42 (2010) 58–65.

Polymeric thin film composite membrane for CO_2 separation

13

Kar Chun Wong, Pei Sean Goh and Ahamad Fauzi Ismail
Universiti Teknologi Malaysia (UTM), Skudai, Johor, Malaysia

Chapter Outline

13.1 Introduction

Carbon dioxide (CO_2) is one of the major greenhouse gases that contributes to global warming whereby the drastic increase of the globe temperature changes the world's climate patterns and upsets the health of our ecological system. This scenario will ultimately lead to the loss of crops production and natural biodiversity [1]. Anthropogenic activities such as biogas anaerobic digestion, coal gasification, fossil fuel combustion, and natural gas exploration are the sources of emission of large amount of CO_2 into the atmosphere. Since it is anticipated that the global population will expand close to 10 billion inhabitants by year 2050 [2], increase in the world resources and energy demand is inevitable and will only bring about more greenhouse gases emission [3]. Apart from the aforementioned impact of CO_2 at global scale, a return to natural gas processing presents a serious problem because it can lower the heating value of the gas stream. Additionally, CO_2 acidic nature decreases of the gas stream pH and causes corrosion to the pipelines, which

Hybrid Polymer Composite Materials: Processing. DOI: http://dx.doi.org/10.1016/B978-0-08-100789-1.00013-7

severely impairs the transportation system [4]. Therefore, removal of CO_2 has become a hot research topic recently and CO_2 capture and storage (CCS) is widely accepted as a viable solution to mitigate this issue [5–7]. However, considering that CO_2 separation alone constitutes nearly 80% of the total cost involved in CO_2 sequestration [8–11], developing cost- and energy-efficient technologies for capturing CO_2 from the source of emission plays a crucial part in the implementation of CCS [8,9,12].

Chemical absorption, pressure swing adsorption, and cryogenic separation are the most common techniques employed for the aforementioned objectives. Even though these technologies have reached developmental maturity, they are constantly plagued by problems such as flooding, foaming, entrainment, weeping, and absorbents degradation. Additionally, their high capital and maintenance costs, great energy consumption and operation complexity force investors and scientists to look for better alternatives. Membrane technology prevails as a feasible candidate because it is a relatively economic, energy efficient, and is environmentally benign compared to the conventional approaches [12–15]. Table 13.1 contrasted membrane separation with the chemical adsorption and cryogenic separation methods.

Table 13.1 Comparison between different CO_2 separation technologies [4,10,16]

Technologies	Advantages	Disadvantages
Chemical absorption	• >95% CO_2 purity • Mature	• Absorbents (amines) are corrosive and susceptible to degradation • High capital cost • High energy consumption: 4–6 MJ/kg CO_2 • High operating and maintenance cost: \$40–70/ ton CO_2 • Complicated and large equipment • Risk from chemical handling
Cryogenic separation	• >99.95% purity • no chemical adsorbents • produce liquid CO_2 (easily transported)	• Required costly conditioning of feed gas to remove water, O_2, NO_x and SO_x content • High operation cost: \$32.7/ton CO_2 (lab-scale) • High energy consumption • Solid CO_2 deposition reduces efficiency
Membrane	• Simple process design (almost no mechanical part) • Energy efficient • Modular construction (easy scale up and small footprint)	• Relative low purity compared to matured technologies • Performance degrade over long-term operation

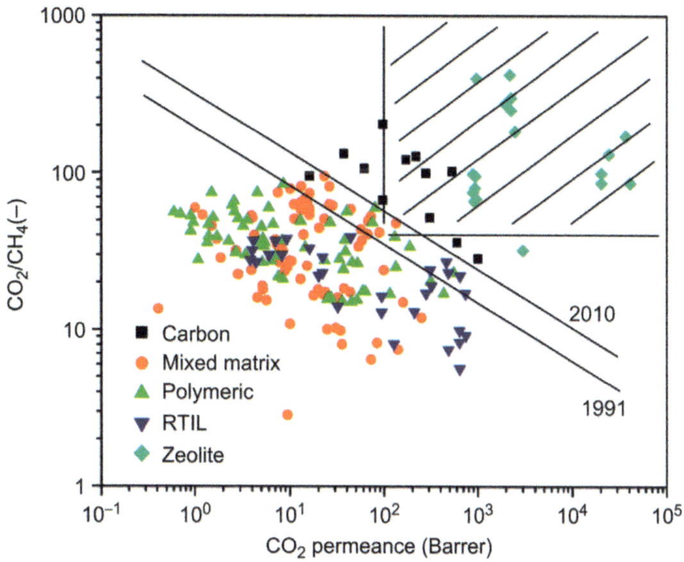

Figure 13.1 Graphical illustration of CO_2/CH_4 selectivity versus CO_2 permeance of different membrane systems including inorganic membranes for comparison [20].

It has been reported that a combination of oxygen-enriched biogas production and membrane separation was able to achieve separation target of 90% CO_2 purity and 90% CO_2 capture ratio using only a quarter of the energy needed by amine adsorption technology [8]. Similarly, Hussain *et al.* [17] suggested that the deployment of the a N_2-selective/CO_2-selective hybrid membrane system can capture CO_2 from flue gas with a low energy requirement of 1 GJ/ton which cost about $20.5/ton. Furthermore, the membrane simple design and ease of operation reduces the reliance on trained labors while its compactness and modularity allows it to be retrofitted into the existing industrial system without taking up much valuable space [18,19].

Attributed by their low material cost and high processability, polymeric membrane is widely used in the gas separation industry. However, the selectivity-permeance tradeoff (Robeson tradeoff) of polymeric material limits the performance achievable by this class of membrane. Fig. 13.1 illustrates the advancement in membrane technology over the past two decades as represented by the shift of the upper boundary from year 1991 to year 2010. Clearly, the performance of pure polymeric membrane is lagging behind other class of membranes such as mixed matrix membrane (MMM) and inorganic membrane.

In order to understand the performance limitation discussed above, let us first understand the mechanism involved in the separation process using polymeric membrane. Transport of gas molecules relies on the availability of free volume in the polymer matrix and the driving force supplied so that the molecules have adequate energy to overcome the interchain forces [21]. Table 13.2 depicts the general mechanisms governing the transport of gas across the polymeric membrane. Since

Table 13.2 Summary of membrane transport mechanisms [22]

Membrane morphology	Transport mechanism	Gas selectivity
Macroporous (pore diameter, $d_p > 50$ nm)	Convective pore flow	None ($\alpha = 1$)
Mesoporous ($2 \text{ nm} < d_p < 50$ nm)	Knudsen diffusion	$\alpha_{Knudsen\ (a/b)} = \sqrt{M_b}/\sqrt{M_a}$
Microporous ($d_p < 2$ nm)	where: Knudsen number, $K_n = \lambda/d_p >> 1$; λ = gas mean free path length	where: M_i = molecular weight of gas i
Dense	Solution–diffusion	$\alpha_{(a/b)} = P_a/P_b = D_a\ S_a/D_b\ S_b$ where: P_i = permeability of i; D_i = diffusivity coefficient of i; S_i = solubility coefficient of i

the skin layer of polymeric membrane has a dense structure, the transport of gas across the thin film follows the solution–diffusion mechanism where the permeability, P of a gas species is defined as the product of solubilty, S, and diffusivity, D across the polymeric chains of the membrane. On the other hand, pore flow mechanism predominates in the porous support layer due to its low resistivity toward mass transfer that has a negligible impact on the gas permeance. If $K_n \geq 1$ separation is based on the difference in gas molecular mass, as the gas molecules interact with the pore wall and other gases [22], but molecules with similar kinetic diameter is not distinguishable [21]. In this condition, Knudsen diffusion dominates.

The flow rate of gas transported via Knudsen diffusion can be calculated using Eq. (13.1):

$$N_t = nN_p = \frac{n\pi d_p{}^3}{3(2\pi RTM)^{0.5}} \frac{\Delta P}{L} \tag{13.1}$$

where N_t = total molar flow rate (mol s^{-1}), N_p = molar flow rate through single pore (mol s^{-1}), n = number of pores, d_p = pore diameter (m), ΔP = pressure differential (bar), R = universal gas constant (bar m^3 mol^{-1} K^{-1}), T = temperature (K), M = molar mass of gas (kg mol^{-1}), and L = pore length (m) [22].

Striking a balance between the penetrant permeability and selectivity is a challenging task for traditional membranes because the transport rate of penetrant across the membrane is inversely correlated with the thickness [23] and cross-linking degree [24] while reducing the cross-linking could seriously impairs the molecular sieving property of the membrane. In order to enhance the performance of the polymeric membrane, two or more polymers with different properties have been combined to give a composite membrane, which exhibits better separation performance and stability. Generally, the properties of a composite membrane can be tuned by manipulating the portion of glassy and rubbery materials used during fabrication whereby the rubbery segment provides high gas permeation while the glassy segment regulates the selectivity of the membrane.

13.2 Thin film composite

Composite membrane can be fabricated via a number of techniques such as blending, coating and copolymerization but this chapter focuses on thin film composite (TFC) fabricated via interfacial polymerization (IP). IP occurs at the interface between two immiscible aqueous and organic phases near the support surface whereby the dissolved monomers, usually polyamines and polyacyl halides, undergo polycondensation reaction [21,25] to form the polyamides (PA) skin layer. Since the reaction is instantaneous [26] and self-terminating, IP is capable of forming ultrathin (0.1–1.0 μm) defect-free selective layer [15,27–29] and the wide variety of monomers enables the thin film chemistry to be freely tuned [25,30]. Furthermore, this technique is extremely scalable [13,31,32] and is already widely adopted in the industry. These advantages make IP more appealing than other methods. IP usually involves three major steps which are:

1. impregnation of support or substrate layer with aqueous phase (amine solution)
2. removal of excess amine solution by draining in a vertical position [15,31], dapping with tissue paper [33,34] or filter paper [35], rolled off with soft rubber brayer [36–40] or blown off with air-knife [41–43].
3. reaction with acyl chloride solution for a certain duration by bringing the impregnated substrate into contact with organic phase.

Alternately, solutions used in step (1) and (3) can be exchanged.

As the two phases are brought into contact, amine molecules diffuse more readily across the liquid–liquid interface solutions to embark in a complex mix of mass transport, polycondensation reaction and film growth processes with the acid chlorides [26,29,44,45] (Fig. 13.2A). The PA chains formed quickly band together into loose clumps, which are suspended above the support pores. As the amine monomer continuously diffuse out of the support pores and reacts instantaneously with the acid chloride, the tufts grow larger resulting in grainy polyamide structure formation atop the pores (Fig. 13.2B). This grainy structure hinders the upward diffusion of amine monomer leading to the preference of lateral diffusion that resulted in the formation of belt-like structure (Fig. 13.2C). As the reaction proceeds, the support pores are covered by the growing thin film and the amine monomer is forced to permeate through the grains resulting in continuous growth of the grains. Eventually, the growing grains become too large and gradually merge with neighboring grains to form a dense structure that overshadows the belt-like structure (Fig. 13.2D). This morphology is generally referred as ridge–valley [46] or corrugate structures

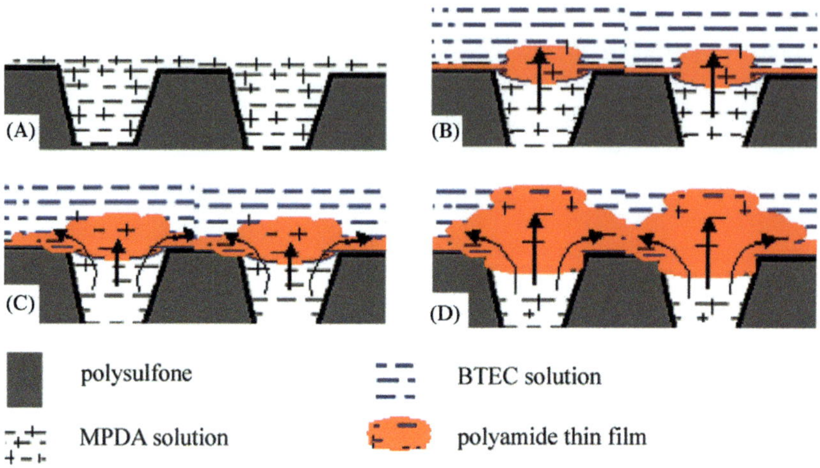

Figure 13.2 Model of thin film growth during interfacial polymerization [50].

[47,48]. As the polymerization proceeds, the reaction zone shifts perpendicularly away from the liquid–liquid interface into the organic phase and the film grows in the same direction [25,29,45,49]. At the same time, the degree of cross-linking of the film continues to increase until finally the diffusion of monomers are restricted by the dense incipient film which terminated the reaction (considering the monomers supply is not exhausted).

Another growth model was proposed by Yuan and coworkers [45] who have visually inspected the formation of thin film surrounding a pendant drop of amine solution immersed into an organic phase. They pointed out that pinhole defects tend to develop evenly on the incipient PA film during IP due to the enormous surface energy per volume of the film which makes it highly unstable (Fig. 13.3A). Amine molecules diffuse through these pinholes into the organic phase as tiny droplets of aqueous solution which later coalesce to form larger droplets due to instability of the droplets partitions. New layer of bubble-like film is formed as freshly synthesized PA precipitated at the droplet–organic interface and halts the coalescing process (Fig. 13.3B). With each successive PA layer formation, the film becomes more stable with larger bubble overlapping it smaller predecessor (Fig. 13.3C). This process continues to propagates towards the organic phase until the diffusion of amine is restricted by the nascent film (Fig. 13.3D). When the membrane is dried, these hierarchical bubble films will shrink and solidify into corrugated structures with peaks and valleys. Their observation is in line with the findings of Freger [29] and Ji [44] where a highly crosslinked region is established within the interior of the less crosslinked region of the PA film, and it is believed that this dense inner region plays an important role in the membrane separation performance [24,50]. The subsequent topics discuss about the suitable properties of nanocomposite film for gas separation and the IP parameters that can be tweaked to obtain desired film characteristics.

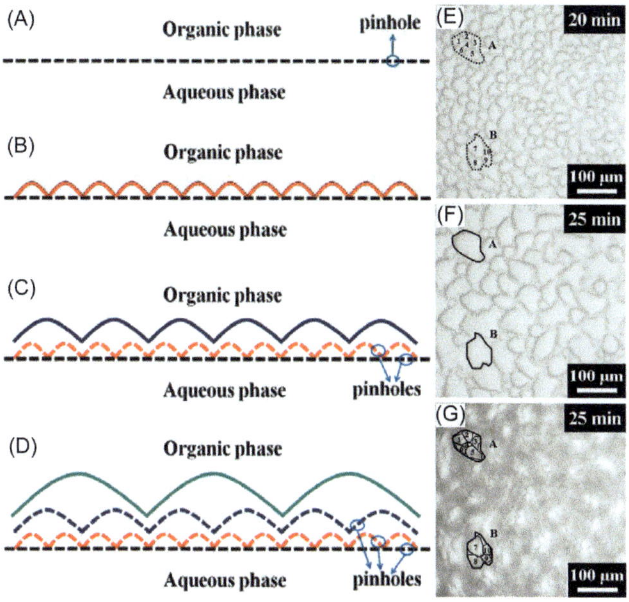

Figure 13.3 Schematic of film growth: (A) formation of nascent film with uniformly distributed pinholes; (B, C, and D) formation of second, third, and fourth generation of bubble-like films. The surface morphology of film after (E) 20 min and (F) 25 min of reaction as well as (G) internal morphology are shown [45].

13.3 Parameters of interfacial polymerization

Generally, IP is governed by the mass transfer of monomer across the interface of the immiscible phases and the polymerization between the reactive amines and acid chlorides [44]. Setting the appropriate boundary variables to control the diffusion rate of the monomer is vital for designing the PA layer [31,36,51]. Even the slightest variation in IP parameter can have tremendous impact on the formation of the ultrathin film [52]. Additionally, the properties of support layer such as pore size, pore distribution, porosity, and hydrophilicity can also affect the characteristics of the final composite membrane [43,45,53–58]. In view of the large degree of freedom in the fabrication process, optimizing the TFC membrane is a challenging task. Thus, an in-depth discussion on the critical factors affecting thin film formation in presented in the following subtopics to get to the heart of TFC fabrication. Currently, polysulfone (PSf) is the most employed material [18,35,43,54–56,59,60] to produce ultraporous support [15,61] because it is strong, thermal, and chemically stable, cheap and easily accessible [54], whereas m-phenylenediamine (MPD) and trimesoyl chloride (TMC) are the two most comprehensively studied monomers used to produce the PA thin film [62].

13.3.1 Monomer and solvent

Monomer diffusion determines the rate of polycondensation reaction, which affects both the polymerization degree and thickness of PA film [26,29]. When amine diffusivity (D_{amine}) is much greater than the amine solubility (S_{amine}) in organic phase, amines react vigorously with acid chloride to produce large amount of PA chains, instantly forming a dense film, which bring the reaction towards self-termination. Consequently, a thin but highly crosslinked PA film is formed. On the contrary, when $S_{amine} >> D_{amine}$, polycondensation proceeds slowly, which gives more time for the amine to diffuse deep into the organic phase and result in a thick but less crosslinked PA film [27]. Additionally, the low reaction rate allows acid chloride to diffuse close to the liquid–liquid interface and hydrolyze upon contact with water [63]. Hydrolysis of acid chloride can impair the formation of amide linkages and result in a defected polymer network [18]. In order to prevent this phenomenon, a high amine to acid chloride ratio is generally used. The excessive amine can promote amine diffusion and drive the reaction zone into the organic phase [29].

The thickness of the resultant film increases with the monomers concentration because a greater initial concentration gradient encourages the monomers to diffuse faster and farther away from the liquid–liquid boundary which expand the polymerization zone. Hence, more PA molecules can be synthesized to generate greater film mass [31,41,61,64,65]. This condition generally holds true before self-termination. Further raise in monomer concentration will not affect the film thickness, as the diffusion of monomers is restricted [25,49]. Besides, formation of thin film at low acid chloride concentration range usually results in more cross-linked structures than those formed at high acid chloride concentration range [27]. At low acid chloride concentration, polymerization process is driven by the diffusion of amine and the acid chloride react completely with the amine to form a defect-free amide network. At high acid chloride concentration, excess acid chloride could diffuse into the aqueous phase, react with water and introduce defects in the nascent film matrix. Similar condition could also happen if the amine concentration is too low whereby polymerization reaction proceeds at the aqueous phase due to lower diffusion rate of amine compared to the acid chloride [61,66]. In this case, competition between polycondensation and hydrolysis of acid chloride resulted in a thin and highly defect film formation.

Generally, a thin and dense PA layer is preferred for gas separation, as thinner skin layer provides a shorter flow path for gases across the membrane whereas highly crosslinked network enhances the membrane sieving property [31,36,67]. As mentioned in Section 13.1, the transport of CO_2 can be improved by tuning the chemistry of the thin film. While this is achievable by choosing monomers (especially amine) with high content of functional groups that is CO_2 soluble and reactive [68], the presence of large amount of functional groups can increase the monomers molecular size thus altering their diffusion coefficient [26,47]. Work done by Li *et al.* [47] revealed that the use of diaminopolyethylene glycol (DAmPEG, molecular weight, MW: 1000 g/mol) increased the CO_2 permeance of the resultant membrane compared to the counterpart that is produced using diethylene glycol bis(3-aminopropyl) ether

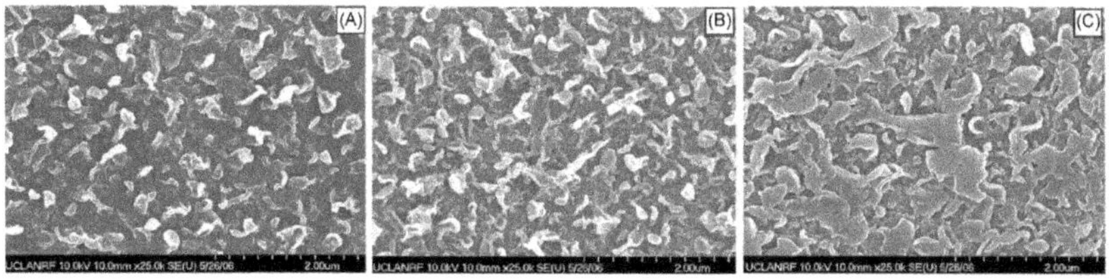

Figure 13.4 SEM and TEM images of TFC prepared using (A) hexane, (B) heptane, and (C) cyclohexane as organic solvent [24].

(DGBAmE, MW: 220 g/mol). The greater permeance enhancement was attributed to the higher EO content in DAmPEG (21 EO units) compared to DGBAmE (3 EO units), which increases the membrane overall CO_2 solubility. Yet, the TFC produced from DAmPEG suffered from low selectivity due to loss of molecular sieving property of PA film. The long chain of DAmPEG reduced its diffusivity and impeded the transport of monomer across the incipient film [29,44,45]. This limited the amine supply into the reaction zone thus, producing a thin PA layer with low cross-linking degree. Since the characteristics of monomers mutually affect one another, the concentration of amine and acid chloride should be optimized simultaneously in order to obtain an appropriate monomer ratio which endows desirable thickness and cross-linking degree [25,29,33].

The type of solvent used can influence the ability of the amine to solubilize, diffuse into the organic phase and subsequently take part in the polymerization process. Ghosh *et al.* [24] found that the surface morphology of TFC changed from nodular to more leaf-like structures when an organic solvent was used in the order of hexane < heptane < cyclohexane (Fig. 13.4). Considering the facts that the diffusivity of amine on an organic solvent decreased in the order of hexane > heptane > cyclohexane whereas its solubility decreased in the opposite order [24,45], when hexane ($S_{amine} << D_{amine}$) was used, the diffusion of amine was limited by the nascent film which yield low mass PA layer with small nodular structures. In contrast, when cyclohexane ($S_{amine} >> D_{amine}$) was used, amine supply was not instantaneously interrupted hence allowed the nodular structures to grow in size and resulted in the leaf-like morphology. Yuan *et al.* [27] suggested that the cross-linking of thin film increased in the order of cyclohexane < heptane < hexane due to the lower solubility of the amine in hexane. Their results also indicated that the use of hexane as an organic solvent could promote the CO_2-induced plasticization resistance of the membrane, probably due to the higher degree of cross-linking that has hampered the polymeric chain mobility. Therefore, the selection of an organic solvent should be carefully performed while preparing a composite via IP.

13.3.2 Support characteristics

Since the support layer hosts the polymerization solutions and provides mechanical support for the thin PA layer, the substrate properties is also an important factor in IP [15,54,69]. Ideally, an asymmetric substrate consists of a thin sponge-like layer with pore sizes smaller than the film thickness [70] (or molecular weight cut-off <300 kDa [53]) atop of a porous polymeric matric with finger-like macrovoids [59] is preferred for IP. The dense surface structure of the substrate ensures that the PA film formed remained intact under high pressure while the porous layer underneath provides the overall mechanical strength for the membrane with minimal mass transfer resistance. Li *et al.* [58] found that the PA film formed on the substrate with small pore size was relatively smooth and less compact compared to those formed on substrate with the larger pore (Fig. 13.5). In a small pore, diffusion and simple convection govern amine transport and lead to a slow supply of amine monomer into the reaction zone. Besides, the small pore is easily clogged, which limits the supply of reactant and resulted in a low mass PA film. In contrast, large pore allows the amine to rapidly transported into the organic phase. Mixing between the phases can occur at the interface due to turbulence from the fast mass transport, which leads to a vigorous polycondensation reaction. Additionally, the turbulence can swirl, twist, and band the initial PA clumps into larger globules, which results in a rough film surface with a leaf-like structure [43].

The effects of pore size have intricate links with lag time and support chemistry. The introduction of lag time or drying before bringing the impregnated support into contact with the second reaction solution greatly impacts the film formation mechanisms. An impregnation solution is able to seep deep into the substrate pores with a long lag time or drying duration. This effect is more prominent in bigger pores, which could result in the formation of PA inside the pores (Fig. 13.5) [33,56]. On the other hand, the chemical property of the support determines the wettability of the membrane by an impregnation solution. A hydrophilic support

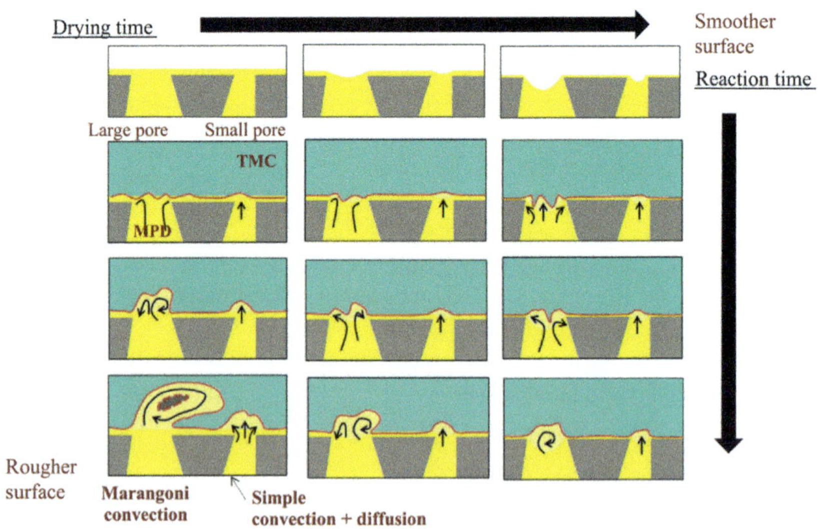

Figure 13.5 Thin film formation at varying drying time, reaction duration, and pore size [33].

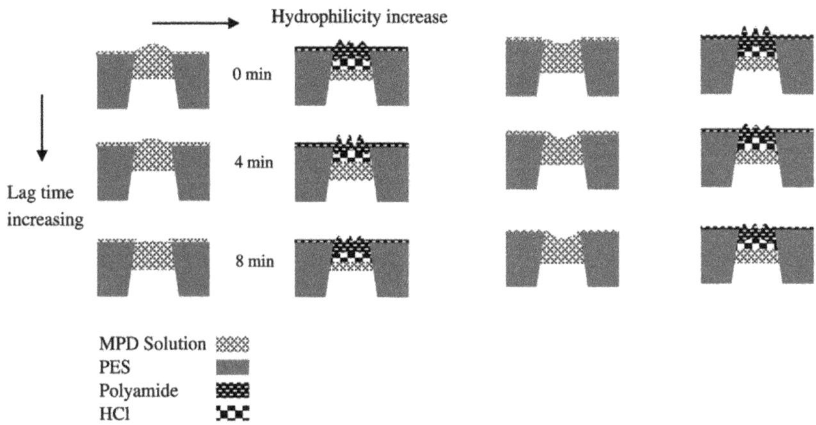

Figure 13.6 Illustration of the effect of support hydrophilicity and lag time [43].

favors wetting by the aqueous phase, leading to faster diffusion of the solution into the pores with a concave meniscus. Additionally, the presence of polar additives (residue of common pore former such as PEG and polyvinylpyrrolidone (PVP) used to increase membrane hydrophilicity) can attract amines to the pore wall, which inhibits the monomer from leaving the pore into the reaction zone during IP [71]. As such, PA formation on better wetted support is similar to that of produced with increased lag time [72]. Conversely, if an organic phase is used, a hydrophilic support surface might not be well-wetted and a convex meniscus that slightly protrudes out of surface is formed. This prevents penetration of the solution and formation of PA within the pores (Fig. 13.6) [43,54].

Although the formation of PA within the pores can prevent delamination of skin layer by anchoring it to the support layer, this risks blocking the pore, which could hamper the membrane permeability [73,74]. This problem can be overcome by using a support with a small pore size and employing pre- or posttreatment. Pretreatment, such as coating the substrate with a gutter layer of polydimethylsiloxane (PDMS), can avoid PA penetration and improves skin-support adhesion (Fig. 13.7) [18,27,75]. Since the PDMS layer function by controlling the diffusion rate of monomers, hydrolysis of acid chloride can be reduced by optimizing the supply of acid chloride [18]. Posttreatment, such as cleaning the membrane with solvents (water, ethanol and hexane) to remove residue reactants [21,22,47,52] and loose PA, can free up clogged pores [76] and increase the membrane free volume [77].

13.3.3 Additives

An acid acceptor is usually added into the aqueous phase to neutralize hydrochloric acid (by-product generated during polycondensation reaction) and catalyze the polymerization [2,24,78] to obtain highly cross-linked film layer. However, several studies suggested that the addition of trimethylamine (TEA) could lead to formation of

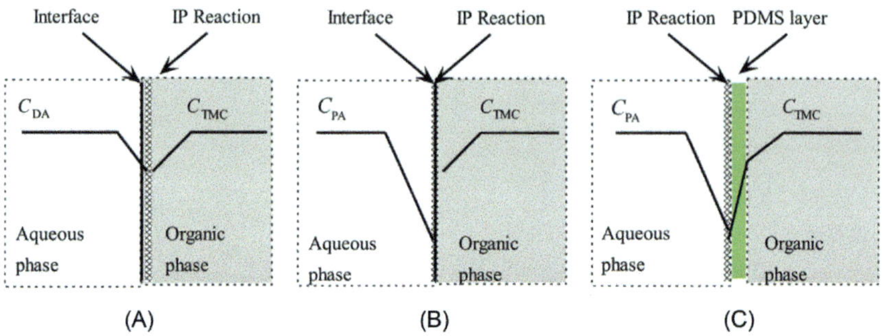

Figure 13.7 Monomer concentration profile at different interfacial polymerization processes; (A) small diamine on support without PDMS coating, (B) long chain polyamine on support without PDMS coating, (C) long-chain amine on support with PDMS coating. C_{DA} refers to the concentration of small diamine; C_{PA} is concentration of long-chain polyamine and C_{TMC} in concentration of TMC [18].

defects [61], as this basic additive can interrupt the amine diffusion and reaction [24]. Moreover, the addition of TEA increases the pH (>10) of aqueous phase, which promotes the hydrolysis of acid chloride thus reducing the PA network cross-linking degree [36]. Klaysom *et al.* [33] suggested that TEA should be used in combination with SDS (surfactant) in order to benefit from the acid acceptor addition.

A surfactant is a well-known wetting agent used to reduce the surface tension of liquid. In IP, dilute surfactant solution is used to increase the wettability of the substrate layer, which encourages the impregnation solution to be absorbed into the substrate pores [24,33,79]. Furthermore, the addition of a surfactant into the reaction solution can reduce the liquid–liquid interface tension of the immiscible phases and facilitates monomers transport across the interface, which results in abetter polymerization reaction [62,80]. For this reason, when a surfactant is used with a small quantity of TEA, the acid acceptor can be effortlessly transferred into the reaction zone to catalyze the polycondensation. The catalyzed reaction rapidly yields a large mass of PA and results in a thin but highly cross-linked film structure [29,33,81]. Yet, a surfactant concentration that is too high (>0.1 wt% in reaction solution [33]) can lead to the formation of surfactant-PA micelles, which introduces large amount of voids in the PA network [81].

13.3.4 Preparation conditions

The condition of which IP is conducted can influence the final film properties. It has been reported that impregnating the support for a long period of time can result in the formation of thicker film [65]. A long impregnation time ensures that the support pores, which act as the reactant reservoirs, are completely saturated with monomer. If the amine content in the reservoir is insufficient, supply of the monomer can be interrupted, yielding low PA mass. In addition, this effect is dependent on the support layer chemical and structural properties (see Section 13.3.2). On the other hand, IP duration or reaction duration exhibited the same effects on thin film formation as that of monomer concentration. Increased reaction duration gives more time for the polycondensation process to proceed towards completion thus more PA linkages can be formed to give thicker and highly crosslinked PA network [36,82].

Heat treatment is widely employed to improve membrane stability under high pressure by constricting the voids within the membrane to increase the matrix density and encourage the completion of polymer formation [13,67,83,84]. This enhances the sieving property of the membrane leading to higher selectivity towards small gases [85,86]. In IP, subjecting the thin film layer to heat treatment helps to remove residual solvent, dehydrate the hydrolyzed acid chloride and encourage cross-linking of the unreacted monomers [36]. However, prolonged heating of a membrane or heating at high temperature can degrade the structural integrity and introduce defects to the thin film [57].

13.4 Polyethylene oxide for membrane with high CO_2 solubility

Recently, it has been demonstrated that the performance of a composite membrane can be enhanced by fabricating a membrane with high CO_2 diffusivity, solubility, and reactivity [87–90]. Highly CO_2-philic materials such as polyethylene oxide (PEO) can be used to overcome the limitation of size-based separation discussed in Section 13.1 [91]. The polar ether oxygen groups in PEO have high affinity toward CO_2 [47] which enable CO_2 to be selectively separated from smaller gases [1,19,68,75,92] via quadrupole–dipole interaction between the CO_2 with the ether groups [19]. Low molecular weight PEO with low crystallinity exists in liquid form at ambient environment which limits its direct application for gas separation membrane. As such, PEO is often incorporated into membrane as copolymers containing ethylene oxide blocks, additives, and cross-linking agents. Blending of polyethylene glycol (PEG) into the polymer matrix is a simple yet robust method that enables the membrane voids and polymer chain flexibility to be tweaked according to requirements. PEO-based membranes have been reported to show a reverse–selective character whereby the membranes could achieve higher selectivity during plasticization. This behavior contradicts from that of glassy polymer during CO_2-induced plasticization because the higher segmental mobility results in higher free volume and increases interaction between the polymer and CO_2 [93].

PEO-containing membranes is suitable for low pressure and temperature application such as conditioning precombustion feed and can potentially be used for post-combustion CO_2 capture. Li *et al.* [47] have produced TFC membranes containing ethylene oxide (EO) groups. They interfacial polymerized diethylene glycol bis(3-aminopropyl) ether (DGBAmE) with trimesoyl chloride atop of a PDMS/PSf supports and tested the TFC using mixed gases. By optimizing the concentration of the monomers used (DGBAmE: 0.0115 mol/L, TMC: 0.0104 mol/L), the fabricated membranes show high CO_2 permeance (815GPU in CO_2/H_2 gas test, 727GPU in CO_2/CH_4 gas test, 973GPU in CO_2/N_2 gas test) and good selectivity (CO_2/H_2

selectivity: 10, CO_2/CH_4 selectivity: 31, CO_2/N_2 selectivity: 84) at a feed pressure of 0.11 MPa. Here, selection of the PEO properties such as length of polymer chain and end functional groups is of great importance. The addition of PEO needs to be carefully controlled to obtain appropriate crystallinity and free volume of the membrane matrix [94,95].

13.5 CO_2-facilitated transport behavior of amine-based membrane

The incorporation of CO_2 reactive carriers such as amine groups into a polymer matrix is the gist of a facilitated transport membrane (FTM) [87,89,96–104]. Dissolved CO_2 from the feed stream undergoes reversible reaction with the carriers to form complexes that are then transported through the membrane via a concentration difference and finally released at the low pressure side while other gas species that do not react with the carrier slowly permeated across the membrane via the solution–diffusion mechanism. This carrier-mediated transport is also known as the CO_2-facilitated transport mechanism, which makes fabricating a membrane with high selectivity without forfeiting the permeance possible [32,68,103,105]. In the presence of water, CO_2 forms bicarbonate ions (HCO_3-) and complexes (R_xNHCOO^-, $x = 1,2$) with primary and secondary amines via the reactions shown below [13,102].

$$2RNH_2 + 2CO_2 + H_2O \rightleftharpoons RHNCOOH + RNH_3^+ + HCO_3^- \tag{R1}$$

$$2R_2NH + 2CO_2 + H_2O \rightleftharpoons R_2NCOOH + R_2NH_2^+ + HCO_3^- \tag{R2}$$

It is suggested that tertiary amine too play the role of catalyst and reacts with CO_2 as follows [31,96,106,107]:

$$R_3N + CO_2 + H_2O \rightleftharpoons R_3NH^+ + HCO_3^- \tag{R3}$$

CO_2 is then transported across the membrane in the form of HCO_3^- ions until finally the bicarbonate ions decompose at the permeated and CO_2 is released [87] (Fig. 13.8). Apart from aiding the hydration reaction to promote ion transfer, water can slightly increase the chain mobility of the polymer matrix hence reducing the resistance on gas diffusion [84,106,107]. Although primary and secondary amines can facilitate CO_2 transport in the absence of water (Reaction 4 and 5) [105], tertiary amine demonstrates greater CO_2 absorption capacity [106], higher catalytic efficiency [32] and greater stability in air. Furthermore, the high crystallinity of primary and secondary amines can impair the permselectivity and structural integrity of the membrane [108]. This makes tertiary amine, which has a lower crystallinity more appealing for gas separation application.

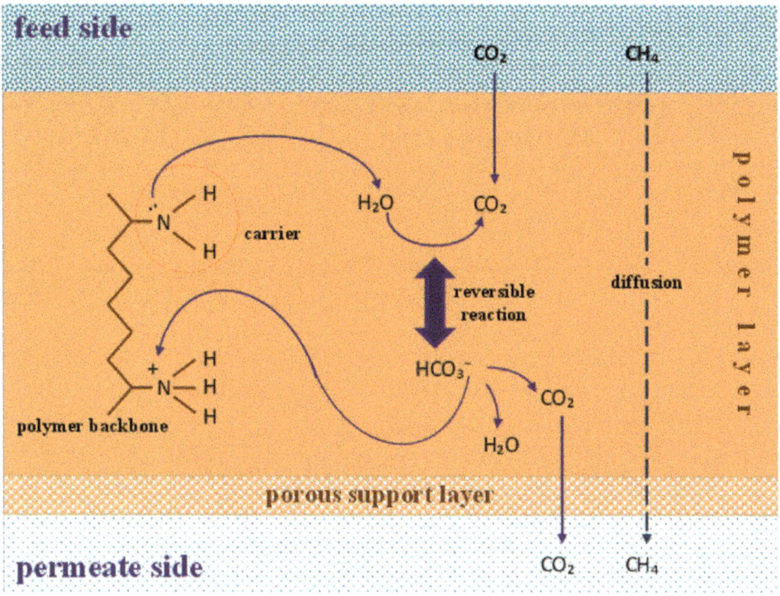

Figure 13.8 Schematic of transport mechanism in fixed carrier facilitated transport membrane.

$$2RNH_2 + CO_2 \rightleftharpoons RHNCOO^- + RNH_3^+ \tag{R4}$$

$$2R_2NH + CO_2 \rightleftharpoons R_2NCOO^- + R_2NH_2^+ \tag{R5}$$

The total flux of CO_2 in a membrane that is transported with facilitated mechanism is described by Eq. (13.2),

$$J_{CO_2} = \frac{D_{CO_2}}{l}\left(C_{CO_2,0} - C_{CO_2,l}\right) + \frac{D_{complex}}{l}\left(C_{complex,0} - C_{complex,l}\right) \tag{13.2}$$

which comprises the Fickian diffusion (left term) and carrier-mediated diffusion (right term) where:

D_x = diffusion coefficient, $x = CO_2$, CO_2-carrier complex
$C_{x,0}$ = concentration in feed side of membrane
$C_{x,l}$ = concentration in permeate side of membrane
l = thickness of membrane

$C_{CO_2,0}$varies proportionally to feed pressure and feed CO_2 concentration whereas $C_{complex,0}$ is determined by the CO_2-carrier reaction and limited by the saturation of carriers. Once the carrier is saturated, $C_{complex,0}$ stops increasing with the feed pressure. Since the hydration of CO_2 at low pressure before carriers reach saturation is fast, $C_{complex,0}$ increases more rapidly than $C_{CO_2,0}$. Hence, CO_2 transport predominantly follows the facilitated transport. Considering that facilitated transport is negatively affected by membrane thickness [87], thinner membrane benefits more from the presence of the reactive carriers, leading to higher permselectivity. Meanwhile, at high pressure condition, the concentration of free CO_2 could become higher than $C_{complex,0}$ when the all the carriers are saturated and cannot bind with fresh CO_2 molecules present in the membrane [84,106]. At this point, Fickian diffusion dominates, resulting in decrease in CO_2 permeance due to the loss of facilitated effect which lead to loss of CO_2/N_2 [13] selectivity. Besides, the diffusion of N_2 is more comparable with that of CO_2 when at high pressure and without facilitated effect, which can also account for the loss of selectivity [87]. As such, separation of CO_2 using carrier-mediated membrane is best done under lower pressure condition. It has been reported that an increase in operation temperature could increase the CO_2 hydration and decomplexation rates, which liberates the bounded carriers and favors the desorption of CO_2 at the permeate side whereas increasing the amine content could provide more reaction sites. Yet, it should be cautioned that overly high operation temperature could degrade the membrane structure and sieving property whereas membrane containing too much amine could lower the CO_2 solubility and impedes the HCO_3^- diffusion [106].

Andrew Lee *et al.* [89] found that wetted commercial TFC exhibits FTM-like properties due to the presence of amide and residue unreacted amine within the PA layer which functioned as CO_2 carriers. In year 2010, Yu *et al.* [31] fabricated TFC containing tertiary amino, which showed a high CO_2 permeance of 173 GPU with CO_2/N_2 selectivity of 70. They described the high performance of the composite membrane as resulting from the facilitated transport of CO_2 by the amino groups and the thin defect-free skin layer produced by IP. The preference on amine monomer for producing CO_2-selective TFC has changed over time in favor of long chain amines rather than the commonly used aromatic m-phenylenediamine (MPD). The employment of long chain amine enables a more flexible amide network to be formed which gives higher gas permeability. Additionally, the presence of functional units (CO_2 soluble and reactive groups) along the amine chain enhances the selectivity of the thin film. In year 2012, Li *et al.* [90] demonstrated that the synergy between a highly CO_2 soluble monomer (contains ethylene oxide) and a highly CO_2 reactive monomer (contains tertiary amino) gives TFC with superior CO_2 separation performance capable of achieving CO_2 permeance of 1612 GPU and CO_2/N_2 selectivity of 138. All these highlighted the potential of IP technique and the importance of amine monomers selection to prepare CO_2-selective composite membrane.

13.6 Nanomaterials for the ultimate membrane enhancement

While it is not the intent of this chapter to discuss nanomaterials, the benefits of nanofiller incorporation on membrane performance is briefly discussed in this section. Incorporation of inorganic nanofillers into the polymer matrix will disturbs the polymer segments packing which increases the free volumes within the polymer matrix. This, reduces the resistance on penetrant species and leads to improvement

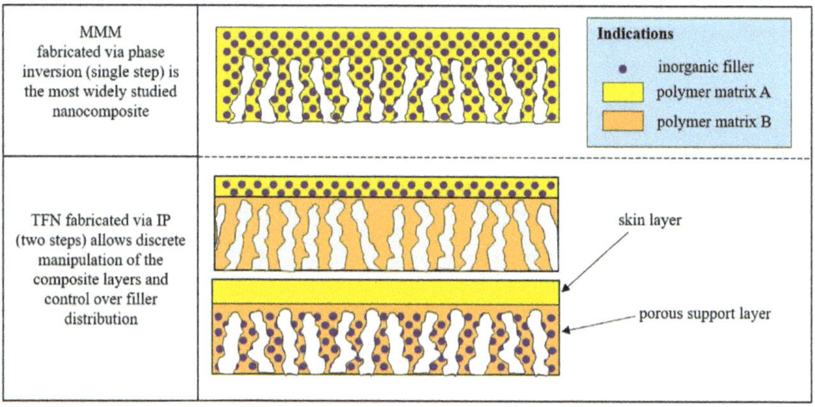

Figure 13.9 Schematic diagram of MMM and TFN.

in gas diffusion. Additionally, polar penetrant species might interact with the functional groups on the surface of the inorganic particles which enhances the solubility of the gases [109]. For instance, Ahn and coworkers [110] demonstrated that mixing of 20 volume percent of silica particles into polysulfone matrix increases the membrane void volume by 2.6% which increase the total free volume leading to increase in the permeance of CO_2 from 6.3 Barrer to 19.7 Barrer and CH_4 from 0.22 Barrer to 1.10 Barrer. Furthermore, porous inorganic particles usually exhibit performance beyond the Robeson tradeoff attributed by their precise pore structure. Therefore, addition of porous fillers can theoretically enhance not only the permeance but selectivity of the membrane as well. In 2003, Vu *et al.* [111] demonstrate that their MMM can achieve improvement in both CO_2 permeance (from 10 to 12.6 Barrer) and selectivity (from 35.3 to 51.7) by embedding 36 vol percent of primed carbon molecular sieve (CMS 800-2) in Matrimid matrix. Thin film nanocomposite (TFN) is the alternative of MMM that is fabricated via IP technique which inherent all the advantages of TFC. Fig. 13.9 depicts the typical schematics of MMM and TFN.

Table 13.3 depicted some of the works done on TFC and nanocomposites for CO_2 separation. application. Although embedding fillers, especially functionalized nanoparticles, into the polymer matrix can generally bring about at least 50% increment in CO_2 permeance while maintaining or, in most cases, elevating the selectivity of the nanocomposite (both MMM and TFN), the overall performance of TFN is under par compared to MMM mainly due to lack of optimization. As such, there are still lots of room for the improvement of TFN.

13.7 Current challenges in TFC development

Owing to its low energy requirement, which can be driven by pressure as low as one bar, TFC with its distinguished ultrathin PA active layer is well suited for

Table 13.3 **Contemporary progress in composite and nanocomposite membranes for CO_2 separation**

Membrane type	Optimum fabrication parameters	Testing conditions & optimum performance	Remark	Ref.
TFC	PSf support (MWCO: 5 kDa) was saturated with aqueous solution of 2 %w/v MPD for 2 min then reacted with 0.5 %w/v isophthaloyl chloride (IPC) in hexane for 30 s. Resultant TFC was heated in oven for 50 min at 130°C.	Tested with pure CO_2, H_2S, CH_4, O_2, and N_2. Temperature: 30°C Pressure 1 MPa CO_2 Permeance: *15 GPU* CO_2/CH_4 Selectivity: 14 H_2S Permeance: *52 GPU* H_2S/CH_4 Selectivity: 49 N_2 Permeance: 0.95 GPU O_2/N_2 selectivity: 5.4	• Moderate thermal annealing increases selectivity of TFC polyamide membrane without significant loss in permeance • Increase in feed pressure resulted in increase in CO_2 permeance and CO_2/CH_4 selectivity due to sorption of CO_2 in polyamide membrane while the nonpolar CH_4 showed no interaction with the polymer matrix	[15]
TFC	PSf support was saturated with aqueous solution containing 2 wt% MPD + 0.15 wt% sodium lauryl sulfate (SLS) and excess solution was removed with filter paper before reacted with 0.1 wt% TMC in hexane for 1 min. Membranes undergone different posttreatment: 1. dried at room temperature (24 hr) then in oven (120 °C, 30 min) 2. immersed in ethanol (5 min) followed by hexane (1 min) then solvent evaporated at room temperature (15 min) 3. immersed in aqueous t-butanol solution (15 min) then freeze dried in pure t-butanol under vacuum (2 h)	Tested with pure He, H_2, CO_2, O_2, N_2, propane (C_3H_8) and sulfur hexafluoride (SF_6). Temperature: 16–200 °C Pressure: 0.25 MPa CO_2 permeance, 150 °C: *178 GPU* CO_2/N_2 selectivity: 14 He permeance, 150 °C: *1248 GPU* He/N_2 selectivity: 99	• Polyamide layer of the composite membranes was not perfectly homogenous, but rather consisted of a dense matrix and highly permeable structures in which gases permeated via Knudsen diffusion and determine the membrane separation performance in dry samples. • While ethanol–hexane drying and freeze drying have better enhancement to gas permeance compared to room temperature–oven drying, the selectivity of He, H_2 and CO_2 toward N_2 were lower	[21]

TFC	PSf support (MWCO: 6 kDa) was dipped in aqueous sodium dodecyl sulfate (SDS) solution for 48 h then washed with RO deionized water and dried. Treated support was impregnated with aqueous phase containing 0.0062 mol/L 3,3′-diamino-N-methyldipropylamine (DNMDAm) + 0.4 wt% Na_2CO_3 for 10 min then reacted with 0.0226 mol/L TMC in hexane for 3 min.	Tested with CO_2/N_2 mix (20/80 by vol) and CO_2/CH_4 mix (10/90 by volume) using H_2 as sweep gas. All gas saturated with water vapor by passing through humidifier. Temperature: Room temperature Pressure: 0.11 MPa CO_2 Permeance: *173 GPU* (in CO_2/N_2 mix) CO_2/N_2 Selectivity: 70 CO_2 Permeance: *118 GPU* (in CO_2/CH_4 mix) CO_2/CH_4 Selectivity: 37	• Water facilitates transport of CO_2 in the presence of tertiary amino groups • Concentration of both diamine and acid chloride play important role in thin film formation with increasing film thickness as concentration of the monomers increases • Polymerization rate and diffusivity of monomers can influence the IP process	[31]
TFC	PSf substrate (MWCO: 6 kDa) was coated with 3 wt% PDMS + 1 wt% tetraethoxysilane (TEOS) + 1 wt% ditin butyl dilaurate (DBD) in heptane. Substrate impregnated with 0.015 mol/L TMC in hexane for 10 min and excess was dripped off before reacted with aqueous solution containing 0.0133 mol/L polyetheramine (PEA, MW: 900) + 0.038 mol/L Na_2CO_3 for 3 min.	Tested with CO_2/N_2 mix (20/80 by vol) and CO_2/CH_4 mix (10/90 by volume) using H_2 as sweep gas. All gas saturated with water vapor by bubbling through water bottles. Temperature: 25°C Pressure: 0.03 MPa CO_2 Permeance: *360 GPU* (in CO_2/N_2 mix) CO_2/N_2 Selectivity: 67 CO_2 Permeance: *275 GPU* (in CO_2/CH_4 mix) CO_2/CH_4 Selectivity: 32	• Increasing PEA molecular weight increases the ether group content which improves the CO_2 solubility and CO_2 permeability in membrane • Increasing PEA molecular weight ($MW > 900$) will negatively affect selectivity because longer PEA molecular chain leads to low cross-linking density • Higher PEA molecular weight leads to lower diffusion of the amine from the bulk aqueous phase to the organic–aqueous interface during IP which could result in incomplete polymerization reaction and formation of defects	[18]

(Continued)

Table 13.3 **(Continued)**

Membrane type	Optimum fabrication parameters	Testing conditions & optimum performance	Remark	Ref.
TFC	PSf substrate (MWCO: 6 kDa) was coated with 2 wt% PDMS + 1 wt% TEOS + 1 wt% DBD in heptane. Treated substrate was impregnated with 0.0104 mol/L TMC in hexane for 10 min and excess was drained before react with aqueous phase containing 0.0115 mol/L diethylene glycol bis(3-aminopropyl) ether (DGBAmE) + 0.0377 mol/L Na_2CO_3 for 3 min. Resultant membrane was washed with RO and dried at 30°C for 12 h at 40% relative humidity.	Tested with CO_2/H_2 mix (40/60 by vol) using Ar as sweep gas while CO_2/CH_4 mix (10/90 by vol) and CO_2/N_2 mix (15/85 by volume) using H_2 as sweep gas. All gas was saturated with water vapor. Temperature: 22 °C Pressure: 0.11 MPa CO_2 Permeance: *815 GPU* (in CO_2/H_2 mix) CO_2/H_2 Selectivity: 10 CO_2 Permeance: *727 GPU* (in CO_2/CH_4 mix) CO_2/CH_4 Selectivity: 31 CO_2 Permeance: *973 GPU* (in CO_2/N_2 mix) CO_2/N_2 Selectivity: 84	Length of monomer molecules can influence the formation of polyamide thin film in term of cross-linking and thickness	[47]
TFC	Polysulfone (PSf) support (MWCO: 6 kDa) was coated with 2 wt% PDMS + 1 wt% TEOS + 1 wt% DBD in heptane. Treated support was impregnated with 0.0250 mol/L trimesoyl chloride (TMC) in hexane for 10 min and excess was blown off with an air-knife before reacted with aqueous solution containing	Tested with CO_2/N_2 mix (15/85 by vol) using H_2 as sweep gas. All gas saturated with water vapor by bubbling through water bottles. Temperature: 25°C Pressure: 0.11 MPa CO_2 Permeance: *1035 GPU* CO_2/N_2 Selectivity: 87	• Higher cross-linking extent of skin layer gives higher total carrier content in the membrane, which facilitates more CO_2 molecules across the membrane • CO_2 permeance is determined by both thickness and cross-linking of the skin layers at the whole	[27]

	0.0615 mol/L N-methyldiethanolamine (MEDA) + 0.0379 mol/L sodium carbonate (Na_2CO_3) for 10 min.		feed pressure range while N_2 permeance is dominated by skin layer thickness and crystallinity at the lower feed pressure but by cross-linking at the higher feed pressure. • Membranes with high CO_2 permeance and high CO_2/N_2 selectivity could be obtained by decreasing TMC concentration in organic phase and increasing MEDA concentration in aqueous phase	
TFC	PDMS coated PSf was immersed in organic phase of 0.0104 mol/L TMC in hexane for 10 min. Excess solution was drained off before reacting the saturated support with aqueous solution containing 0.0115 mol/L DGBAmE + 0.0115 mol/L 3,3′-diamino-*N*-methyldipropylamine (DNMDAm) + 0.0377 mol/L Na_2CO_3 for 3 min at 25°C. Resultant membrane was washed with DI water and stored in artificial climate chamber (30°C, 40% relative humidity) for 12 hrs	Tested with CO_2/H_2 mix (40/60 by vol) using Ar as sweep gas while CO_2/CH_4 mix (10/90 by vol) and CO_2/N_2 mix (15/85 by volume) using H_2 as sweep gas. All gas saturated with water vapor. Temperature: 22°C Pressure: 0.11 MPa CO_2 Permeance: *1612 GPU* (in CO_2/N_2 mix) CO_2/N_2 Selectivity: 138 CO_2 Permeance: *1224 GPU* (in CO_2/CH_4 mix) CO_2/CH_4 Selectivity: 51 CO_2 Permeance: *746 GPU* (in CO_2/H_2 mix) CO_2/H_2 Selectivity: 22	• Appropriate thin film stiffness, CO_2 selectivity and reactivity can be tuned by manipulating the content of DGBAmE (flexible chain, high CO_2 solubility) and DNMDAm (rigid, CO_2 reactive) • Increasing rigidity of polymer chain can improve CO_2-plasticization resistance • Membrane with multiple separation mechanism (i.e., diffusivity−selectivity, solubility−selectivity and reactivity−selectivity) could enhance the removal of polar and acidic gas	[90]

(Continued)

Table 13.3 (Continued)

Membrane type	Optimum fabrication parameters	Testing conditions & optimum performance	Remark	Ref.
MMM	0.1 g PEI with dicarboxylic acid ends was dissolved in 3 mL of N,N′-dimethyacetamide (DMAc) and TiO_2 (12 wt% of total solid weight) was added with a syringe. The dispersion was stirred at room temperature for 30 min before casted onto glass plate. Membrane was heat treated from 50 to 250°C for 1 h in every 50°C rise in temperature and 300°C for 12 h.	Tested with pure N_2 and CO_2 at Temperature: Room temperature Pressure: 0.07 MPa CO_2 Permeance: *199 Barrer* CO_2/N_2 Selectivity: 38	• Compared to neat: ◦ CO_2 permeance increased 153% ◦ CO_2/N_2 selectivity increased 26% • Addition of nanoparticles into glassy polymer membranes results in polymer chain packing rearrangement which increased the free volume	[112]
MMM	Pebax (MH 1657, 2 wt% in solution) was dissolved in ethanol/water (70/30 wt%) at 80°C under reflux with for 2 h. ZIF-7 (8 wt% of total solid weight) was dispersed by sonication for 10 min before the solution was stirred at room temperature. Dope solution was casted onto polytrimethylsilylpropyne (PTMSP) coated polyacrylonitrile (PAN) support and kept at room temperature for 1 day	Tested with pure N_2, CH_4 and CO_2 at Temperature: 20°C Pressure: 0.375 MPa CO_2 Permeance: *291 GPU* CO_2/N_2 Selectivity: 68 CO_2/CH_4 Selectivity: 23	• Compared to neat: ◦ CO_2 permeance increased 1% ◦ CO_2/N_2 selectivity increased 64% ◦ CO_2/CH_4 selectivity increased 100% • ZIF-7 loading beyond 8 wt% could further enhanced the selectivities but reduces CO_2 permeance due to increase membrane rigidity by nanofillers addition	[113]
MMM	Pebax (MH 1657, 4 wt% of total solution) was dissolved in ethanol/water (70/30 wt %) solution at 80°C under reflux for 2 hrs. 20 wt% (solid weight)	Tested with humidified pure N_2, CH_4 and CO_2 and mix gases of CO_2/N_2 (10/90 vol%) and CO_2/CH_4 (30/70 vol%)	• compared to neat: ◦ CO_2 permeance increased 210%	[104]

	PEI-MCM-41 was dispersed into the solution via sonication for 30 min before stirred vigorously for 2 h. Dope solution was casted on a flat glass plate, dried at ambient temperature for 24 h and at 50°C for another 24 h in vacuum oven.	Temperature: 25 °C Pressure: 0.1 MPa CO_2 Permeance: *1521 Barrer* CO_2/N_2 Selectivity: 102 CO_2/CH_4 Selectivity: 41 CO_2 Permeance: *1080 Barrer* (in CO_2/N_2 mix) CO_2/N_2 Selectivity: 102 CO_2 Permeance: *1400 Barrer* (in CO_2/CH_4 mix) CO_2/CH_4 Selectivity: 39	◦ CO_2/N_2 selectivity increased 96% ◦ CO_2/CH_4 selectivity increased 116% • Improvement in CO_2 selectivity and permeability of MMM containing PEI-MCM-41 compare to MMM containing MCM-41 was attributed to the presence of amine carriers (PEI) on the filler surface which facilities the transport of CO_2 and increases filler-polymer compatibility	
MMM	0.1 wt% (solid weight) beta-cyclodextrin functionalized MWNTs (β-CD-MWNTs) were dispersed in acetic acid (CH_3COOH)/H_2O (70:30 wt%) mix via sonication for 20 min followed by 4 h stirring. Cellulose acetate (CA, 10 wt% of total solution) was gradually added with stirring at 55°C for 3 h. Dope solution was cooled to room temperature, stirred for another 12 h and degassed for 10 min before casted using automatic film applicator. The casted dope was immersed into coagulation bath for 5 min then transferred and immersed in another water bath for 24 h. Next MMM was immersed in ethanol (4 h) followed by n-hexane (1 h) and dried for 24 h.	Tested with pure N_2 and CO_2 at Temperature: Room temperature Pressure: 0.3 MPa CO_2 Permeance: *742 GPU* CO_2/N_2 Selectivity: 40	• Compared to neat: ◦ CO_2 permeance increased 85% ◦ CO_2/N_2 selectivity increased 52% • Functionalized MWNTs are less prone to aggregation compared to pristine counterpart which ensure good filler-polymer interaction and defect-free membrane	[114]

(Continued)

Table 13.3 **(Continued)**

Membrane type	Optimum fabrication parameters	Testing conditions & optimum performance	Remark	Ref.

TFN	PSf (MWCO: 6 kDa) was placed in aqueous solution containing 0.0615 mol/L DNMDAm + 0.038 mol/L Na_2CO_3 + 0.0017 mol/L SDS + 0.00363 g/mL LUDOX silica for 10 min then reacted with 0.01 mol/L TMC in hexane for 3 min at 25°C.
	Resultant membrane was rinsed with pure hexane, heat treated in a 70°C hot-air circulation oven for 12 min, washed with RO water and finally kept in room temperature for 12 h.
		Tested with CO_2/N_2 mix (20/80 by vol) using H_2 as sweep gas which was humidified with moistener.
		Temperature: Room temperature
		Pressure: 0.1 MPa
		CO_2 Permeance: *59 GPU* (in CO_2/N_2 mix)
		CO_2/N_2 Selectivity: 85
			• Compared to neat:
			  ◦ CO_2 permeance increased 136%
			  ◦ CO_2/N_2 selectivity increased 12%
			• −OH groups on the silica could react with TMC
			• Addition of silica changed the surface morphology of polyamide layer.
			• Permeance of CO_2 initially increased then decreased with increasing silica content.
			• Separation selectivities of rubbery polymers are mainly contributed by solubility selectivities.
				[32]

TFN	PSf support was coated with 2 wt% PDMS solution containing 0.5 g/L of modified MWNTs. Then, support was saturated with organic phase of 0.28 %w/v TMC in hexane for 10 min before reacting with aqueous solution containing 0.35 % w/v DGBAmE + 0.4 %w/v Na_2CO_3 for 3 min.
	Resultant membrane was dried under ambient conditions for 12 h.
		• Tested with pure N_2, CH_4 and CO_2 at
		• Temperature: Room temperature
		• Pressure: 0.2 MPa
		• CO_2 Permeance: *71 GPU*
		• CO_2/N_2 Selectivity: 67
		• CO_2/CH_4 Selectivity: 29
			• Compared to neat:
			  ◦ CO_2 permeance increased 29%
			  ◦ CO_2/N_2 selectivity increased 47%
			  ◦ CO_2/CH_4 selectivity increased 9%
			• Incorporating fillers in coating layer ensure well-adhesion of nanotubes and formation of defect-free skin layer
				[115]

1 GPU = 1.0×10^{-6} cm^3 (STP) /(cm^2 s cmHg) = 3.35×10^{-10} mol /(m^2 s Pa) [20]
1 Barrer = 1.0×10^{-10} cm^3 (STP) cm/(cm^2 s cmHg)

postcombustion CO_2 capture application [116]. However, attention needs to be paid on the membrane stability and separation factor at high pressure, as the thin skin is susceptible to both CO_2-induced plasticization and carrier saturation. Although a plasticization effect can be effectively circumvented by blending stiff polymeric or compatible inorganic materials into the active layer, highly CO_2 soluble or reactive compounds often need to be added side-by-side to accommodate the negative impact that results from matrix rigidification. For this reason, ever CO_2-philic materials are constantly being sought after. Recently, a special class of salt known as ionic liquid (IL), which possess incredibly high CO_2-sorption capability has been vigorously developed. This material has been immobilized in membrane pores, blended, homo-polymerized, copolymerized, and coated on the fillers to produce membrane with high CO_2 separation efficiency [117–121]. Continuous advancement in material science and nanotechnology is expected to propel the performance of TFC to greater height.

In regards to membrane performance, enhancement brings about by nanofiller incorporation, since the micron-thick skin layer of the TFN is more prone to defects compared to MMM, the issues of fillers aggregation and uneven distribution still persist. Depending on the nature of the bulk host polymer, membrane application, and method of fillers incorporation, different types of modification such as mechanical treatment and physicochemical treatment can be engaged to resolve these issues [122]. Some researchers fabricated multilayered PA film through several IP cycles but the resulting skin layer is usually thicker than the normal TFN, which compromises gas permeability [46,60]. Currently, nanofiller functionalization is the most widely adopted and well-documented technique to improve the nanoparticle dispersibility and avoid agglomeration due it robustness and capability to alter the chemical properties of the particles [104,123–127]. Ansaloni *et al.* [101] pointed out that the formation of interface voids in their nanocomposite is subjugated through excellent bonding between the amino moieties on the modified CNTs surface and the polyvinylalcohol-polysiloxane (PVA-POS) matrix, which gave membrane with CO_2 permeance of 957 barrer, CO_2/CH_4 selectivity of 264 and CO_2/N_2 selectivity of 384. Peyravi *et al.* [37] and Khan *et al.* [128] also revealed that their modified fillers (amine grafted TiO_2 and PEG functionalized CNTs, respectively) have better dispersion in polymer matrix, which gives more homogeneous and defect-free membranes compared to pristine counterparts. Furthermore, they also prove that the incorporation of rigid nanofillers into the polymer system can improve the overall mechanical integrity of the membrane leading to greater solvent, thermal, and pressure stability. Likewise, Jadav and coworkers [129] found that the glass-transition temperature (T_g) of silica-filled membrane is higher than the neat membrane and suggested that the strong silica–polymer interaction, which apparently fused the two distinct phases into a homogeneous matrix, is responsible for the enhanced feature. Therefore, ensuring good filler–polymer interaction through filler modification is not only crucial to produce membranes with superior separation performance but nanocomposites with excellent strength as well [130–133].

13.8 Conclusion

In light of the extreme tunability of an ultrathin skin layer for high CO_2 permselectivity and scalability of IP, TFC makes a very attractive candidate for commercialization. It is anticipated that, with the continuous expansion in monomer varieties, enhancements in filler incorporation and optimization of fabrication techniques, high-performing TFC can be developed to resolve the contemporary global warming issue.

References

[1] C.E. Powell, G.G. Qiao, Polymeric CO_2/N_2 gas separation membranes for the capture of carbon dioxide from power plant flue gases, J. Memb. Sci. 279 (2006) 1–49. Available from: http://dx.doi.org/10.1016/j.memsci.2005.12.062.
[2] B.S. Lalia, V. Kochkodan, R. Hashaikeh, N. Hilal, A review on membrane fabrication: structure, properties and performance relationship, Desalination. 326 (2013) 77–95. Available from: http://dx.doi.org/10.1016/j.desal.2013.06.016.
[3] J.K. Adewole, A.L. Ahmad, S. Ismail, C.P. Leo, Current challenges in membrane separation of CO_2 from natural gas: a review, Int. J. Greenh. Gas Control 17 (2013) 46–65. Available from: http://dx.doi.org/10.1016/j.ijggc.2013.04.012.
[4] Y. Zhang, J. Sunarso, S. Liu, R. Wang, Current status and development of membranes for CO_2/CH_4 separation: a review, Int. J. Greenh. Gas Control 12 (2013) 84–107. Available from: http://dx.doi.org/10.1016/j.ijggc.2012.10.009.
[5] J. Gibbins, H. Chalmers, Carbon capture and storage, Energy Policy. 36 (2008) 4317–4322. Available from: http://dx.doi.org/10.1016/j.enpol.2008.09.058.
[6] T. Dixon, K. Romanak, S. Neades, A. Chadwick, Getting science and technology into international climate policy: carbon dioxide capture and storage in the UNFCCC, Energy Procedia, Elsevier B.V, 2013, pp. 7590–7595. Available from: http://dx.doi.org/10.1016/j.egypro.2013.06.703.
[7] J.C.M. Pires, F.G. Martins, M.C.M. Alvim-Ferraz, M. Simões, Recent developments on carbon capture and storage: an overview, Chem. Eng. Res. Des. 89 (2011) 1446–1460. Available from: http://dx.doi.org/10.1016/j.cherd.2011.01.028.
[8] E. Favre, R. Bounaceur, D. Roizard, Biogas, membranes and carbon dioxide capture, J. Memb. Sci. 328 (2009) 11–14. Available from: http://dx.doi.org/10.1016/j.memsci.2008.12.017.
[9] E. Favre, Membrane processes and postcombustion carbon dioxide capture: challenges and prospects, Chem. Eng. J. 171 (2011) 782–793. Available from: http://dx.doi.org/10.1016/j.cej.2011.01.010.
[10] D. Aaron, C. Tsouris, Separation of $CO_2$from flue gas: a review, Sep. Sci. Technol. 40 (2005) 321–348. Available from: http://dx.doi.org/10.1081/SS-200042244.
[11] S.-H.H. Choi, J.-H.H. Kim, Y. Lee, Pilot-scale multistage membrane process for the separation of CO_2 from LNG-fired flue gas, Sep. Purif. Technol. 110 (2013) 170–180. Available from: http://dx.doi.org/10.1016/j.seppur.2013.03.016.
[12] M. Czyperek, P. Zapp, H.J.M. Bouwmeester, M. Modigell, K. Ebert, I. Voigt, et al., Gas separation membranes for zero-emission fossil power plants: MEM-BRAIN, J. Memb. Sci. 359 (2010) 149–159. Available from: http://dx.doi.org/10.1016/j.memsci.2010.04.012.

[13] J. Zhao, Z. Wang, J. Wang, S. Wang, Influence of heat-treatment on CO_2 separation performance of novel fixed carrier composite membranes prepared by interfacial polymerization, J. Memb. Sci. 283 (2006) 346–356. Available from: http://dx.doi.org/10.1016/j.memsci.2006.07.004.
[14] A. Brunetti, F. Scura, G. Barbieri, E. Drioli, Membrane technologies for CO_2 separation, J. Memb. Sci. 359 (2010) 115–125. Available from: http://dx.doi.org/10.1016/j.memsci.2009.11.040.
[15] S. Sridhar, B. Smitha, S. Mayor, B. Prathab, T.M. Aminabhavi, Gas permeation properties of polyamide membrane prepared by interfacial polymerization, J. Mater. Sci. 42 (2007) 9392–9401. Available from: http://dx.doi.org/10.1007/s10853-007-1813-5.
[16] T.C. Merkel, H. Lin, X. Wei, R. Baker, Power plant post-combustion carbon dioxide capture: an opportunity for membranes, J. Memb. Sci. 359 (2010) 126–139. Available from: http://dx.doi.org/10.1016/j.memsci.2009.10.041.
[17] A. Hussain, S. Farrukh, F.T. Minhas, Two-stage membrane system for post-combustion CO_2 capture application, Energy Fuels 29 (2015) 6664–6669. Available from: http://dx.doi.org/10.1021/acs.energyfuels.5b01464.
[18] A.A.M.M. Salih, C. Yi, H. Peng, B. Yang, L. Yin, W. Wang, Interfacially polymerized polyetheramine thin film composite membranes with PDMS inter-layer for CO_2 separation, J. Memb. Sci. 472 (2014) 110–118. Available from: http://dx.doi.org/10.1016/j.memsci.2014.08.025.
[19] S. Feng, J. Ren, K. Hua, H. Li, X. Ren, M. Deng, Poly(amide-12-b-ethylene oxide)/polyethylene glycol blend membranes for carbon dioxide separation, Sep. Purif. Technol. 116 (2013) 25–34. Available from: http://dx.doi.org/10.1016/j.seppur.2013.05.002.
[20] D.D. Iarikov, S. Ted Oyama, Review of CO_2/CH_4 separation membranes, Inorganic, Polym. Compos. Membr. Struct. Funct. Other Correl, 1st ed., Elsevier B.V, 2011, pp. 91–115. Available from: http://dx.doi.org/10.1016/B978-0-444-53728-7.00005-7.
[21] J. Albo, J. Wang, T. Tsuru, Gas transport properties of interfacially polymerized polyamide composite membranes under different pre-treatments and temperatures, J. Memb. Sci. 449 (2014) 109–118. Available from: http://dx.doi.org/10.1016/j.memsci.2013.08.026.
[22] J.S. Louie, I. Pinnau, M. Reinhard, Gas and liquid permeation properties of modified interfacial composite reverse osmosis membranes, J. Memb. Sci. 325 (2008) 793–800. Available from: http://dx.doi.org/10.1016/j.memsci.2008.09.006.
[23] A.P. Rao, S.V. Joshi, J.J. Trivedi, C.V. Devmurari, V.J. Shah, Structure-performance correlation of polyamide thin film composite membranes: effect of coating conditions on film formation, J. Memb. Sci. 211 (2003) 13–24. Available from: http://dx.doi.org/10.1016/s0376-7388(02)00305-8.
[24] A.K. Ghosh, B.H. Jeong, X. Huang, E.M.V. Hoek, Impacts of reaction and curing conditions on polyamide composite reverse osmosis membrane properties, J. Memb. Sci. 311 (2008) 34–45. Available from: http://dx.doi.org/10.1016/j.memsci.2007.11.038.
[25] Y. Song, P. Sun, L.L. Henry, B. Sun, Mechanisms of structure and performance controlled thin film composite membrane formation via interfacial polymerization process, J. Memb. Sci. 251 (2005) 67–79. Available from: http://dx.doi.org/10.1016/j.memsci.2004.10.042.
[26] A.V. Berezkin, A.R. Khokhlov, Mathematical modeling of interfacial polycondensation, J. Polym. Sci. Part B Polym. Phys 44 (2006) 2698–2724. Available from: http://dx.doi.org/10.1002/polb.20907.

[27] F. Yuan, Z. Wang, S. Li, J. Wang, S. Wang, Formation-structure-performance correlation of thin film composite membranes prepared by interfacial polymerization for gas separation, J. Memb. Sci 421-422 (2012) 327–341. Available from: http://dx.doi.org/10.1016/j.memsci.2012.07.035.

[28] J. Petersen, K.-V. Peinemann, Novel polyamide composite membranes for gas separation prepared by interfacial polycondensation, J. Appl. Polym. Sci. 63 (1997) 1557–1563. Available from: http://dx.doi.org/10.1002/(SICI)1097-4628(19970321)63:12 < 1557::AID-APP6 > 3.0.CO;2-N.

[29] V. Freger, Kinetics of film formation by interfacial polycondensation, Langmuir. 21 (2005) 1884–1894. Available from: http://dx.doi.org/10.1021/la048085v.

[30] J. Espeso, A.E. Lozano, J.G. de la Campa, J. de Abajo, Effect of substituents on the permeation properties of polyamide membranes, J. Memb. Sci. 280 (2006) 659–665. Available from: http://dx.doi.org/10.1016/j.memsci.2006.02.023.

[31] X. Yu, Z. Wang, Z. Wei, S. Yuan, J. Zhao, J. Wang, et al., Novel tertiary amino containing thin film composite membranes prepared by interfacial polymerization for CO_2 capture, J. Memb. Sci. 362 (2010) 265–278. Available from: http://dx.doi.org/10.1016/j.memsci.2010.06.043.

[32] X. Yu, Z. Wang, J. Zhao, F. Yuan, S. Li, J. Wang, et al., An effective method to improve the performance of fixed carrier membrane via incorporation of CO_2-selective adsorptive silica nanoparticles, Chinese Journal of Chemical Engineering, Chemical Industry and Engineering Society of China (CIESC) and Chemical Industry Press (CIP), 2011, pp. 821–832. Available from: http://dx.doi.org/10.1016/S1004-9541(11)60062-1.

[33] C. Klaysom, S. Hermans, A. Gahlaut, S. Van Craenenbroeck, I.F.J.J. Vankelecom, Polyamide/polyacrylonitrile (PA/PAN) thin film composite osmosis membranes: film optimization, characterization and performance evaluation, J. Memb. Sci. 445 (2013) 25–33. Available from: http://dx.doi.org/10.1016/j.memsci.2013.05.037.

[34] W.S. Hung, C.L. Lai, Q. An, M. De Guzman, T.J. Shen, Y.H. Huang, et al., A study on high-performance composite membranes comprising heterogeneous polyamide layers on an electrospun substrate for ethanol dehydration, J. Memb. Sci. 470 (2014) 513–523. Available from: http://dx.doi.org/10.1016/j.memsci.2014.07.066.

[35] N. Widjojo, T.S. Chung, M. Weber, C. Maletzko, V. Warzelhan, The role of sulphonated polymer and macrovoid-free structure in the support layer for thin-film composite (TFC) forward osmosis (FO) membranes, J. Memb. Sci. 383 (2011) 214–223. Available from: http://dx.doi.org/10.1016/j.memsci.2011.08.041.

[36] M. Liu, S. Yu, J. Tao, C. Gao, Preparation, structure characteristics and separation properties of thin-film composite polyamide-urethane seawater reverse osmosis membrane, J. Memb. Sci. 325 (2008) 947–956. Available from: http://dx.doi.org/10.1016/j.memsci.2008.09.033.

[37] M. Peyravi, M. Jahanshahi, A. Rahimpour, A. Javadi, S. Hajavi, Novel thin film nanocomposite membranes incorporated with functionalized TiO_2 nanoparticles for organic solvent nanofiltration, Chem. Eng. J. 241 (2014) 155–166. Available from: http://dx.doi.org/10.1016/j.cej.2013.12.024.

[38] M. Fathizadeh, A. Aroujalian, A. Raisi, M. Fotouhi, Preparation and characterization of thin film nanocomposite membrane for pervaporative dehydration of aqueous alcohol solutions, Desalination. 314 (2013) 20–27. Available from: http://dx.doi.org/10.1016/j.desal.2013.01.001.

[39] S. Yu, M. Ma, J. Liu, J. Tao, M. Liu, C. Gao, Study on polyamide thin-film composite nanofiltration membrane by interfacial polymerization of polyvinylamine (PVAm) and isophthaloyl chloride (IPC), J. Memb. Sci. 379 (2011) 164–173. Available from: http://dx.doi.org/10.1016/j.memsci.2011.05.061.
[40] D. Emadzadeh, W.J. Lau, T. Matsuura, A.F. Ismail, M. Rahbari-Sisakht, Synthesis and characterization of thin film nanocomposite forward osmosis membrane with hydrophilic nanocomposite support to reduce internal concentration polarization, J. Memb. Sci. 449 (2014) 74–85. Available from: http://dx.doi.org/10.1016/j.memsci.2013.08.014.
[41] M.L. Lind, A.K. Ghosh, A. Jawor, X. Huang, W. Hou, Y. Yang, et al., Influence of zeolite crystal size on zeolite-polyamide thin film nanocomposite membranes, Langmuir. 25 (2009) 10139–10145. Available from: http://dx.doi.org/10.1021/la900938x.
[42] D. Hu, Z. Xu, C. Chen, Polypiperazine-amide nanofiltration membrane containing silica nanoparticles prepared by interfacial polymerization, Desalination. 301 (2012) 75–81. Available from: http://dx.doi.org/10.1016/j.desal.2012.06.015.
[43] M. Fathizadeh, A. Aroujalian, A. Raisi, Effect of lag time in interfacial polymerization on polyamide composite membrane with different hydrophilic sub layers, Desalination. 284 (2012) 32–41. Available from: http://dx.doi.org/10.1016/j.desal.2011.08.034.
[44] J. Ji, M. Mehta, Mathematical model for the formation of thin-film composite hollow fiber and tubular membranes by interfacial polymerization, J. Memb. Sci 192 (2001) 41–54. Available from: http://dx.doi.org/10.1016/S0376-7388(01)00496-3.
[45] F. Yuan, Z. Wang, X. Yu, Z. Wei, S. Li, J. Wang, et al., Visualization of the formation of interfacially polymerized film by an optical contact angle measuring device, J. Phys. Chem. C. 116 (2012) 11496–11506. Available from: http://dx.doi.org/10.1021/jp210209v.
[46] D. Wu, Y. Huang, S. Yu, D. Lawless, X. Feng, Thin film composite nanofiltration membranes assembled layer-by-layer via interfacial polymerization from polyethyleneimine and trimesoyl chloride, J. Memb. Sci. 472 (2014) 141–153. Available from: http://dx.doi.org/10.1016/j.memsci.2014.08.055.
[47] S. Li, Z. Wang, C. Zhang, M. Wang, F. Yuan, J. Wang, et al., Interfacially polymerized thin film composite membranes containing ethylene oxide groups for CO_2 separation, J. Memb. Sci. 436 (2013) 121–131. Available from: http://dx.doi.org/10.1016/j.memsci.2013.02.038.
[48] M.N.A. Seman, M. Khayet, N. Hilal, Nanofiltration thin-film composite polyester polyethersulfone-based membranes prepared by interfacial polymerization, J. Memb. Sci. 348 (2010) 109–116. Available from: http://dx.doi.org/10.1016/j.memsci.2009.10.047.
[49] J.N. Shen, C.C. Yu, H.M. Ruan, C.J. Gao, B. Van der Bruggen, Preparation and characterization of thin-film nanocomposite membranes embedded with poly(methyl methacrylate) hydrophobic modified multiwalled carbon nanotubes by interfacial polymerization, J. Memb. Sci. 442 (2013) 18–26. Available from: http://dx.doi.org/10.1016/j.memsci.2013.04.018.
[50] Y. Liu, B. He, J. Li, R.D. Sanderson, L. Li, S. Zhang, Formation and structural evolution of biphenyl polyamide thin film on hollow fiber membrane during interfacial polymerization, J. Memb. Sci. 373 (2011) 98–106. Available from: http://dx.doi.org/10.1016/j.memsci.2011.02.045.

[51] H. Wang, L. Li, X. Zhang, S. Zhang, Polyamide thin-film composite membranes prepared from a novel triamine 3,5-diamino-N-(4-aminophenyl)-benzamide monomer and m-phenylenediamine, J. Memb. Sci. 353 (2010) 78–84. Available from: http://dx.doi.org/10.1016/j.memsci.2010.02.033.
[52] A. Jezowska, T. Schipolowski, G. Wozny, Influence of simple pre-treatment methods on properties of membrane material, Desalination. 189 (2006) 43–52. Available from: http://dx.doi.org/10.1016/j.desal.2005.06.011.
[53] L. Shi, S.R. Chou, R. Wang, W.X. Fang, C.Y. Tang, A.G. Fane, Effect of substrate structure on the performance of thin-film composite forward osmosis hollow fiber membranes, J. Memb. Sci. 382 (2011) 116–123. Available from: http://dx.doi.org/10.1016/j.memsci.2011.07.045.
[54] A.K. Ghosh, E.M. V. Hoek, Impacts of support membrane structure and chemistry on polyamide-polysulfone interfacial composite membranes, J. Memb. Sci. 336 (2009) 140–148. Available from: http://dx.doi.org/10.1016/j.memsci.2009.03.024.
[55] P. Ji, Y. Cao, X. Jie, M. Li, Q. Yuan, Impacts of coating condition on composite membrane performance for CO_2 separation, Sep. Purif. Technol. 71 (2010) 160–167. Available from: http://dx.doi.org/10.1016/j.seppur.2009.11.015.
[56] N. Misdan, W.J. Lau, A.F. Ismail, T. Matsuura, Formation of thin film composite nanofiltration membrane: effect of polysulfone substrate characteristics, Desalination. 329 (2013) 9–18. Available from: http://dx.doi.org/10.1016/j.desal.2013.08.021.
[57] R. Zhang, J. Vanneste, L. Poelmans, A. Sotto, X. Wang, B. Van der Bruggen, Effect of the manufacturing conditions on the structure and performance of thin-film composite membranes, J. Appl. Polym. Sci. 125 (2012) 3755–3769. Available from: http://dx.doi.org/10.1002/app.36542.
[58] X. Li, K.Y. Wang, B. Helmer, T.S. Chung, Thin-film composite membranes and formation mechanism of thin-film layers on hydrophilic cellulose acetate propionate substrates for forward osmosis processes, Ind. Eng. Chem. Res 51 (2012) 10039–10050. Available from: http://dx.doi.org/10.1021/ie2027052.
[59] A. Tiraferri, N.Y. Yip, W.A. Phillip, J.D. Schiffman, M. Elimelech, Relating performance of thin-film composite forward osmosis membranes to support layer formation and structure, J. Memb. Sci 367 (2011) 340–352. Available from: http://dx.doi.org/10.1016/j.memsci.2010.11.014.
[60] H. Wu, B. Tang, P. Wu, Optimization, characterization and nanofiltration properties test of MWNTs/polyester thin film nanocomposite membrane, J. Memb. Sci 428 (2013) 425–433. Available from: http://dx.doi.org/10.1016/j.memsci.2012.10.042.
[61] S. Hermans, H. Mariën, E. Dom, R. Bernstein, I.F.J. Vankelecom, Simplified synthesis route for interfacially polymerized polyamide membranes, J. Memb. Sci. 451 (2014) 148–156. Available from: http://dx.doi.org/10.1016/j.memsci.2013.10.005.
[62] W.J. Lau, A.F. Ismail, N. Misdan, M.A. Kassim, A recent progress in thin film composite membrane: a review, Desalination. 287 (2012) 190–199. Available from: http://dx.doi.org/10.1016/j.desal.2011.04.004.
[63] N.K. Saha, S.V. Joshi, Performance evaluation of thin film composite polyamide nanofiltration membrane with variation in monomer type, J. Memb. Sci. 342 (2009) 60–69. Available from: http://dx.doi.org/10.1016/j.memsci.2009.06.025.
[64] S.H. Chen, D.J. Chang, R.M. Liou, C.S. Hsu, S.S. Lin, Preparation and separation properties of polyamide nanofiltration membrane, J. Appl. Polym. Sci. 83 (2002) 1112–1118. Available from: http://dx.doi.org/10.1002/app.2282.

[65] C.L. Li, S.H. Huang, D.J. Liaw, K.R. Lee, J.Y. Lai, Interfacial polymerized thin-film composite membranes for pervaporation separation of aqueous isopropanol solution, Sep. Purif. Technol. 62 (2008) 694–701. Available from: http://dx.doi.org/10.1016/j.seppur.2008.03.031.

[66] T.D. Matthews, H. Yan, D.G. Cahill, O. Coronell, B.J. Mariñas, Growth dynamics of interfacially polymerized polyamide layers by diffuse reflectance spectroscopy and Rutherford backscattering spectrometry, J. Memb. Sci. 429 (2013) 71–80. Available from: http://dx.doi.org/10.1016/j.memsci.2012.11.040.

[67] Y. Li, T. Chung, Y. Xiao, Superior gas separation performance of dual-layer hollow fiber membranes with an ultrathin dense-selective layer, J. Memb. Sci. 325 (2008) 23–27. Available from: http://dx.doi.org/10.1016/j.memsci.2008.08.018.

[68] S. Yuan, Z. Wang, Z. Qiao, M. Wang, J. Wang, S. Wang, Improvement of CO_2/N_2 separation characteristics of polyvinylamine by modifying with ethylenediamine, J. Memb. Sci. 378 (2011) 425–437. Available from: http://dx.doi.org/10.1016/j.memsci.2011.05.023.

[69] G.Z. Ramon, M.C.Y. Wong, E.M. V. Hoek, Transport through composite membrane, Part 1: Is there an optimal support membrane? J. Memb. Sci 415-416 (2012) 298–305. Available from: http://dx.doi.org/10.1016/j.memsci.2012.05.013.

[70] R.D. Offeman, C.N. Ludvik, A novel method to fabricate high permeance, high selectivity thin-film composite membranes, J. Memb. Sci. 380 (2011) 163–170. Available from: http://dx.doi.org/10.1016/j.memsci.2011.07.005.

[71] M.F. Jimenez-Solomon, P. Gorgojo, M. Munoz-Ibanez, A.G. Livingston, Beneath the surface: Influence of supports on thin film composite membranes by interfacial polymerization for organic solvent nanofiltration, J. Memb. Sci. 448 (2013) 102–113. Available from: http://dx.doi.org/10.1016/j.memsci.2013.06.030.

[72] S.H. Huang, W.S. Hung, D.J. Liaw, H.A. Tsai, G.J. Jiang, K.R. Lee, et al., Positron annihilation study on thin-film composite pervaporation membranes: correlation between polyamide fine structure and different interfacial polymerization conditions, Polymer. 51 (2010) 1370–1376. Available from: http://dx.doi.org/10.1016/j.polymer.2010.01.064.

[73] C.L. Chapman, D. Bhattacharyya, R.C. Eberhart, R.B. Timmons, C.J. Chuong, Plasma polymer thin film depositions to regulate gas permeability through nanoporous track etched membranes, J. Memb. Sci. 318 (2008) 137–144. Available from: http://dx.doi.org/10.1016/j.memsci.2008.02.030.

[74] J.T. Arena, B. McCloskey, B.D. Freeman, J.R. McCutcheon, Surface modification of thin film composite membrane support layers with polydopamine: enabling use of reverse osmosis membranes in pressure retarded osmosis, J. Memb. Sci. 375 (2011) 55–62. Available from: http://dx.doi.org/10.1016/j.memsci.2011.01.060.

[75] W. Yave, A. Car, J. Wind, K.-V. Peinemann, Nanometric thin film membranes manufactured on square meter scale: ultra-thin films for CO_2 capture, Nanotechnology. 21 (2010) 395301. Available from: http://dx.doi.org/10.1088/0957-4484/21/39/395301.

[76] M.F. Jimenez Solomon, Y. Bhole, A.G. Livingston, High flux membranes for organic solvent nanofiltration (OSN)—interfacial polymerization with solvent activation, J. Memb. Sci 423-424 (2012) 371–382. Available from: http://dx.doi.org/10.1016/j.memsci.2012.08.030.

[77] V.K. Agrawal, P.S. Singh, Alcohol-induced change of "reverse osmosis" polyamide membrane surface, J. Appl. Polym. Sci. 124 (2012) E290–E299. Available from: http://dx.doi.org/10.1002/app.36561.

[78] A. Idris, F. Kormin, M.Y. Noordin, Application of response surface methodology in describing the performance of thin film composite membrane, Sep. Purif. Technol. 49 (2006) 271–280. Available from: http://dx.doi.org/10.1016/j.seppur.2005.10.010.
[79] C. Zhou, Y. Shi, C. Sun, S. Yu, M. Liu, C. Gao, Thin-film composite membranes formed by interfacial polymerization with natural material sericin and trimesoyl chloride for nanofiltration, J. Memb. Sci. 471 (2014) 381–391. Available from: http://dx.doi.org/10.1016/j.memsci.2014.08.033.
[80] Y. Mansourpanah, S.S. Madaeni, A. Rahimpour, Fabrication and development of interfacial polymerized thin-film composite nanofiltration membrane using different surfactants in organic phase; study of morphology and performance, J. Memb. Sci. 343 (2009) 219–228. Available from: http://dx.doi.org/10.1016/j.memsci.2009.07.033.
[81] Y. Mansourpanah, K. Alizadeh, S.S. Madaeni, A. Rahimpour, H. Soltani Afarani, Using different surfactants for changing the properties of poly(piperazineamide) TFC nanofiltration membranes, Desalination. 271 (2011) 169–177. Available from: http://dx.doi.org/10.1016/j.desal.2010.12.026.
[82] A. Ahmad, B. Ooi, Properties–performance of thin film composites membrane: study on trimesoyl chloride content and polymerization time, J. Memb. Sci. 255 (2005) 67–77. Available from: http://dx.doi.org/10.1016/j.memsci.2005.01.021.
[83] D. Bastani, N. Esmaeili, M. Asadollahi, Polymeric mixed matrix membranes containing zeolites as a filler for gas separation applications: a review, J. Ind. Eng. Chem. 19 (2013) 375–393. Available from: http://dx.doi.org/10.1016/j.jiec.2012.09.019.
[84] J. Zou, W.S.W. Ho, CO_2-selective polymeric membranes containing amines in crosslinked poly(vinyl alcohol), J. Memb. Sci. 286 (2006) 310–321. Available from: http://dx.doi.org/10.1016/j.memsci.2006.10.013.
[85] L.Y. Jiang, T.S. Chung, S. Kulprathipanja, An investigation to revitalize the separation performance of hollow fibers with a thin mixed matrix composite skin for gas separation, J. Memb. Sci. 276 (2006) 113–125. Available from: http://dx.doi.org/10.1016/j.memsci.2005.09.041.
[86] K. Vanherck, G. Koeckelberghs, I.F.J. Vankelecom, Crosslinking polyimides for membrane applications: a review, Prog. Polym. Sci. 38 (2013) 874–896. Available from: http://dx.doi.org/10.1016/j.progpolymsci.2012.11.001.
[87] L. Deng, T.J. Kim, M.B. Hägg, Facilitated transport of CO_2 in novel PVAm/PVA blend membrane, J. Memb. Sci. 340 (2009) 154–163. Available from: http://dx.doi.org/10.1016/j.memsci.2009.05.019.
[88] S. Saedi, S.S. Madaeni, F. Seidi, A.A. Shamsabadi, S. Laki, Synthesis and application of a novel Amino-Starch derivative as a new polymeric additive for fixed facilitated transport of carbon dioxide through an asymmetric polyethersulfone (PES) membrane, Int. J. Greenh. Gas Control. 19 (2013) 126–137. Available from: http://dx.doi.org/10.1016/j.ijggc.2013.06.019.
[89] S. Andrew Lee, G.W. Stevens, S.E. Kentish, Facilitated transport behavior of humidified gases through thin-film composite polyamide membranes for carbon dioxide capture, J. Memb. Sci. 429 (2013) 349–354. Available from: http://dx.doi.org/10.1016/j.memsci.2012.11.047.
[90] S. Li, Z. Wang, X. Yu, J. Wang, S. Wang, High-performance membranes with multi-permselectivity for CO_2 separation, Adv. Mater. 24 (2012) 3196–3200. Available from: http://dx.doi.org/10.1002/adma.201200638.
[91] H. Lin, B.D. Freeman, Materials selection guidelines for membranes that remove CO_2 from gas mixtures, J. Mol. Struct. 739 (2005) 57–74. Available from: http://dx.doi.org/10.1016/j.molstruc.2004.07.045.

[92] A. Car, C. Stropnik, W. Yave, K.V. Peinemann, Tailor-made polymeric membranes based on segmented block copolymers for CO_2 separation, Adv. Funct. Mater. 18 (2008) 2815–2823. Available from: http://dx.doi.org/10.1002/adfm.200800436.
[93] S.L. Liu, L. Shao, M.L. Chua, C.H. Lau, H. Wang, S. Quan, Recent progress in the design of advanced PEO-containing membranes for CO_2 removal, Prog. Polym. Sci. 38 (2013) 1089–1120. Available from: http://dx.doi.org/10.1016/j.progpolymsci.2013.02.002.
[94] W. Yave, A. Car, S.S. Funari, S.P. Nunes, K.V. Peinemann, CO_2-Philic polymer membrane with extremely high separation performance, Macromolecules. 43 (2010) 326–333. Available from: http://dx.doi.org/10.1021/ma901950u.
[95] N.P. Patel, A.C. Miller, R.J. Spontak, Highly CO_2-permeable and -selective membranes derived from crosslinked poly(ethylene glycol) and its nanocomposites, Adv. Funct. Mater. 14 (2004) 699–707. Available from: http://dx.doi.org/10.1002/adfm.200305136.
[96] H. Matsuyama, M. Teramoto, H. Sakakura, Selective permeation of CO_2 through poly {2-(N,N-dimethyl)aminoethyl methacrylate} membrane prepared by plasma-graft polymerization technique, J. Memb. Sci. 114 (1996) 193–200. Available from: http://dx.doi.org/10.1016/0376-7388(95)00318-5.
[97] R. Du, A. Chakma, X. Feng, Interfacially formed poly(N,N-dimethylaminoethyl methacrylate)/polysulfone composite membranes for CO_2/N_2 separation, J. Memb. Sci. 290 (2007) 19–28. Available from: http://dx.doi.org/10.1016/j.memsci.2006.12.010.
[98] L. Deng, M.-B. Hägg, Carbon nanotube reinforced PVAm/PVA blend FSC nanocomposite membrane for CO_2/CH_4 separation, Int. J. Greenh. Gas Control. 26 (2014) 127–134. Available from: http://dx.doi.org/10.1016/j.ijggc.2014.04.018.
[99] J. Zhao, Z. Wang, J. Wang, S. Wang, High-performance membranes comprising polyaniline nanoparticles incorporated into polyvinylamine matrix for CO_2/N_2 separation, J. Memb. Sci. 403- 404 (2012) 203–215. Available from: http://dx.doi.org/10.1016/j.memsci.2012.02.048.
[100] Y. Cai, Z. Wang, C. Yi, Y. Bai, J. Wang, S. Wang, Gas transport property of polyallylamine-poly(vinyl alcohol)/polysulfone composite membranes, J. Memb. Sci. 310 (2008) 184–196. Available from: http://dx.doi.org/10.1016/j.memsci.2007.10.052.
[101] L. Ansaloni, Y. Zhao, B.T. Jung, K. Ramasubramanian, M.G. Baschetti, W.S.W. Ho, Facilitated transport membranes containing amino-functionalized multi-walled carbon nanotubes for high-pressure CO_2 separations, J. Memb. Sci. 490 (2015) 18–28. Available from: http://dx.doi.org/10.1016/j.memsci.2015.03.097.
[102] R. Nasir, H. Mukhtar, Z. Man, B.K. Dutta, M.S. Shaharun, M.Z. Abu Bakar, Mixed matrix membrane performance enhancement using alkanolamine solution, J. Memb. Sci. 483 (2015) 84–93. Available from: http://dx.doi.org/10.1016/j.memsci.2015.02.041.
[103] S. Saedi, S.S. Madaeni, F. Seidi, A.A. Shamsabadi, S. Laki, Fixed facilitated transport of CO_2 through integrally-skinned asymmetric polyethersulfone membrane using a novel synthesized Poly (acrylonitrile-co-N, N-Dimethylaminopropyl acrylamide), Chem. Eng. J. 236 (2014) 263–273. Available from: http://dx.doi.org/10.1016/j.cej.2013.09.092.
[104] H. Wu, X. Li, Y. Li, S. Wang, R. Guo, Z. Jiang, et al., Facilitated transport mixed matrix membranes incorporated with amine functionalized MCM-41 for enhanced gas separation properties, J. Memb. Sci. 465 (2014) 78–90. Available from: http://dx.doi.org/10.1016/j.memsci.2014.04.023.
[105] M. Ostwal, R.P. Singh, S.F. Dec, M.T. Lusk, J.D. Way, 3-Aminopropyltriethoxysilane functionalized inorganic membranes for high temperature CO_2/N_2 separation, J. Memb. Sci. 369 (2011) 139–147. Available from: http://dx.doi.org/10.1016/j.memsci.2010.11.053.

[106] G.J. Francisco, A. Chakma, X. Feng, Membranes comprising of alkanolamines incorporated into poly(vinyl alcohol) matrix for CO_2/N_2 separation, J. Memb. Sci. 303 (2007) 54–63. Available from: http://dx.doi.org/10.1016/j.memsci.2007.06.065.

[107] T.J. Kim, L.I. Baoan, M.B. Hägg, Novel fixed-site-carrier polyvinylamine membrane for carbon dioxide capture, J. Polym. Sci. Part B Polym. Phys. 42 (2004) 4326–4336. Available from: http://dx.doi.org/10.1002/polb.20282.

[108] C. Yi, Z. Wang, M. Li, J. Wang, S. Wang, Facilitated transport of CO_2 through polyvinylamine/polyethlene glycol blend membranes, Desalination 193 (2006) 90–96. Available from: http://dx.doi.org/10.1016/j.desal.2005.04.139.

[109] H. Cong, J. Zhang, M. Radosz, Y. Shen, Carbon nanotube composite membranes of brominated poly(2,6-diphenyl-1,4-phenylene oxide) for gas separation, J. Memb. Sci. 294 (2007) 178–185. Available from: http://dx.doi.org/10.1016/j.memsci.2007.02.035.

[110] J. Ahn, W.-J. Chung, I. Pinnau, M.D. Guiver, Polysulfone/silica nanoparticle mixed-matrix membranes for gas separation, J. Memb. Sci. 314 (2008) 123–133. Available from: http://dx.doi.org/10.1016/j.memsci.2008.01.031.

[111] D.Q. Vu, W.J. Koros, S.J. Miller, Mixed matrix membranes using carbon molecular sieves: I. Preparation and experimental results, J. Memb. Sci. 211 (2003) 311–334. Available from: http://dx.doi.org/10.1016/S0376-7388(02)00429-5.

[112] M. Dinari, H. Ahmadizadegan, Synthesis, structural characterization and properties of novel functional poly(ether imide)/titania nanocomposite thin films, Polymer. 55 (2014) 6252–6260. Available from: http://dx.doi.org/10.1016/j.polymer.2014.10.021.

[113] T. Li, Y. Pan, K. Peinemann, Z. Lai, Carbon dioxide selective mixed matrix composite membrane containing ZIF-7 nano-fillers, J. Memb. Sci. 425- 426 (2013) 235–242. Available from: http://dx.doi.org/10.1016/j.memsci.2012.09.006.

[114] A.L. Ahmad, Z.A. Jawad, S.C. Low, S.H.S. Zein, A cellulose acetate/multi-walled carbon nanotube mixed matrix membrane for CO_2/N_2 separation, J. Memb. Sci. 451 (2014) 55–66. Available from: http://dx.doi.org/10.1016/j.memsci.2013.09.043.

[115] K.C. Wong, P.S. Goh, B.C. Ng, A.F. Ismail, Thin film nanocomposite embedded with polymethyl methacrylate modified multi-walled carbon nanotubes for CO_2 removal, RSC Adv. 5 (2015) 31683–31690. Available from: http://dx.doi.org/10.1039/C5RA00039D.

[116] E. Torralba-Calleja, J. Skinner, D. Gutiérrez-Tauste, CO_2 capture in ionic liquids: a review of solubilities and experimental methods, J. Chem 2013 (2013). Available from: http://dx.doi.org/10.1155/2013/473584.

[117] Z. Dai, R.D. Noble, D.L. Gin, X. Zhang, L. Deng, Combination of ionic liquids with membrane technology: a new approach for CO_2 separation, J. Memb. Sci. 497 (2016) 1–20. Available from: http://dx.doi.org/10.1016/j.memsci.2015.08.060.

[118] H. Seon Bang, S. Jang, Y. Soo Kang, J. Won, Dual facilitated transport of CO_2 using electrospun composite membranes containing ionic liquid, J. Memb. Sci. 479 (2015) 77–84. Available from: http://dx.doi.org/10.1016/j.memsci.2015.01.020.

[119] F. Moghadam, E. Kamio, A. Yoshizumi, H. Matsuyama, An amino acid ionic liquid-based tough ion gel membrane for CO_2 capture, Chem. Commun. 51 (2015) 13658–13661. Available from: http://dx.doi.org/10.1039/C5CC04841A.

[120] X. Sun, M. Zhang, R. Guo, J. Luo, J. Li, CO_2-facilitated transport performance of poly(ionic liquids) in supported liquid membranes, J. Mater. Sci. (2014) 104–111. Available from: http://dx.doi.org/10.1007/s10853-014-8570-z.

[121] J. Zhou, M.M. Mok, M.G. Cowan, W.M. McDanel, T.K. Carlisle, D.L. Gin, et al., High-permeance room-temperature ionic-liquid-based membranes for CO_2/N_2 separation, Ind. Eng. Chem. Res. 53 (2014) 20064–20067. Available from: http://dx.doi.org/10.1021/ie5040682.

[122] H. Wang, D.R. Paul, T.S. Chung, Surface modification of polyimide membranes by diethylenetriamine (DETA) vapor for H_2 purification and moisture effect on gas permeation, J. Memb. Sci. 430 (2013) 223–233. Available from: http://dx.doi.org/10.1016/j.memsci.2012.12.008.

[123] P.C. Ma, N.A. Siddiqui, G. Marom, J.K. Kim, Dispersion and functionalization of carbon nanotubes for polymer-based nanocomposites: a review, Compos. Part A Appl. Sci. Manuf 41 (2010) 1345–1367. Available from: http://dx.doi.org/10.1016/j.compositesa.2010.07.003.

[124] A. Surapathi, H. Chen, E. Marand, J. Karl Johnson, Z. Sedlakova, Gas sorption properties of zwitterion-functionalized carbon nanotubes, J. Memb. Sci. 429 (2013) 88–94. Available from: http://dx.doi.org/10.1016/j.memsci.2012.11.021.

[125] K.C. Wong, P.S. Goh, A.F. Ismail, Gas separation performance of thin film nanocomposite membranes incorporated with polymethyl methacrylate grafted multi-walled carbon nanotubes, Int. Biodeterior. Biodegrad. 102 (2015) 339–345. Available from: http://dx.doi.org/10.1016/j.ibiod.2015.02.007.

[126] M. Namvar-Mahboub, M. Pakizeh, S. Davari, Preparation and characterization of UZM-5/polyamide thin film nanocomposite membrane for dewaxing solvent recovery, J. Memb. Sci. 459 (2014) 22–32. Available from: http://dx.doi.org/10.1016/j.memsci.2014.02.014.

[127] A.D. Kiadehi, M. Jahanshahi, A. Rahimpour, S.A.A. Ghoreyshi, The effect of functionalized carbon nano-fiber (CNF) on gas separation performance of polysulfone (PSf) membranes, Chem. Eng. Process. Process Intensif 90 (2015) 41–48. Available from: http://dx.doi.org/10.1016/j.cep.2015.02.005.

[128] M.M. Khan, V. Filiz, G. Bengtson, S. Shishatskiy, M.M. Rahman, J. Lillepaerg, et al., Enhanced gas permeability by fabricating mixed matrix membranes of functionalized multiwalled carbon nanotubes and polymers of intrinsic microporosity (PIM), J. Memb. Sci. 436 (2013) 109–120. Available from: http://dx.doi.org/10.1016/j.memsci.2013.02.032.

[129] G.L. Jadav, P.S. Singh, Synthesis of novel silica-polyamide nanocomposite membrane with enhanced properties, J. Memb. Sci. 328 (2009) 257–267. Available from: http://dx.doi.org/10.1016/j.memsci.2008.12.014.

[130] Z. Zhao, Z. Yang, Y. Hu, J. Li, X. Fan, Multiple functionalization of multi-walled carbon nanotubes with carboxyl and amino groups, Appl. Surf. Sci. 276 (2013) 476–481. Available from: http://dx.doi.org/10.1016/j.apsusc.2013.03.119.

[131] C.-F. de Lannoy, E. Soyer, M.R. Wiesner, Optimizing carbon nanotube-reinforced polysulfone ultrafiltration membranes through carboxylic acid functionalization, J. Memb. Sci. 447 (2013) 395–402. Available from: http://dx.doi.org/10.1016/j.memsci.2013.07.023.

[132] A.F. Ismail, N.H. Rahim, A. Mustafa, T. Matsuura, B.C. Ng, S. Abdullah, et al., Gas separation performance of polyethersulfone/multi-walled carbon nanotubes mixed matrix membranes, Sep. Purif. Technol. 80 (2011) 20–31. Available from: http://dx.doi.org/10.1016/j.seppur.2011.03.031.

[133] E.S. Kim, B. Deng, Fabrication of polyamide thin-film nano-composite (PA-TFN) membrane with hydrophilized ordered mesoporous carbon (H-OMC) for water purifications, J. Memb. Sci. 375 (2011) 46–54. Available from: http://dx.doi.org/10.1016/j.memsci.2011.01.041.

Index

Note: Page numbers followed by "*f*" and "*t*" refer to figures and tables, respectively.

Q

R

CPI Antony Rowe
Chippenham, UK
2017-06-13 18:23